ELECTRODE PROCESSES AND ELECTROCHEMICAL ENGINEERING

ELECTRODE PROCESSES AND ELECTROCHEMICAL ENGINEERING

Fumio Hine

Nagoya Institute of Technology
Nagoya, Japan

PLENUM PRESS • NEW YORK AND LONDON

Library of Congress Cataloging in Publication Data

Hine, Fumio.
  Electrode processes and electrochemical engineering.

  Includes bibliographical references and index.
  1. Electrochemistry, Industrial. 2. Electrodes. I. Title.
TP255.H56   1985   666.2′97   84-24782
ISBN 978-1-4757-0111-1   ISBN 978-1-4757-0109-8 (eBook)
DOI 10.1007/978-1-4757-0109-8

© 1985 Plenum Press, New York
Softcover reprint of the hardcover 1st edition 1985
A Division of Plenum Publishing Corporation
233 Spring Street, New York, N.Y. 10013

FOREWORD

This book has been planned and written by Dr. Hine with his knowledge and experience in electrochemical science and engineering for over thirty years since he joined with me at Kyoto University in 1948.

This book is unique and is useful for engineers as well as scientists who are going to work in any interdisciplinary field connected with electrochemistry.

Science is sure to clarify the truth of nature as well as bring prosperity and an improvement to the welfare of human beings. The origin of the word "science" is the same as of "conscience," which means the truth of our heart.

When we consider a scientific and technological subject, first we classify it into the components and/or factors involved, and then we clarify them individually. Second, we combine them to grasp the whole meaning and feature of the subject under discussion. Computers may help us greatly, but how to establish the software that will be most desirable for our purposes is of great importance. We need to make these efforts ourselves, and not decorate with borrowed plumes.

With this concept in mind, this book is attractive because the author describes the basic science in electrochemistry and practice, and discusses the electrochemical engineering applications as a combination of science and technology.

Furthermore, Dr. Hine discusses the present situation and future trends in the industry with a number of drawings and photographs. He also tries to compare the electrochemical process to other competitive processes. He does this well because he has experience both in the field of science and in chemical process industries. Also, he is very happy that he has many friends worldwide who give him useful information and ideas.

Dr. Hine attended classes in electrical engineering at the technical college in Osaka, then he came to my laboratory at the Department of Industrial Chemistry, Kyoto University, where he received his doctorate degree. He has studied hard and specialized in electrochemical engineering. This book is the fruit of his efforts.

It is a pleasure to provide these introductory remarks for Fumio Hine, who is so willing to work actively with an international sense.

SHINZO OKADA
Doctor of Engineering
Professor Emeritus, Kyoto University
Advisor, Japan Storage Battery Company, Kyoto

PREFACE

The principal role for the chemical process industry, including electrochemical processes, is to contribute to society through an adequate utilization of limited resources and energy in the world. Although industry should have an adequate size to meet the requirements of society, the scale-up of plant size as well as equipment and apparatus to be used is a general trend depending on the progress of the activity of our society. Since the chemical process industries, especially electrochemical processes, are energy-intensive, how to save process energy as well as raw materials to be processed is the most important subject for the chemical engineer. The electrochemical industries are consuming, surprisingly, almost ten percent of the total electric energy generated in industrialized countries. This is the reason why we must study the concepts of the chemical engineering of electrolysis and related problems, because the concept is based on the optimization of the process of interest from the sides of both technology and economy.

The electrochemical technology has unique features since the electrochemical process is a heterogeneous catalytic reaction on the working electrode assisted by electrical energy. It is generally flexible, and hence applicable to various fields of chemical processing. These features are desirable, and electrochemical technology could be a candidate for assisting today's difficulties in chemical economy. Trials of application of electrochemistry and its technology to new fields such as organic synthesis and solid-state chemistry are examples. Consequently, any systematization of electrochemical technology and engineering based on electrochemical science is important. Electrochemical engineering is concerned with practical subjects related either directly or indirectly to industrial processes with the knowledge of electrochemistry and chemical engineering. Therefore, electrochemical engineering is the bridge between science and practice in the electrochemical field.

Electrochemistry is close to physical chemistry, and has the features of both unit process and unit operation. It is almost a complete science, but its technology has still a number of problems to be solved.

Since the electrochemical industry and its related industries have important positions in the industrialized society of today, first we need to study the present situation. With the concepts described above, this textbook involves three parts: (i) the outline of electrochemical sciences, (ii) the outline of electrochemical industries, and (iii) the design and operation of electrochemical processes and electrolyzers. Of these, the electrochemical processes are outlined from the viewpoints of thermodynamics and kinetics in the first part, probably a minimum requirement for engineers, since there are many publications discussing details of electrode processes.

In the second part, the relationship and comparison between the electrochemical process and the nonelectrochemical process have been emphasized so as to understand the feasibility of the electrochemical method in the field of chemical industry.

I remember that we spoke a new phrase, "engineering concepts on industrial electrochemical processes and electrolyzers", in Mr. MacMullin's home in Niagara Falls in the summer of 1960, at a time when I worked at Case Western Reserve University, Cleveland, Ohio. I have been influenced greatly by MacMullin's philosophy on electrochemical engineering since then. But electrochemical engineering is unpopular even today. The concepts and procedures have not been established as yet. I hope that the last part of this book can be a milestone toward our goal.

In any event, the electrochemical industry has experienced many business ups and downs in the hundred years of its history. However, electrochemistry and its industry have faced a turning point in the last thirty years. It is the "revolution of electrochemistry and its industry." A number of the scientific approaches have been presented; on the other hand, the electrochemical industries have developed extensively. It was good fortune for me that I could have the opportunity of studying electrochemical science and engineering under the directorship of Professor Shinzo Okada at Kyoto University in this period. Professor Okada is ninety years old but still active and interested in our subjects.

I wish to acknowledge with many thanks the helpful advice and assistance of my collaborators and students. This textbook was published originally in Japanese by Kagaku Dojin Publishing Company, Kyoto. The company has approved English translation without any obligation. Some of the data and the figures as well as descriptions in this book have been revised since the Japanese book was published more than five years ago.

I appreciate the many friends and several companies who gave me useful photographs and data. Also several figures were quoted by permission of the publishers. Their names can be found in the captions.

Also thanks to Plenum Publishing Company, who gave me an opportunity for publication.

Finally, I wish to express my appreciation to Dr. Leonard Nanis, Palo Alto, California, for his review and advice on preparation of the manuscript. This book could not have been published without his kind efforts.

FUMIO HINE

CONTENTS

II. ELECTROCHEMICAL INDUSTRIES

III. ELECTROCHEMICAL ENGINEERING

I INTRODUCTION TO ELECTROCHEMISTRY

CONCEPT AND ROLE OF ELECTROCHEMICAL ENGINEERING

1.1. ELECTROCHEMICAL SCIENCE AND TECHNOLOGY

An electrode process is a heterogeneous catalytic reaction accompanied by charge transfer at the electrode surface in contact with the electrolyte. The amount of mass transfer is exactly proportional to the amount of electricity passing through the electrode–electrolyte interface and is governed by the Faraday law. Hence, the current density on the working electrode is the same as the reaction rate. There is a potential difference at the electrode–electrolyte interface, called the *electrode potential*, which is related to the free energy change for the electrode process under discussion. The electrode potential may deviate from its equilibrium state, called the *equilibrium* or *reversible potential*, when the charge transfer reaction takes place. Consequently, analysis of the electrode process of interest can be made from the viewpoints of both thermodynamics and chemical kinetics by using the experimental relationship between the electrode potential (the driving force for the reaction) and the current density (the reaction rate). Advancement of electronic instrumentation has promoted scientific research in these fields.

The electrochemical industry is a process industry in which one or a number of electrolyzers (the electrochemical reactors) are operated to make chemical products. Industrial electrochemistry and electrochemical technology are the engineering concept and technique, respectively, which interpret and utilize useful knowledge from all fields of natural and political sciences, as well as from electrochemistry, for further advancement of the electrochemical industry.

The electrochemical process and its rate occurring on a unit area of the working electrode as part of a commercial scale cell are the same as those of a small electrode in a laboratory cell from the viewpoint of electrochemical science. On the other hand, from the engineering point of view, there are important considerations related to various subjects, such as mass and heat transfer, balances of materials and energy, and materials selection and geometry of the electrolyzer. Selection of the dc power source is also of importance. In general, we may consider that electrochemical engineering is a method for the optimal design and operation of the electrochemical process industry of sufficient size to meet market demand under reasonable conditions of economy and return on investment, using scientific knowledge and technology from many related fields.

For example, the chlor-alkali industries in Western countries have a total capacity of some 25 million tons per year of caustic and chlorine, and the capacity of the aluminum industry in these countries is more than 10 million tons per year. The power consumption in the electrochemical industry in advanced countries is estimated to be about 6–7% of the total generation of electricity. One side of electrochemistry is concerned with microscopic details and is highly scientific while the other side is concerned with similar reactions on a large scale and the related aspects of industrial processes.

The electrochemical reaction is a unit process occurring on the working electrode where either oxidation or reduction takes place without any chemical agent being required for the reaction. Electricity accomplishes the oxidation or reduction so that there are no byproduct species. Today, this is a very important feature of electrochemical processes from the viewpoints of environmental protection and materials conservation. For example, there are several processes and flowsheets for producing chlorine and caustic soda from crude salt via an electrochemical route. Of course, the basic science is universal, but the reactor, i.e., the cell, and the flowsheet are usually modified, and the plant is sometimes operated under unique conditions based on the insights of the designer. Faith is probably paramount in mystical experiences to devout religionists, but it is perhaps less useful in plant design.

A special feature of electrochemical processes is the use of a large amount of electric energy as a driving force for the chemical reaction. Of course, electricity is utilized by other industries as well, but electricity in the electrochemical industry is considered to be a kind of feedstock rather than a utility. Thus, we need to make special efforts for the design of electrochemical processes under the general heading of electrochemical engineering. Electrochemical engineering utilizes the ideas of chemical engineering, which deals with unit operations, but it proceeds with further developments for its own special requirements. Electrochemical engineering involves both unit operations and unit processes. Electrochemical engineering may be considered as the junction between electrochemical science and industry, as illustrated in Fig. 1.1. New developments related to electrochemical engineering are frequent topics for specialized symposia of the Electrochemical Society, the American Institute of Chemical Engineers, and others. Recent symposium topics include "Industrial Electro-Organic Processes," "Fluorocarbon Ion Exchange Membranes," "Modern Methods of

Studying Transport Properties," and "Anodes for the Electrolytic Industries," to mention but a few. In addition, important industrial areas such as chlor-alkali are frequent topics for symposium sessions.

The electrochemical industry has developed because of the applicability of electrochemical processes, while on the other hand, there is a growing application of electrochemical methods in various fields because of their unique features. Certain electrochemical methods are used in various fields outside of electrochemistry, such as process control and monitoring, because electrochemical methods readily permit the determination of small deviations for phenomena in which charge transfer takes place. The control of pH and process control by measurement of the oxidation–reduction potential are examples. Polarization measurements are commonly used for the fundamental study of the corrosion of metals and alloys, because corrosion is an electrochemical process. Electrochemical methods for corrosion monitoring and control are widely used in many industries.

Corrosion of metals at high temperatures is the chemical reaction of metal with oxygen or other oxidizing gases. However, there is a close connection since mass transfer and charge transfer are the main subjects of today's solid-state electrochemistry.

There is no doubt that environmental protection is an important concern of our industrialized society. For this purpose, an electrochemical

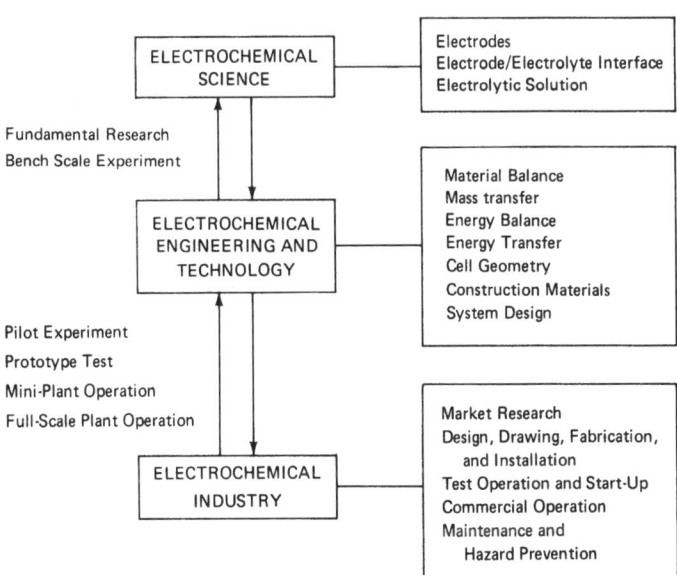

FIGURE 1.1. Relationship and role of electrochemical science, engineering, and industry.

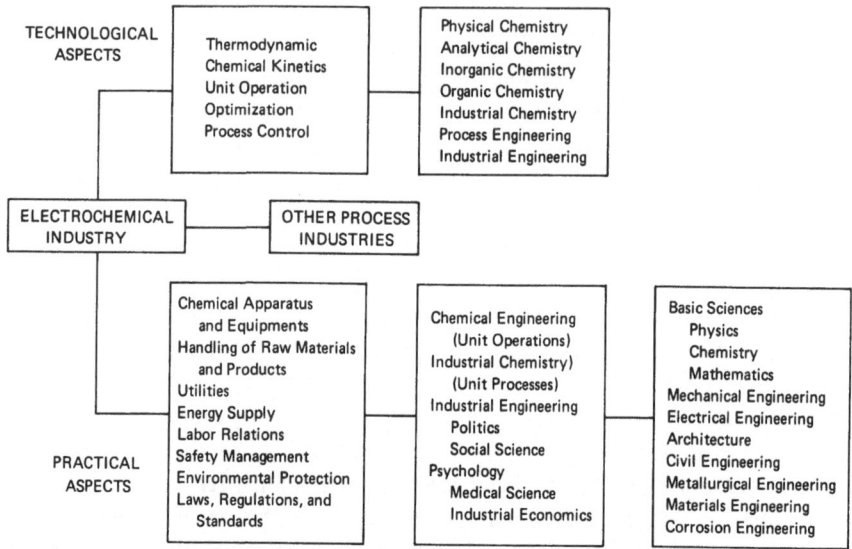

FIGURE 1.2. Background of electrochemical industry.

technique is applied and various products of electrochemistry are consumed. Consumption of chlorine for water treatment in the United States was estimated to be some 80 thousand tons annually in the early 1960s. Electrochemical methods have been proposed for industrial waste metal recovery and for purification of industrial effluents.

Conservation and effective utilization of energy are of importance. The nuclear reactor and also coal are considered to be potential major sources for primary energy in the twenty-first century, and several candidates, such as solar energy, geothermal, and the fuel cell, are being discussed as subsidiary sources. Electrochemical concepts and electrochemical processes are also related closely to these fields. Load leveling of electric energy by the storage of off-peak generated electricity may be achieved through electrochemical methods. New load-leveling battery systems are being studied intensively under government and utility sponsorship in the United States and elsewhere.

As indicated by Bockris,[1] electrochemistry is related to various fields of science, such as thermodynamics, electronics, physics, surface physics and chemistry, metallurgy, crystallography, quantum mechanics, physical chemistry, optics, spectroscopy, statistical mechanics, and hydrodynamics. The electrochemical industry is based on electrochemical science with all of the above-mentioned features, plus various aspects of engineering and technology, as shown in Fig. 1.2.

1.2. FEATURES OF ELECTROCHEMICAL PROCESSES

Although the electrochemical industry is one of the typical process industries, it has several unique features

1. It requires a large amount of dc electric power of low voltage and large amperage for driving the chemical reaction;
2. The mass transfer is exactly proportional to the amount of dc power passing through the electrolyzer as reactor, by the Faraday law;
3. The electrochemical reaction takes place only at the two electrodes, the anode and the cathode, and is evidently a heterogeneous, catalytic reaction at the interface between electrode and electrolyte; and
4. The electrochemical process and its rate can be controlled by the constituents of the system, such as the working electrode and the electrolyte, and by adjustment of the operating conditions, such as the electrode potential or the current density over a relatively wide range.

Although electricity is expensive compared to other energies, there are many unique possibilities with electrochemical science and technology. Therefore, the design engineer should confirm the soundness of the process from the viewpoints of both science and economy, before the decision-making state for the project is reached.

Water electrolysis, to make hydrogen for ammonia, was active in Japan about thirty years ago, but it was completely shut down when the petrochemical complex started to supply cheap hydrogen.

An electrochemical technology developed by Monsanto for making adiponitrile via acrylonitrile with an amalgam cathode has a bright future, and has generated new interest in organic electrochemical processes. [2-4]

The Nalco Chemical Company has electrolyzed a Grignard reagent with a lead anode to produce tetra-alkyl lead, [4,5] and Dupont has synthesized quinone by means of anodic oxidation of benzene. [6] But the Nalco tetra-alkyl lead process has been shut down because of the restriction against leaded gasoline in the United States. Although many scientific reports and reviews of organic electrochemistry have been published, only few applications have been announced in practice. [7-13] Recently, commercialization of organic synthesis via the electrochemical route is being watched with interest for several reasons. [12] It is evident that the electrochemical method for synthesizing organic chemicals is based on either anodic oxidation or cathodic reduction, but organic reactions such as substitution, addition, cleavage, and coupling can also be carried out with suitable procedures in electrolysis. The current efficiency for organic electro-synthesis has been greatly improved, thus keeping the energy consumption low. The production cost via the electrochemical route becomes comparatively low because cheap materials can be used as feedstock in some cases. Ease of environmental protection is also an important factor.

Development of technology on fused salt electrolysis is slow and only a few papers are found in journals. Electrochemical production of titanium to replace the Kroll process of Mg reduction of $TiCl_4$ has not been commercialized as yet, although considerable research has been conducted.[14–16] The amalgam process for producing metallic sodium developed by Kyoto University and Tekkosha Co. has lost its standing because the amalgam-type chlorine cell plants have been designated for shutdown as a result of environmental politics in Japan. Greatly decreased production of tetra-alkyl lead, which is an antiknocking agent for gasoline, has reduced the demand for sodium. Tetra-alkyl lead production had been a major market for sodium.

Innovation of technology for the aluminum industry is still active. Approaches to reduce energy consumption for electrolysis with the Hall–Heroult process have included operation of large-scale electrolyzers with prebaked anodes, decrease of the electrode gap, and addition of Li_2CO_3 to the cryolite melt to increase electrical conductivity of the molten salt.[17–20]

Alcoa has announced a new process using aluminum chloride instead of the cryolite used in the conventional Hall–Heroult process. The energy consumption is reported to be only 9,000–11,000 kW h/t-Al, which is about 30% less than that of the conventional process.

Generally, an industrial process is composed of several steps, such as treatment of raw materials, reaction, and products polishing, so that the mass transfer and the energy balance throughout the entire system must be considered. Of course, the electrochemical process industry must take these concepts into account. The mass transfer and the energy balance in the electrolytic cell should be considered simultaneously because the electrolyzer consumes an amount of electricity as a kind of feedstock than as energy, and the mass transfer for electrochemical processing is closely related to the amount of electricity passed.

Performance of the electrochemical system is affected greatly by the design and operating procedure for the electrolytic cell. Configuration and construction materials used are of importance. Table 1.1 is an example of a checklist for an adequate selection and design of an electrolytic cell, which consists of the cell or vessel, separator or diaphragm, anode, and cathode. These components must be selected and designed carefully from the viewpoints of various fields of science and engineering. The final target is operation of the process plant under optimum conditions so as to keep the production cost low. Indeed, this is the role of electrochemical engineering.

TABLE 1.1
A Checklist for the Design of Electrochemical Systems

		Cell	Diaphragm	Anode	Cathode
Materials selection	Corrosion resistance	Metallic or nonmetallic / Uniform or localized corrosion and rate / Corrosion prevention	Metallic or nonmetallic / Corrosion resistance to both anolyte and catholyte	Soluble anode: dissolution rate, uniform dissolution / Durable anode: degradation, service life	Corrosion resistance at shutdown time / Possibility of use of cell body as cathode
	Mechanical strength and machining	Mechanical strength / Effect of temperature / Machinability, cold working and weldment	Elongation and shrinkage / Deformation under service conditions / Fitting	Mechanical strength—initial and final stage of service / Machinability / Fitting and electric lead-in / Withstand emergency such as short-circuiting	
	Other properties	Electric insulation / Electric conductivity / Thermal conductivity	Porosity and pore size / Permeability / Tortuosity	Activity for proposed electrode reaction / Polarization behavior / Electric conductivity	
	Standardized materials	Use of standardized material to save initial cost	Size of constituents / Composition of materials	Pretreatment and polishing / Surface treatment	
Mechanical and geometrical design	Layout	Number of cells and amperage / Size of cell room / Cell foundation	Indoor or outdoor plant / Connection, series and parallel / Handling of feed and product	Ratio of volume to electrode area / Procedure of energy supply / Type of control system	
	Geometry and configuration	Horizontal or vertical / Monopolar or bipolar / Size and shape / Open or covered / Continuous or batch	Fabric or microporous sheet / Vacuum deposited layer / Ceramics, metallic, organic / Ion exchange membrane	Size and shape / Method of fitting and lead-in / Current distribution / Anode-to-cathode spacing, or electrode-to-diaphragm / Brine flow and bubble effect	
	Operation and maintenance	Start-up and shutdown procedures / Control of operating conditions, manual or automatic / Hazard warning and prevention—flow, heat, electricity, mechanical failures, lightning, earthquake, etc.		Standby and spares / Maintenance / Clean-up and washing / Lighting and ventilation	
Material balance and energy balance	Material balance and rate	Feed incomings and product outgoings / Recovery of raw material / Losses / Water balance	Mass transfer of ionic and molecular species	Current efficiency and factors of current loss / Main reaction vs. side reaction / Brine concentration, pH, independent salt, etc. / Effects of impurities	
	Energy balance and rate	Heating and cooling / Recovery of heat / Ohmic voltage drops	Ohmic voltage drop and change due to plugging	Reversible potential and static potential / Overvoltage and current density / Bubble effect	

REFERENCES

1. J. O'M. Bockris and A. K. N. Reddy, *Modern Electrochemistry*, Vol. 1, p. 26, Plenum Press, New York (1970).
2. Anonymous, *Chem. Eng.*, p. 80 (June 21, 1965).
3. I. H. Prescott, *Chem. Eng.*, p. 238 (November 8, 1965).
4. D. E. Danly, *Hydrocarbon Process.*, p. 159 (June 1969).
5. L. L. Bott, *Hydrocarbon Process.* **44**, 115 (1965).
6. J. L. Fitzjohn, *Chem. Eng. Prog.* **71**(2), 85 (1975).
7. R. B. MacMullin, *Electrochem. Technol.* **2**, 106 (1964).
8. Anonymous, *Chem. Eng.*, p. 56B (April 20, 1971).
9. Anonymous, *Chem. Eng.*, p. 78B (May 17, 1971).
10. Anonymous, *Chem. Ing. Techn.* **46**, 708 (1974).
11. M. Fleischmann and D. Pletcher, *Chem. Br.* **11**(2), 50 (1975).
12. J. C. Davis, *Chem. Eng.*, p. 44 (July 7, 1975).
13. *Symposium on Electrochemical Reaction Engineering*, AIChE, August 18–21, 1974, Salt Lake City, Utah.
14. T. Hashino, M. Kawane, and S. Okada, *Denki Kagaku* (J. Electrochem. Soc. of Japan) **25**, 63 (1957).
15. S. Okada, M. Kawane, and T. Hashino, *Kogyo Kagaku Zasshi* (J. Chem. Soc. of Japan, Industrial Chemistry Section) **63**, 48 (1960).
16. *Kagaku Binran, Oyo-hen* (Handbook of Chemistry, Applied Chemistry Section), p. 298, Maruzen, Tokyo (1973).
17. Anonymous, *Chem. Eng.*, p. 3 (June 9, 1975).
18. L. C. Fuhrmeister and A. T. Emery, *J. Electrochem. Soc.* **120**, 7C (1973).
19. Anonymous, *Wall Street Journal*, January 12 (1973).
20. U.S. Patent 3,725,222 (1973).

Chapter 2

THERMODYNAMICS OF ELECTROCHEMICAL PROCESSES

2.1. ELECTROMOTIVE FORCE AND DECOMPOSITION VOLTAGE

According to the theory of chemical kinetics, the chemical reaction represented by

$$A + B + \cdots \rightleftarrows M + N + \cdots + Q$$

can proceed in both directions, with respective rates, $\vec{v}$ and $\overleftarrow{v}$. The overall reaction rate v, which can be determined directly by experiment, is shown by a difference:

$$v = \vec{v} - \overleftarrow{v}$$

Thus, the reaction may proceed in the right-hand direction (the forward direction) preferentially, if $\vec{v} > \overleftarrow{v}$. Here Q represents the formation energy of the species located on the right-hand side of the reaction.

It is clear that $v = 0$ at $\vec{v} = \overleftarrow{v}$, but both $\vec{v}$ and $\overleftarrow{v}$ may not be equal to zero. This is called the *equilibrium state*, which is a subject of the theory of thermodynamics, and is important as the basis for a discussion of chemical kinetics.

The concept is valid, of course, for electrochemical processes, whereas we need to add up the terms for charge transfer. In other words, a change of the free energy for the electrochemical reaction of interest is shown by a change of the electric energy or by the terms of electricity. We consider these concepts with several examples.

2.1.1. Example 1: Formation and Electrolysis of Hydrochloric Acid

When hydrogen gas is burned with chlorine gas in a reactor made of impregnated carbon, hydrogen chloride gas forms with a large amount of reaction heat. The gas is brought to an absorption column, also made of impregnated carbon, to produce hydrochloric acid solution.

TABLE 2.1

Formation Energy of HCl at 25 °C

$\frac{1}{2}H_2(g) + \frac{1}{2}Cl_2(g) = HCl(g)$ $\frac{1}{2}H_2(g) + \frac{1}{2}Cl_2(g) = HCl(aq)$

	Gas	Aqueous solution
(a) ΔH^0, kcal/mol	−22.063	−40.023
ΔG^0, kcal/mol	−22.769	−31.350
ΔS^0, cal/mol K	44.617	13.2

	ΔH^0, kcal/mol	S^0, cal/mol K
(b) $H_2(g)$	0	31.211
$Cl_2(g)$	0	53.286
HCl(aq)	−40.023	13.2

NOTE: The activity of species is assumed to be unity.

The reaction

$$\frac{1}{2}H_2(g) + \frac{1}{2}Cl_2(g) = HCl(aq) \tag{2.1}$$

is exothermic. Table 2.1 shows the formation energy at 25°C.

Now we may recover the formation energy completely as electric energy under the equilibrium state of reaction (2.1) at constant temperature and pressure; thus we have an equation:

$$-\Delta G = nFE_f \tag{2.2}$$

where E_f is the electromotive force or emf for the electrochemical reaction consisting of H_2 and Cl_2 as reactants, F is the Faraday constant, and n is the number of charge transfer. Since the units of ΔG and E_f are kcal/mol and volts, respectively, and n has units of equivalents per mole,

$$F = 23.06 \text{ kcal/V g-eq.} \tag{2.3}$$

Thus, from Table 2.1, we have

$$E_f^0 = -\frac{\Delta G^0}{nF} = \frac{31.350}{1 \times 23.06} = 1.359 \text{ (V)}$$

Since free energy data such as shown in Table 2.1 are conveniently tabulated in reference books such as the JANAF and *International Critical Tables*, E_f can be calculated directly. If ΔG cannot be found, then we may calculate it with the well-known thermodynamic equation:

$$\Delta G = \Delta H - T\Delta S \tag{2.4}$$

For reaction 2.1, for example, ΔS^0 can be obtained from the data in Table 2.1(b), where the superscript ^{0} represents the standard state for unit activity of species.

The computational details for obtaining ΔS^0 are shown as follows:

	S^0 (cal/g mol K)
$H_2(g)$	$-31.211 \times \frac{1}{2} = -15.606$
$Cl_2(g)$	$-53.286 \times \frac{1}{2} = -26.643$
$HCl(aq)$	$13.2 \times 1 = +13.2$
$\frac{1}{2}H_2(g) + \frac{1}{2}Cl_2(g) = HCl(aq)$	-29.049

Now we obtain ΔG^0 using Eq. (2.4) for the standard state together with the above computed value for ΔS^0:

$$\Delta G^0 = \Delta H^0 - T\Delta S^0 = -40.023 - 298(-29.049)10^{-3}$$

$$= -31.366 \ (\text{kcal/gmol})$$

This calculated value is in good agreement with the data in Table 2.1.

According to thermodynamics, the free energy change of the reverse direction is of the same magnitude as the forward reaction, but the sign should be opposite. It means that we need to supply to the system heat energy of at least 40.023 kcal/mol (enthalpy change, ΔH^0) when we wish to decompose hydrogen chloride to hydrogen and chlorine. Because of the increase in entropy when gas is produced, the free energy required is FE_f kcal/g mol of electric energy, or the terminal voltage of 1.359 V should be applied to the electrochemical cell as a minimum voltage for decomposing HCl; that is, the theoretical decomposition voltage E_d is shown by the equation:

$$E_d = -E_f \tag{2.5}$$

2.1.2. Example 2: Electrolysis of Water

Oxygen and hydrogen are generated at the anode and the cathode, respectively, in a caustic soda solution, say 20% by weight in concentration. The overall reaction is represented by the equation:

$$H_2O(l) = H_2(g) + \frac{1}{2}O_2(g) \tag{2.6}$$

Both ΔG and E_d may vary with the NaOH concentration and the temperature because the activities of the chemical species concerned are functions of the concentration and the temperature in general.

The decomposition voltage E_d, under given conditions, is represented by the equation:

$$E_d = E_d^0 + \frac{RT}{2F} \ln \left[\frac{(p_{H_2})(p_{O_2})^{1/2}}{a_{H_2O}} \right] \tag{2.7}$$

where E_d^0 is the standard decomposition voltage in volts, p_{H_2} and p_{O_2} are the partial pressure of H_2 and O_2, respectively, in atmospheres, and a_{H_2O} is the activity of H_2O in moles. Assume that $p_{H_2} = 1$ atm, $p_{O_2} = 1$ atm, and $a_{H_2O} = 1M$; we then have $E_d = E_d^0$. The second term of the right-hand side in Eq. (2.7) represents the deviation from the standard state. Note that the term in the brackets should be dimensionless because RT/nF has a unit of voltage. Therefore, the partial pressure and the activities of the chemical species in the equation must be represented by the reduced form such as

$$\bar{p} = p/p^0 \quad \text{and} \quad \bar{a} = a/a^0$$

Generally, we use $p^0 = 1$ atm and $a^0 = 1M$ as standard, and hence we have $\bar{p} = p$ and $\bar{a} = a$. In Eq. (2.7), the bar on p and a is dropped for simplicity. E_d^0 is a function of the temperature:

$$E_d^0 = -\frac{\Delta G^0}{nF} = \frac{RT \ln K_p}{nF} \tag{2.8}$$

where ΔG^0 is the standard Gibbs free energy change, and K_p is the equilibrium constant at constant pressure. Now we may calculate ΔG or E_d at given conditions of the NaOH concentration and temperature. According to thermodynamics

$$\Delta G_T^0 = \Delta H_T^0 - T \Delta S_T^0 \tag{2.9}$$

$$\Delta H_{T_2}^0 = \Delta H_{T_1}^0 + \int_{T_1}^{T_2} \Delta C_p \, dT \tag{2.10}$$

$$\Delta S_{T_2}^0 = \Delta S_{T_1}^0 + \int_{T_1}^{T_2} \left(\frac{\Delta C_p}{T} \right) dT \tag{2.11}$$

where C_p is the specific heat at constant pressure in cal/mol K, and ΔC_p is as follows[1,2]:

$$\Delta C_p = C_p(H_2) + \tfrac{1}{2} C_p(O_2) - C_p(H_2O) \tag{2.12}$$

$$C_p(H_2) = 6.62 + 0.00081 \, T \tag{2.13}$$

$$C_p(O_2) = 8.27 + 0.000258 \, T - \frac{187{,}700}{T^2} \tag{2.14}$$

$$C_p(H_2O) = 1.5928493 - 0.0038828 \, T + 0.000006335 \, T^2 \tag{2.15}$$

Therefore

$$\Delta C_p = 9.16_2 + 0.004821_8\, T - 0.000006335\, T^2 - \frac{93,850}{T^2} \qquad (2.16)$$

Thus, we have

$$\int_{T_1}^{T_2} \Delta C_p\, dT = 9.16_2(T_2 - T_1) + 0.002411(T_2^2 - T_1^2)$$

$$- 0.000002111(T_2^3 - T_1^3) - 93,850\left(\frac{1}{T_2} - \frac{1}{T_1}\right) \qquad (2.17)$$

$$\int_{T_1}^{T_2} \frac{\Delta C_p}{T}\, dT = 9.16_2(\ln T_2 - \ln T_1) + 0.004821_8(T_2 - T_1)$$

$$- 0.000003168(T_2^2 - T_1^2) + 46,925\left(\frac{1}{T_2^2} - \frac{1}{T_1^2}\right) \quad (2.18)$$

Data on ΔH, ΔG, and ΔS can be obtained from Ref. 3:

$$\Delta H_{298}^0 = -68.31 \text{ kcal/mol}$$

$$\Delta G_{298}^0 = -56.72 \text{ kcal/mol}$$

$$S_{298}^0(H_2) = 31.23 \text{ cal/mol deg}$$

$$S_{298}^0(O_2) = 49.02 \text{ cal/mol deg}$$

$$S_{298}^0[H_2O(1)] = 15.9 \text{ cal/mol deg}$$

and hence

$$\Delta S_{298}^0 = 15.9 - 31.23 - \tfrac{1}{2} \times 49.02 = -39.84 \text{ cal/mol deg}$$

Substituting them into Eqs. (2.9)–(2.11), we have

$$\Delta S_{353}^0 = -41.192 \text{ cal/mol deg}$$

$$\Delta H_{353}^0 = -67.80 \text{ kcal/mol}$$

$$\Delta G_{353}^0 = \Delta H_{353}^0 - 353.15\, \Delta S_{353}^0$$

$$= -67.80 - 353.15 \times (41.192) \times 10^{-3} = -53.25 \text{ (kcal/mol)}$$

$$E_{d(353)}^0 = -\frac{\Delta G_{353}^0}{2F} = \frac{53.25}{2 \times 23.06} = 1.15_4 \text{ (V)}$$

We may discuss the logarithmic term of the right-hand side of Eq. (2.7).

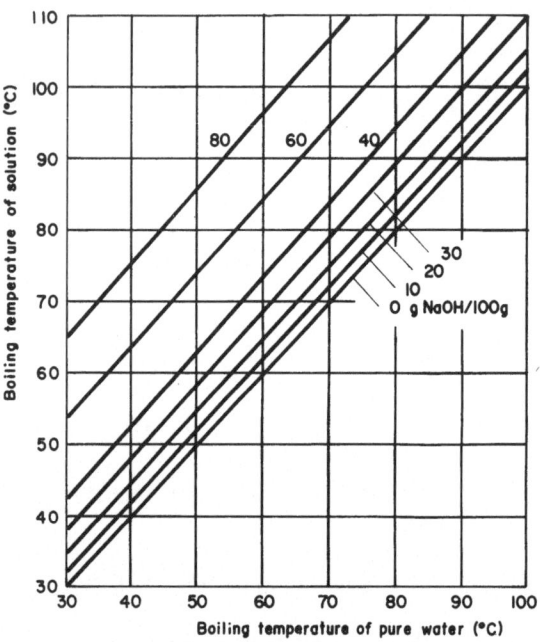

FIGURE 2.1. Dühring diagram of NaOH solution.

It is clear that the activity of water, a_{H_2O}, is shown by the ratio of the water vapor pressure over the NaOH solution, p_{H_2O}, to that of pure water, $p^0_{H_2O}$:

$$a_{H_2O} = \frac{p_{H_2O}}{p^0_{H_2O}} \tag{2.19}$$

p_{H_2O} can be evaluated from the Dühring diagram (Fig. 2.1), while $p^0_{H_2O}$ is obtained from standard reference source.[4] At 80°C, $p^0_{H_2O} = 355.1$ mm Hg and $p_{H_2O} = 289.1$ mm Hg $= 0.380$ atm. Therefore,

$$a_{H_2O} = \frac{289.1}{355.1} = 0.814$$

Hydrogen and oxygen are liberated in the water electrolysis cell at the cathode and the anode, respectively, and those gases are separated by a diaphragm. Thus, the catholyte and the anolyte are saturated with hydrogen and oxygen, respectively. Assume that both hydrogen and oxygen are saturated with water vapor and the total pressure is constant at 1 atm. Then we estimate that

$$p_{H_2} = 1 - p_{H_2O} = 0.620 \text{ (atm)}$$
$$p_{O_2} = 1 - p_{H_2O} = 0.620 \text{ (atm)} \tag{2.20}$$

Substituting these data in Eq. (2.7)

$$E_d = 1.154 + \frac{1.987 \times 353.15}{2 \times 23{,}060} \ln \frac{0.620(0.620)^{1/2}}{0.814}$$

$$= 1.154 - 0.008 = 1.146 \ (V) \tag{2.21}$$

2.2. REVERSIBLE POTENTIAL

2.2.1. Example 1: Anodic and Cathodic Reactions of HCl Electrolysis

Chlorine and hydrogen may be generated at the graphite anode and the cathode, respectively, when a hydrochloric acid solution is electrolyzed in a cell having a diaphragm separator to avoid mixing of both gases, as shown in Fig. 2.2. In other words, in a HCl solution the transfer of charge of H^+ to form H_2 occurs at the cathode, while at the anode, the charge is transferred to form Cl_2. The free energy for the electrode reaction in the HCl cell is as follows:

$$\Delta G^0, \text{kcal/mol}$$

		ΔG^0, kcal/mol	
At the anode:	$Cl^- = \frac{1}{2} Cl_2(g) + e$	-31.350	(2.22)
At the cathode:	$\frac{1}{2}H_2(g) = H^+ + e$	0	(2.23)
Overall reaction:	$H^+ + Cl^- = \frac{1}{2}H_2(g) + \frac{1}{2}Cl_2(g)$	-31.350	(2.24)

Thus we have the Gibbs free energy change of -31.350 kcal/mole for the Cl^-/Cl_2 electrode reaction, and hence the chlorine electrode potential, E_{Cl}^0, is calculated as follows:

$$E_{Cl}^0 = \frac{31.350}{23.06} = 1.359 \ (V)$$

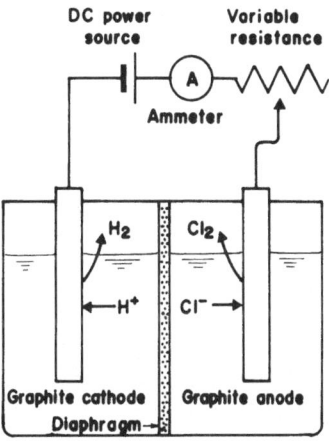

FIGURE 2.2. HCl electrolyzer.

The physical meaning of this potential is the potential difference between the electrode and the solution in contact with each other in a HCl solution saturated with chlorine at 1 atm partial pressure, in which the activity of Cl^- is unity in the equilibrium state.

The Gibbs free energy change with respect to reaction (2.22) is

$$\Delta G^0_{Cl^-/Cl_2} = \mu^0_{Cl^-} - \tfrac{1}{2}\mu^0_{Cl_2} - \mu_e \tag{2.25}$$

where μ_i is the chemical poential of species i. For the hydrogen electrode reaction (2.23)

$$\Delta G^0_{H_2/H^+} = \tfrac{1}{2}\mu^0_{H_2} - \mu^0_{H^+} - \mu_e = 0 \tag{2.26}$$

in a solution in which $a_{H^+} = 1\,M$ and $p_{H_2} = 1$ atm, by convention. This is a constitution of the basis of electrode potential and, hence, we call it the *standard hydrogen electrode* (SHE). From Eqs. (2.25) and (2.26)

$$\Delta G^0_{Cl^-/Cl_2} - \Delta G^0_{H_2/H^+} = \Delta G^0_{Cl^-/Cl_2} = \mu^0_{Cl^-} - \tfrac{1}{2}\mu^0_{Cl_2} - \tfrac{1}{2}\mu^0_{H_2} + \mu^0_{H^+}$$

Now we have set the Gibbs free energy change and the equilibrium potential, or reversible potential, of the chlorine electrode reaction to be -31.350 kcal/mol and 1.359 V, respectively.

The preceding explanations are valid for the special condition that the activity of a species is unity. But we can obtain ΔG for the electrode reaction

$$M = M^{n+} + ne \tag{2.27}$$

under arbitrary conditions of activity with the well-known relationship:

$$\Delta G = -RT \ln K = \Delta G^0 - RT \ln \left(\frac{a_{M^{n+}}}{a_M}\right) \tag{2.28}$$

where K is the equilibrium constant. Substituting Eq. (2.28) into Eq. (2.2)

$$E = \frac{1}{nF}\left[-\Delta G^0 + RT \ln \left(\frac{a_{M^{n+}}}{a_M}\right)\right] = E^0 + \frac{RT}{nF} \ln \left(\frac{a_{M^{n+}}}{a_M}\right) \tag{2.29}$$

This equation is named the *Nernst equation*, one of the most important relationships in electrochemistry. When the activity of all species concerned is unity, the logarithmic term of the right-hand side of this equation becomes zero. E^0 is the specific potential of the reaction under discussion and is only a function of temperature. E^0 is called the *standard potential*.

The basis of the electrode potential is the standard potential of the hydrogen electrode reaction at the same temperature. Of course, the standard potential of hydrogen electrode varies with temperature, but the hydrogen electrode potential should be the basis in any case, by convention.

Since ΔG^0 and E^0 can be obtained easily from tabulations, we should keep in mind the coefficient of the second term of the right-hand side of the Nernst equation for convenience in further discussion:

$$R = 1.987 \text{ cal/mol K}$$

$$F = 23,060 \text{ cal/V g-eq.}$$

$$T = 298 \text{ K}$$

$$\ln x = 2.303 \log x$$

and hence,

$$(RT/F) \times 2.303 = 0.059 \text{ V at } 25°C$$

For example, we may calculate the reversible potential of the chlorine electrode [Eq. (2.22)] in $6\,M$ HCl at 25°C. Assume that $p_{Cl_2} = 0.8$ atm, $\gamma_{\pm} = 3.22$, $E^0_{Cl} = 1.359$ V vs. SHE. Therefore,

$$a_{Cl^-} = C_{Cl^-} \cdot \gamma_{\pm} = 6 \times 3.22 = 19.32$$

Substituting these data into the Nernst equation:

$$E_{Cl} = E^0_{Cl} + \frac{RT}{F} \ln \frac{(p_{Cl_2})^{1/2}}{a_{Cl^-}}$$

we have

$$E_{Cl} = 1.359 + 0.059(\tfrac{1}{2} \log 0.8 - \log 19.32) = 1.280 \text{ V vs. SHE}$$

where V vs. SHE means the potential in volts referred to the standard hydrogen electrode.

One might ask how nonconductive hydrogen gas takes part in an electrochemical reaction. Chlorine gas is also nonconductive. In the case of gas electrodes, an auxiliary metallic electrode, such as platinum, plays a part, as shown in Fig. 2.3. To some extent hydrogen dissolves in solution, and the concentration of dissolved hydrogen, C_{H_2}, is proportional to the partial pressure, p_{H_2}, in the gas phase in equilibrium with solution:

$$C_{H_2} = Hp_{H_2} \tag{2.30}$$

Equation (2.30) is named Henry's law, and H is the Henry coefficient, which is a function of the solution composition and the temperature. Since the surface of the platinum electrode in the solution is catalytically active, a part of dissolved hydrogen is adsorbed on it in atomic form, named the *adatom*, written as H-M or H_{ad}. The adsorbed hydrogen is an intermediate state and

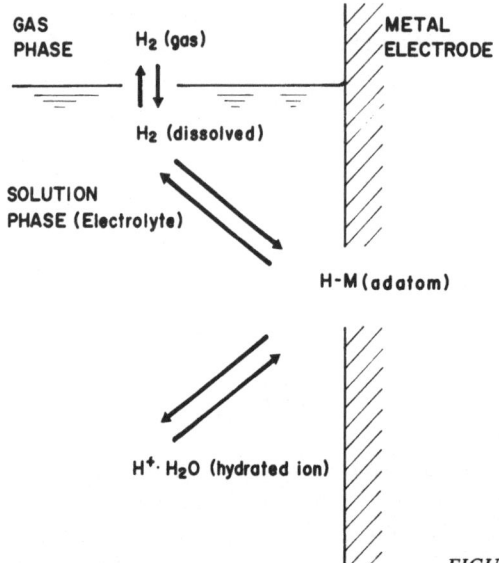

GAS
PHASE   H₂ (gas)   METAL
                                                      ELECTRODE

H₂ (dissolved)

SOLUTION
PHASE (Electrolyte)

H-M (adatom)

H⁺· H₂O (hydrated ion)

FIGURE 2.3. Model for hydrogen electrode.

can react with water molecule in the solution. Thus, we have several steps between gaseous H_2 and H^+ in the solution:

$$H_2(\text{gas}) \rightleftarrows H_2(\text{dissolved}) \rightleftarrows H\text{-M(adsorbed)} \rightleftarrows H^+ \cdot H_2O$$

The reversible electrochemical reaction may consist of

$$H_2(\text{dissolved}) \rightleftarrows 2H^+ + 2e$$

Since determination of the activity of H_2 (dissolved) is difficult, we use the more convenient form

$$H_2(\text{gas}) \rightleftarrows 2H^+ + 2e \tag{2.31}$$

since a state of equilibrium is still indicated, using Eq. (2.30).

The hydrogen electrode is useful for electrochemistry because the reaction attains equilibrium quickly and exactly, its potential is located near the center of the series of electrode potentials, and the hydrogen electrode process is a common one, being sometimes in competition with the electrode process of interest in aqueous solution.

Again, we may consider the Nernst equation for the hydrogen electrode, Eq. (2.23) (with squared terms for convenience):

$$E_H = E_H^0 + \frac{RT}{2F} \ln \frac{(a_{H^+})^2}{p_{H_2}} \tag{2.32}$$

By convention, $E_H^0 = 0$. Assume that $p_{H_2} = 1$ atm. Then we have

$$E_H = -\frac{2.303\,RT}{F}\,(\text{pH}) \tag{2.33}$$

At 25°C

$$E_H = -0.059\,(\text{pH}) \tag{2.34}$$

The hydrogen electrode potential is proportional to the solution pH with a coefficient of -0.059 V/pH at 25°C.

2.2.2. Example 2: Hydrogen Electrode Reaction versus Oxygen Electrode Reaction

When a dilute sulfuric acid solution is electrolyzed with lead electrodes used for both anode and cathode, oxygen evolves at the anode and hydrogen at the cathode.

		ΔG^0, kcal/mol	
At the anode:	$H_2O = 2H^+ + \frac{1}{2}O_2 + 2e$	-56.690	(2.35)
At the cathode:	$H_2 = 2H^+ + 2e$	0.000	(2.36)
Overall reaction:	$H_2O = H_2 + \frac{1}{2}O_2$	-56.690	(2.37)

With the assumption of unit activity of species in this equation, the decomposition voltage at 25°C is

$$E_d = \frac{56.690}{2 \times 23.06} = 1.229\ (\text{V})$$

or the standard potentials of both anodic and cathodic processes are

$$E_A^0: \qquad 1.229\ \text{V vs. SHE}$$
$$E_C^0: \qquad 0.000\ \text{V}$$
$$E_d = E_A^0 - E_C^0: \qquad 1.229\ \text{V}$$

We can use carbon steel as the construction material for cell parts, including electrodes, if the electrolyte is caustic soda solution of adequate concentration and temperature instead of lead electrodes in sulfuric acid, because steel is corrosion resistant to caustic soda. This is of great advantage in industrial application compared with the cell using an acid electrolyte in which only lead and lead alloy can be used. Accordingly, caustic soda or potash are used in conventional water electrolyzers. Now we may calculate the reversible potentials of both anode and cathode processes and the decomposition voltage of the alkaline water electrolyzer.

$$\Delta G^0, \text{kcal/mol}$$

		ΔG^0, kcal/mol	
At the anode:	$2OH^- = H_2O + \frac{1}{2}O_2 + 2e$	-18.500	(2.38)
At the cathode:	$H_2 + 2OH^- = 2H_2O + 2e$	38.190	(2.39)
Overall reaction:	$H_2O = H_2 + \frac{1}{2}O_2$	-56.690	(2.40)

The overall reaction (2.40) is, of course, the same as that of the acid cell shown by Eq. (2.37); thus, the Gibbs free energy change and the decomposition voltage should be the same.

In an acid solution, abundant hydrogen ions may come to the cathode region and become reduced to hydrogen molecules. On the other hand, anodic oxidation of SO_4^{2-} occurs with difficulty under the operating conditions of a water electrolyzer.[†] In this case, H_2O is the only candidate for anodic oxidation to form oxygen.

On the other hand, in an alkaline solution OH^- can be oxidized at the anode to produce oxygen, but neither Na^+ nor K^+ is reduced at the iron cathode. Thus, H_2O is cathodically reduced to hydrogen.

Of course, both reactions (2.35) and (2.38) are called the *oxygen electrode process*, and reactions (2.36) and (2.39) are called the *hydrogen electrode process*. However, the respective reactions differ, so the reversible potentials and the free energy changes differ, as shown in Table 2.2.

Consider the Nernst equation for Reaction 2.35:

$$E = E^0 + \frac{RT}{2F} \ln \frac{(a_{H^+})^2 (p_{O_2})}{a_{H_2O}}$$

[†] SO_4^{2-} can be oxidized on the platinum anode at very high current density or anodic potential to form $S_2O_8^{2-}$. Addition of a small amount of HF is favorable for the reaction. It is the basis for producing H_2O_2 via $S_2O_8^{2-}$ by an electrochemical route.

TABLE 2.2

Standard Potentials of Oxygen and Hydrogen
Electrode Reactions at 25°C

Electrode processes	ΔG^0, kcal/mol	E^0, V vs. SHE
$H_2O = 2H^+ + \frac{1}{2}O_2 + 2e$	-56.690	1.229
$2OH^- = H_2O + \frac{1}{2}O_2 + 2e$	-18.500	0.401
$H_2 = 2H^+ + 2e$	0.000	0.000
$H_2 + 2OH^- = 2H_2O + 2e$	38.190	-0.828

For convenience sake, the oxygen partial pressure p_{O_2} is assumed to be 1 atm, and the H_2O activity is kept constant. Thus we have

$$E = E^0 + \frac{RT}{F} \ln a_{H^+} \qquad (2.41)$$

Since

$$K_w = a_{H^+} \cdot a_{OH^-} = 10^{-14} \text{ at } 25°C$$

a_{H^+} in the alkaline solution of $a_{OH^-} = 1\,M$ is $10^{-14}\,M$. Substituting it into Eq. (2.41)

$$E = 1.229 + 0.059 \log(10^{-14})$$

$$= 1.229 - 0.059 \times 14 = 0.403 \text{ (V vs. SHE at } 25°C)$$

This corresponds to E^0 for reaction (2.38). In other words, reaction (2.38) is a special case of reaction (2.35) at $a_{H^+} = 10^{-14}\,M$.

The Nernst equation for reaction (2.38) is

$$E_B = E_B^0 + \frac{RT}{2F} \ln \frac{a_{H_2O} p_{O_2}}{(a_{OH^-})^2}$$

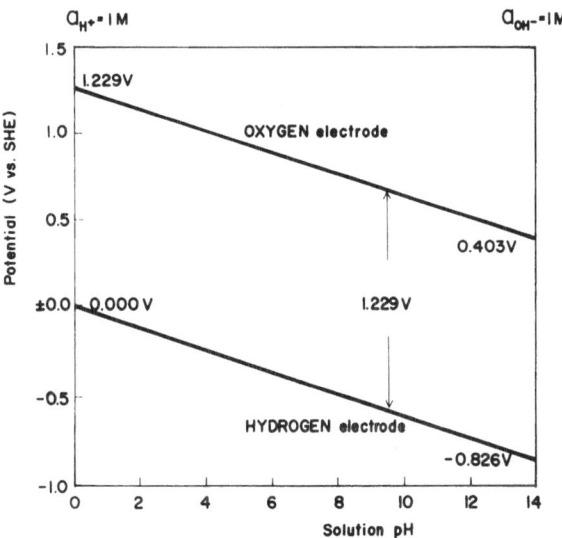

FIGURE 2.4. Relationship between the equilibrium potential and the solution pH of the oxygen electrode and the hydrogen electrode at 25°C.

and for simplicity,

$$E_B = E_B^0 - \frac{RT}{F} \ln a_{OH^-}$$

$$E_B^0 = 0.401 \text{ V vs. SHE at } 25°C$$

(2.42)

where E_B represents the potential in basic solution.

Equations (2.41) and (2.42) are conveniently represented in the potential–pH diagram shown in Fig. 2.4. Similar discussion applies to the hydrogen electrode [reactions (2.36) and (2.39)], where the potential is a linear function of solution pH with slope 0.059 V/pH, as shown in Fig. 2.4. It is clear that the potential vs. pH curves for the oxygen electrode reaction and for the hydrogen electrode reaction are parallel to each other. The difference between the two lines is always 1.229 V at 25°C, independent of the solution pH because the overall reaction of water electrolysis [Eqs. (2.37) and (2.40)] does not change with the pH.

2.2.3. Example 3: Chlorine Electrode Process

The reversible potential of the chlorine electrode process

$$2Cl^- = Cl_2 + 2e$$

(2.43)

is as follows:

$$E_{Cl} = E_{Cl}^0 + \frac{RT}{F} \ln \frac{(p_{Cl_2})^{1/2}}{a_{Cl^-}}$$

(2.44)

where a_{Cl^-} is the Cl^- activity, and p_{Cl_2} the partial pressure of chlorine.

We may calculate E_{Cl} in 4.8 M NaCl at 80°C. The thermodynamic data at arbitrary temperatures are difficult to find in tables, thus we need to evaluate them with adequate equations of thermodynamics. The reversible potential is a function of temperature, t°C or T K, so that we may expand $E(t)$ as a Taylor's series [5]:

$$E(t)^0 = E(25)^0 + (t - 25)\frac{\partial E^0}{\partial T} + \frac{1}{2}(t - 25)^2 \frac{\partial^2 E^0}{\partial T^2}$$

(2.45)

From the table

$$E_{Cl}(25)^0 = +1.3595 \text{ V vs. SHE}$$

$$\frac{\partial E_{Cl}^0}{\partial T} = -1.260 \text{ mV/deg}$$

$$\frac{\partial^2 E_{Cl}^0}{\partial T^2} = -5.454 \ \mu\text{V/deg}^2$$

Substituting these values into Eq. (2.45), we have

$$E_{Cl}(80)^0 = +1.2902 \text{ V vs. SHE}$$

Now we may consider

$$\Delta H = \Delta G + T \Delta S \tag{2.46}$$

and

$$E = -\frac{\Delta G}{nF} \tag{2.47}$$

Differentiating Eq. (2.47) by T at constant pressure gives

$$\frac{\partial E}{\partial T} = -\frac{1}{nF} \left[\frac{\partial \Delta H}{\partial T} - T \frac{\partial \Delta S}{\partial T} - \Delta S \right]_p$$

By definition, we have

$$\left[\frac{\partial \Delta H}{\partial T} \right]_p = \Delta C_p \tag{2.48}$$

$$\left[\frac{\partial \Delta S}{\partial T} \right]_p = \frac{\Delta C_p}{T} \tag{2.49}$$

and, hence, substituting (2.48) and (2.49) in the expression for $\partial E/\partial T$

$$\frac{\partial E}{\partial T} = \frac{\Delta S}{nF} \tag{2.50}$$

$$\frac{\partial^2 E}{\partial T^2} = \frac{1}{nF} \left[\frac{\partial \Delta S}{\partial T} \right]_p = \frac{1}{nF} \left[\frac{\Delta C_p}{T} \right]_p \tag{2.51}$$

From Eqs. (2.46), (2.47), and (2.50), we have

$$\Delta H = -nF \left[E - T \left(\frac{\partial E}{\partial T} \right) \right] \tag{2.52}$$

This equation is important and is named the *Gibbs–Helmholtz equation*.

Since the first temperature derivation of the reversible potential of the chlorine electrode process is negative, ΔS is negative from Eq. (2.50). Therefore, reaction (2.43) would not occur spontaneously. When reaction (2.43) proceeds to the right-hand side with infinitesimal or negligibly small current, the Gibbs free energy change ΔG is larger than the enthalpy change ΔH. Thus, we need to supply external heat $w = \Delta G - \Delta H$ so as to keep the temperature constant. It is clear that $(\partial E/\partial T)_p$, ΔS, and ΔC_p are important with regard to the direction and possibility of reaction.

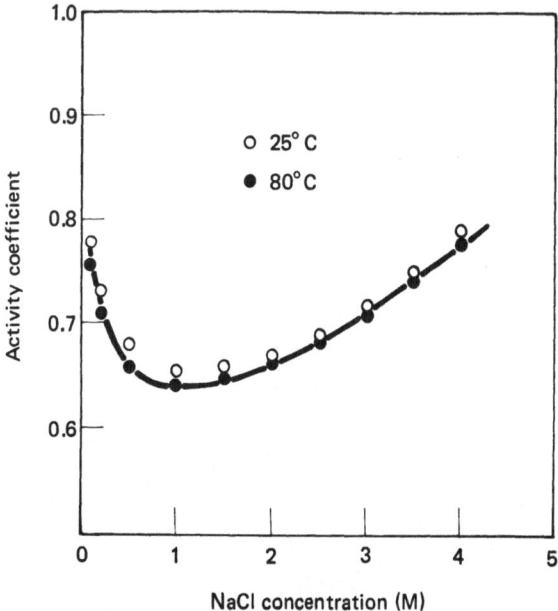

FIGURE 2.5. The mean activity coefficient of NaCl solution.

Figure 2.5 shows the mean activity coefficient $\gamma_\pm$ of NaCl solution.[6] It is clear that $\gamma_\pm$ is a weak function of temperature. From this figure, $\gamma_\pm$ of 4.8 M NaCl is about 0.84 at 80°C. The water vapor pressure under the same conditions is obtained from the Dühring diagram to be about 447 mm Hg; hence, p_{Cl_2} is estimated to be 0.412 atm. Using Eq. (2.45) and substituting those data into Eq. (2.44), we have

$$E_{Cl} = 1.2902 + \frac{1.987 \times 353.15}{23,060} \ln \frac{(0.412)^{1/2}}{4.032} = 1.2342 \text{ V}$$

2.2.4. Example 4: Sodium Amalgam Electrode

The cathodic reaction of the amalgam-type chlorine cell is as follows:

$$Na(Hg) = Na^+ + (Hg) + e \qquad (2.53)$$

Therefore, the reversible potential for this reaction can be shown by Eq. (2.54):

$$E_{Na(Hg)} = E^0_{Na(Hg)} + \frac{RT}{F} \ln \frac{a_{Na^+}}{a_{Na(Hg)}} \qquad (2.54)$$

Let us consider $E^0_{Na(Hg)}$, the standard potential for reaction (2.53). Amalgam is a mercury solution of sodium:

$$Na + (Hg) = Na(Hg) \qquad (2.55)$$

$$\underline{Na(Hg) = Na^+ + (Hg) + e} \qquad (2.53)$$

$$Na + (Hg) = Na^+ + (Hg) + e \qquad (2.56)$$

Reaction (2.55) is dissolution of sodium into mercury, and reaction (2.53) is clearly a charge transfer process between Na and Na^+. Therefore, we can obtain ΔG for reaction (2.53) as a difference of ΔG's of reactions (2.55) and (2.56).

For convenience sake, we may consider a different method. When sodium amalgam having the activity of $[a_{Na(Hg)}]_0$ is in contact with a sodium chloride solution of unit activity, $a_{Na^+} = 1$, in equilibrium, the reversible potential of reaction (2.53) is shown by Eq. (2.57) or (2.58):

$$[E_{Na(Hg)}]_0 = E^0_{Na(Hg)} + \frac{RT}{F} \ln \frac{1}{[a_{Na(Hg)}]_0} \qquad (2.57)$$

$$E_{Na(Hg)} = [E_{Na(Hg)}]_0 + \frac{RT}{F} \ln \frac{a_{Na^+}}{Y} \qquad (2.58)$$

where

$$Y = \frac{a_{Na(Hg)}}{[a_{Na(Hg)}]_0} \qquad (2.59)$$

That is, Eq. (2.58) shows the reversible potential of the sodium amalgam electrode of any concentration or activity referred to the electrode consisting of a specific concentration or activity. For example, the potential of the 0.1 w/o Na-amalgam at 80°C is

$$[E_{Na(Hg)}]_0 = -1.83_4 \text{ V vs. SHE} \qquad (2.60)$$

The relative activity of sodium amalgam Y, referred to 0.1 w/o Na–Hg at 80°C, is shown in Fig. 2.6.[7] In a practical cell, the amalgam concentration is in the range 0.1–0.3 w/o Na, mostly 0.2 w/o. At 0.2 w/o Na, $Y \simeq 3.8$. Thus

$$E_{Na(Hg)} = -1.834 + \frac{1.987 \times 353.15}{23,060} \ln \frac{4.032}{3.8} = -1.832 \text{ V vs. SHE}$$

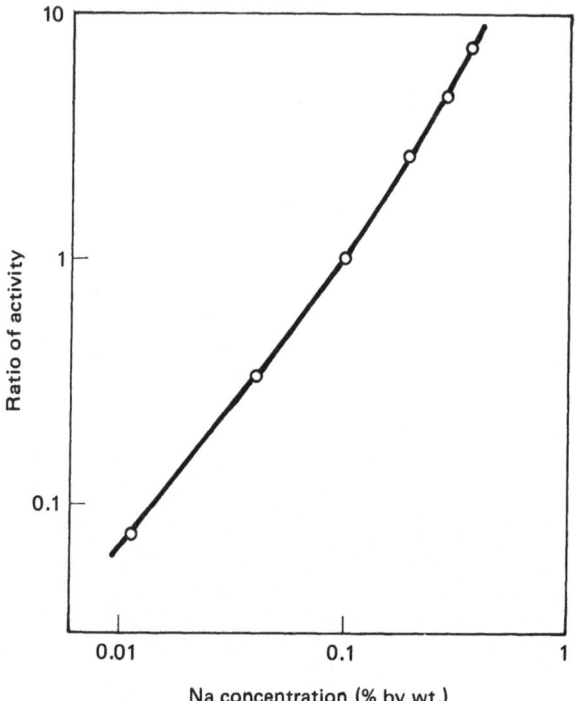

FIGURE 2.6. The ratio of the Na activity in sodium amalgam at 80°C referred to 0.1% Na-amalgam.

2.2.5. Example 5: Solid-Phase Electrodes of Mercury Compounds

The calomel electrode is frequently used as reference for potential measurements. This electrode consists of

$$Hg/Hg_2Cl_2(s)/KCl(aq) \qquad (2.61)$$

The solubility of Hg_2Cl_2 in a KCl solution is very small. The electrode reaction and the reversible potential can be written as follows:

$$Hg + Cl^- = \tfrac{1}{2}Hg_2Cl_2 + e \qquad (2.62)$$

$$E = E^0 - \frac{RT}{F}\ln(a_{Cl^-}) \qquad (2.63)$$

Why does the solid compound behave as the active material for the electrode process? In a KCl solution, a very small amount of Hg_2^{2+} from Hg_2Cl_2 may exist, and Hg_2^{2+} forms a reversible electrode with Hg as indicated in reaction (2.65):

$$Hg_2Cl_2 = Hg_2^{2+} + 2Cl^- \qquad (K_s = 2 \times 10^{-18} \text{ at } 25°C) \qquad (2.64)$$

$$Hg = \tfrac{1}{2}Hg_2^{2+} + e \qquad (2.65)$$

The solubility product of Hg_2Cl_2, K_s, and the Nernst equation for reaction (2.65) are as follows:

$$K_s = (a_{Hg_2^{2+}})(a_{Cl^-})^2 \qquad (2.66)$$

$$E = E_1 + \frac{RT}{2F}\ln(a_{Hg_2^{2+}}) \qquad (2.67)$$

Substituting Eq. (2.66) into Eq. (2.67)

$$E = \underbrace{E_1 + \frac{RT}{2F}\ln(K_s)}_{E^0} - \frac{RT}{F}\ln(a_{Cl^-}) \qquad (2.68)$$

K_s is a function of the temperature, so that we may put E^0 for the sum of the first and the second terms of the right-hand side of Eq. (2.68), which thus agrees with Eq. (2.63).

Other compounds of mercury such as HgO and Hg_2SO_4 can be used in the same way. The electrode reaction and the Nernst equation of the system

$$Hg/HgO/KOH \qquad (2.69)$$

are represented as follows:

$$Hg + 2OH^- = HgO(s) + H_2O + 2e \qquad (2.70)$$

$$E = E^0 - \frac{RT}{F}\ln(a_{OH^-}) \qquad (2.71)$$

For the Hg_2SO_4 electrode in H_2SO_4:

$$Hg/Hg_2SO_4(s)/H_2SO_4 \qquad (2.72)$$

the reaction and the potential are

$$2Hg + SO_4^{2-} = Hg_2SO_4 + 2e \qquad (2.73)$$

$$E = E^0 - \frac{RT}{2F}\ln(a_{SO_4^{2-}}) \qquad (2.74)$$

The electrode process consisting of Cu in acidified $CuSO_4$ solution is

$$Cu = Cu^{2+} + 2e \tag{2.75}$$

An electrode process of this kind is called the *metal–metal ion electrode*. We have studied in Examples 1–3 the *gas–ion electrode* such as

$$H_2 = 2H^+ + 2e \tag{2.76}$$

and

$$2Cl^- = Cl_2 + 2e \tag{2.77}$$

in which case, gaseous material behaves on the auxiliary metallic electrode as an active material. In this section, we have discussed the *solid-phase electrode*. There is also another type of electrode process with ionic species having different valences such as

$$Fe^{2+} = Fe^{3+} + e \tag{2.78}$$

It is named the *redox-type electrode*.

2.2.6. Example 6: The Pourbaix Diagram for the Iron–Water System

Since H^+, OH^-, and/or H_2O are concerned in the electrode process occurring in aqueous solutions, either directly or indirectly, the Nernst equation for the reversible potential of the electrode process involves the activity of those species, so that in many cases the potential varies with the solution pH, as in Examples 1 and 2.

The solution pH is an important factor for an electrode process, and therefore the pH dependency of the reversible potential is discussed frequently, especially in the corrosion science of metals in aqueous solutions.

It is well known that dissolved oxygen is also a major factor for metallic corrosion because it is a strong oxidant, and the oxygen electrode process is affected directly by pH. The reversible potential for the oxygen electrode, of course, depends on the pH. The reversible potential of the metal oxide electrode also varies with the solution pH.

On the other hand, the reversible potential of the electrode process

$$Fe = Fe^{2+} + 2e \tag{2.79}$$

is independent of the solution pH while it varies with the activity of Fe^{2+}

$$E_{Fe/Fe^{2+}} = E^0_{Fe/Fe^{2+}} + \frac{RT}{2F} \ln(a_{Fe^{2+}}) \tag{2.80}$$

In an alkaline solution, ferrous ion (dihypoferrite), $HFeO_2^-$, forms. The electrode process and the Nernst equation in this case are as follows:

$$Fe + 3OH^- = HFeO_2^- + H_2O + 2e \qquad (2.81)$$

$$E_{Fe}(B) = E_{Fe}(B)^0 - 0.089\ pH + 0.030\ \log(a_{HFeO_2^-}) \qquad (2.82)$$

It is evident that this potential is a function of the pH.

These examples show that the reversible potential of the electrode process depends on the solution pH. Consequently, the reversible potentials for various metals have been plotted as a function of the solution pH by Pourbaix.[8] The Pourbaix diagrams contain the curves for both hydrogen and oxygen electrodes because they are important for a discussion of the corrosion of metals in an aqueous solution, as stated above.

As shown in Eqs. (2.80) and (2.82), for example, the reversible potential is also affected by the activity of the ionic species in the solution. Let us consider that a piece of iron is immersed in an acidified and stagnant solution. Iron is corroded, and Fe^{2+} leaves from the metal surface to the bulk of solution by diffusion. The diffusion rate varies with the physicochemical properties of the solution and the operating conditions such as the temperature and flow pattern of the solution. The metal is thought to be resistant enough if its corrosion rate is 5×10^{-3} in./yr (ipy) or less,[9] which corresponds to a current density of about 10^{-7} A/cm^2. The ionic species formed are brought to the bulk of stagnant solution by diffusion, so that the concentration of ions at the surface of a metal is estimated to be about 10^{-5} M as Fe^{2+} in the steady state, with some assumptions. Consequently, it is thought that Fe is in contact with a solution containing 10^{-5} M Fe^{2+} under a quasi-equilibrium state. Now we can calculate the Nernst equation fully. Pourbaix diagrams are usually constructed with curves not only for 10^{-5} M but for various concentrations, say 10^{-6}, 10^{-7}, and 10^{-8} M. Figure 2.7 is an example of the Pourbaix diagram for the iron–water system.

We may calculate the reversible potentials shown by various lines labelled in Fig. 2.7.

Line 1. The electrode process and the reversible potential at $a_{Fe^{2+}} = 10^{-5}$ M are as follows:

$$Fe = Fe^{2+} + 2e \qquad E_1^0 = -0.440\ V\ vs.\ SHE$$

$$E_1 = E_1^0 + \frac{RT}{2F}\ \ln(a_{Fe^{2+}}) = -0.440 + \frac{0.059}{2}\ \log 10^{-5} = -0.590\ V$$

Since E_1 is independent of the pH, the straight line is parallel to the ordinate. All potentials are referred to the SHE at 25°C.

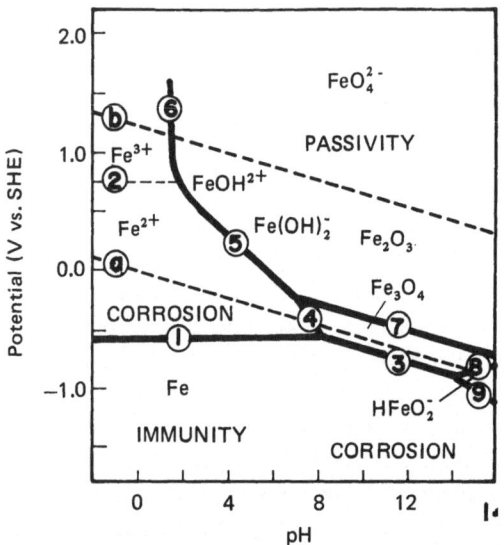

FIGURE 2.7. Pourbaix diagram for the iron–water system at 25°C.

Line 2.

$$Fe^{2+} = Fe^{3+} + e$$

$$E_2^0 = +0.771 \text{ V}$$

$$E_2 = E_2^0 + \frac{RT}{F} \ln \frac{a_{Fe^{3+}}}{a_{Fe^{2+}}}$$

It is clear that $E_2 = E_2^0$ when $a_{Fe^{2+}} = a_{Fe^{3+}} = 10^{-5} M$. The line is parallel to the ordinate.

Line 3.

$$3Fe + 4H_2O = Fe_3O_4 + 8H^+ + 8e$$

$$E_3^0 = -0.085 \text{ V}$$

$$E_3 = E_3^0 + \frac{RT}{F} \ln(a_{H^+}) = -0.085 - 0.059 \text{ pH}$$

where the activities of Fe, H_2O, and Fe_3O_4 are assumed to be kept constant. The E_3 vs. pH curve has a slope of -0.059 V/pH.

Line 4.

$$3Fe^{2+} + 4H_2O = Fe_3O_4 + 8H^+ + 2e$$

$$E_4^0 = +0.980 \text{ V}$$

$$E_4 = E_4^0 + \frac{RT}{2F} \ln \frac{(a_{H^+})^8}{(a_{Fe^{2+}})^3} = +0.980 - 0.236 \text{ pH} - 0.089 \log(a_{Fe^{2+}})$$

That is, the potential is a function of both the solution pH and the Fe^{2+} activity.

Line 5.

$$2Fe^{2+} + 3H_2O = Fe_2O_3 + 6H^+ + 2e$$

$$E_5^0 = +0.728 \text{ V}$$

$$E_5 = E_5^0 + \frac{RT}{F} \ln \frac{(a_{H^+})^3}{a_{Fe^{2+}}} = +0.728 - 0.177 \text{ pH} - 0.059 \log(a_{Fe^{2+}})$$

Line 6.

$$2Fe^{3+} + 3H_2O = Fe_2O_3 + 6H^+$$

This reaction is nonelectrochemical. The equilibrium coefficient K_6 is shown by the next equation:

$$K_6 = \frac{(a_{H^+})^3}{a_{Fe^{3+}}}$$

It is related to Lines 2 and 5, and Line 6 is a curve at $a_{Fe^{3+}} = 10^{-5} M$, which is perpendicular to the pH coordinate.

Line 7.

$$2Fe_3O_4 + H_2O = 3Fe_2O_3 + 2H^+ + 2e$$

$$E_7 = +0.221 \text{ V}$$

$$E_7 = E_7^0 + \frac{RT}{F} \ln a_{H^+} = +0.221 - 0.059 \text{ pH}$$

Line a.   Hydrogen electrode reaction.

$$H_2 = 2H^+ + 2e$$

$$E_a = -0.059 \text{ pH}$$

Line b.   Oxygen electrode reaction.

$$2H_2O = 4H^+ + O_2 + 4e$$

$$E_b^0 = +1.229 \text{ V}$$

$$E_b = E_b^0 + \frac{RT}{F} \ln a_{H^+} = +1.229 - 0.059 \text{ pH}$$

Line 8.   (In alkaline solution.)

$$3HFeO_2^- = Fe_3O_4 + H_2O + OH^- + 2e$$

This reaction corresponds to

$$3HFeO_2^- + H^+ = Fe_3O_4 + 2H_2O + 2e$$

$$E_8^0 = -1.819 \text{ V}$$

$$E_8 = E_8^0 - \frac{RT}{F} \ln(a_{HFeO_2^-})^3 + 0.030 \text{ pH}$$

Line 9.

$$Fe + 3OH^- = HFeO_2^- + H_2O + 2e$$

This corresponds to

$$Fe + 2H_2O = HFeO_2^- + 3H^+ + 2e$$

$$E_9^0 = +0.493 \text{ V}$$

$$E_9 = E_9^0 + \frac{RT}{2F} \ln a_{HFeO_2^-} - 0.088 \text{ pH}$$

In the region surrounded by Lines 1, 3, and 9, Fe is the most stable form. On the other hand, Fe^{2+} and/or Fe^{3+} are stable in a region surrounded by the abscissa and Lines 1, 4, 5, and 6. It means that Fe might be corroded when it is immersed in an acidified solution at potentials more noble than -0.6 V.

The surface of the iron piece would be covered with the oxide film consisting of Fe_3O_4 or Fe_2O_3 in the upper corner of the right-hand side divided by Lines 3–6, and 8, so that the substrate is protected from corrosion. Such an oxide layer is called a *passive film*, which is important for practice. Pourbaix has named those regions in his diagram, *immunity*, *corrosion*, and *passivation*.

The well-known book entitled *Atlas of Electrochemical Equilibria in*

Aqueous Solutions by Pourbaix[8] contains a number of diagrams of various electrochemical systems.

It is important to note that the Pourbaix diagram is based on thermodynamics, but not on kinetics. Therefore, it suggests only a possibility of occurence of the electrochemical process under discussion but does not explain the kinetic aspects of the reaction.

REFERENCES

1. J. H. Perry, *Chemical Engineers' Handbook*, 4th ed., pp. 3–116, McGraw-Hill, New York (1963).
2. *Jitsuyo Kagaku Binran, Jōkan* (Handbook of Practical Chemistry), p. 493, Chem. Ind. Ass'n (1948).
3. Kagaku Binran (Handbook of Chemistry, New Edition), pp. 724 and 741, Maruzen, Tokyo (1958).
4. S. Kamei, *Kagaku-Kikai no Riron to Keisan* (Theory and Calculation for Chemical Process Equipments, New Edition), p. 544, Sangyo Tosho Publishers, Tokyo (1959).
5. G. R. Salvi and A. J. deBethune, *J. Electrochem. Soc.* **108**, 672 (1961).
6. D. Dobos, *Electrochemical Data*, p. 208, Elsevier, New York (1975).
7. S. Yoshizawa and E. Muto, *J. Chem. Soc. Jpn, Ind. Chem. Section* **56**, 387 (1953).
8. M. Pourbaix, *Atlas of Electrochemical Equilibria in Aqueous Solutions*, Pergamon, New York (1969).
9. H. H. Uhlig, ed., *The Corrosion Handbook*, p. 748, Wiley, New York (1969).

KINETICS OF ELECTROCHEMICAL PROCESSES

3.1. THE RATE OF ELECTROCHEMICAL PROCESS

We consider the electrochemical process shown by the formula

$$M \rightleftarrows M^{n+} + ne \tag{3.1}$$

where $\rightleftarrows$ indicates reversibility. According to the theory of chemical processes, the reaction rate which can be measured from the outside is the difference between the forward reaction rate $\vec{v}$ and the backward reaction rate $\bar{v}$, i.e.,

$$v = \vec{v} - \bar{v} \tag{3.2}$$

The reaction will proceed in the forward direction when $v > 0$, and vice versa. It is clear that

$$\vec{v} = \bar{v} = v_0, \qquad \text{at } v = 0 \tag{3.3}$$

v_0 is the exchange reaction rate at equilibrium.

Faraday's law is well known

$$\omega = \frac{M}{nF} I\theta \tag{3.4}$$

where ω is the amount of mass transfer (g), M is the atomic weight (g/mol), n is the number of charge transfer (eq./mol), F is Faraday's constant (26.8 A h/g-eq.), I is the current passed (A), θ is the time of electrolysis (hrs). From Eq. (3.4)

$$i = \frac{I}{A} = nF \frac{\omega}{M} \frac{1}{\theta} \frac{1}{A} = nFv \tag{3.5}$$

where i is the amperage passing through a unit area of electrode surface, and is called the *current density*, and A is the area of the electrode surface. Equation (3.5) shows clearly that the current density is equivalent to the reaction rate of the electrochemical process. Thus, we sometimes call it the electrochemical reaction rate as well as the current density.

Now we may consider the reaction rate as a function of the activity of species of interest a_i, and the free energy change ΔG_i

$$v_i = k_i a_i \exp\left(-\frac{\Delta G_i}{RT}\right) \tag{3.6}$$

where k_i is the specific rate of reaction. The free energy changes for both directions, $\Delta \vec{G}$ and $\Delta \bar{G}$, differ from ΔG in the equilibrium state when the electrode is polarized from the reversible potential E to the potential E' (the potential to which the electrode is polarized) by an outside source of electric power

$$\Delta \vec{G} = \Delta \vec{G}_1 - \alpha n F E' \tag{3.7}$$

$$\Delta \bar{G} = \Delta \bar{G}_1 + (1 - \alpha) n F E' \tag{3.8}$$

where $\Delta \vec{G}_1$ and $\Delta \bar{G}_1$ are terms independent of the potential, and α is named the transfer coefficient to be described later in detail.

The second term of the right-hand side of Eqs. (3.7) and (3.8) is the contribution of polarization to the potential barrier for the reaction. Since $E' = E$ in the equilibrium state, we have

$$\Delta \vec{G}_{eq} = \Delta \vec{G}_1 - \alpha n F E \tag{3.7a}$$

$$\Delta \bar{G}_{eq} = \Delta \bar{G}_1 + (1 - \alpha) n F E \tag{3.8a}$$

where the subscript "eq" refers to equilibrium.

From Eqs. (3.5)–(3.8) for reaction (3.1), there results

$$i = nFk_0 \left[a_M \exp\left(-\frac{\Delta \vec{G}_1 - \alpha n F E'}{RT} \right) \right.$$
$$\left. - a_{M^{n+}} \exp\left(-\frac{\Delta \bar{G}_1 + (1 - \alpha) n F E'}{RT} \right) \right] \tag{3.9}$$

where k_0 is the specific rate of reaction.

At the net current $i = 0$ and the potential $E' = E$, that is, for equilibrium conditions, Eq. (3.9) becomes

$$a_M \exp\left(-\frac{\Delta \vec{G}_1 - \alpha n F E}{RT} \right) = a_{M^{n+}} \exp\left(-\frac{\Delta \bar{G}_1 + (1 - \alpha) n F E}{RT} \right)$$

or

$$E = E^0 + \frac{RT}{nF} \ln \left(\frac{a_{M^{n+}}}{a_M} \right) \tag{3.10}$$

With Eqs. (3.7a), (3.8a), and (3.10)

$$E^0 = \frac{\Delta \vec{G}_1 - \Delta \bar{\bar{G}}_1}{nF} \tag{3.11}$$

where E^0 is the standard potential for reaction (3.1).

Equation (3.10) is in the form of the Nernst equation which we have studied previously [Eq. (2.29)]. With Eqs. (3.9) and (3.10), we have

$$i = i_0 \left\{ \exp \left(\frac{\alpha nF\eta}{RT} \right) - \exp \left[-\frac{(1-\alpha)nF\eta}{FT} \right] \right\} \tag{3.12}$$

$$\eta = E' - E \tag{3.13}$$

and

$$i_0 = nFk_0 a_M \exp \left(-\frac{\Delta \vec{G}_1 - \alpha nFE}{RT} \right) \tag{3.14}$$

It is clear that η is a deviation of the electrode potential from the reversible potential at $i = i$, and is called the *overvoltage*. It is a voltage drop due to the hindrance or difficulty of sustaining the electrode reaction under discussion at the rate or current density i. The term i_0 is the rate of the exchange reaction in the equilibrium state, and is named the *exchange current density*. It is a measure of the reversibility of the reaction concerned, since a large value of i_0 indicates reversible behavior and, according to Eq. (3.12), large currents may be passed with small values of η. Small i_0 values are associated with electrode polarization (large η) and irreversible electrode reactions.

3.2. ELECTRODE PROCESSES CONTROLLED BY A REACTION STEP

3.2.1. Example 1: Hydrogen Electrode Process

The overall reaction of the hydrogen electrode in acidic solution is

$$H_2 = 2H^+ + 2e \tag{3.15}$$

which contains several steps of elementary reactions. The oxidation of H_2 into H^+ is the forward direction and the reduction of H^+ is the backward direction. Here we may consider that the backward reaction (hydrogen formation) is preferred. The direction of the net reaction is shown by the arrow.

Figure 3.1(a) illustrates the overall process of hydrogen electrode from H^+ to H_2. Hydrogen ions in the bulk of the solution are transported to the outside of the double layer near the electrode surface by either diffusion or migration, and are attached and neutralized at the active sites on the electrode to form the activated complex, presumably combined with the metal atom. It might be a form of chemisorbed atoms of hydrogen, although details are unclear, and is named the *adsorbed atom* or simply *adatom*, designated H_{ad} or H-M.

Since the electrode surface is catalytically active, two adatoms of hydrogen, probably neighboring each other, combine to form a H_2 molecule, i.e., the hydrogen formation reaction consists of the three major steps:

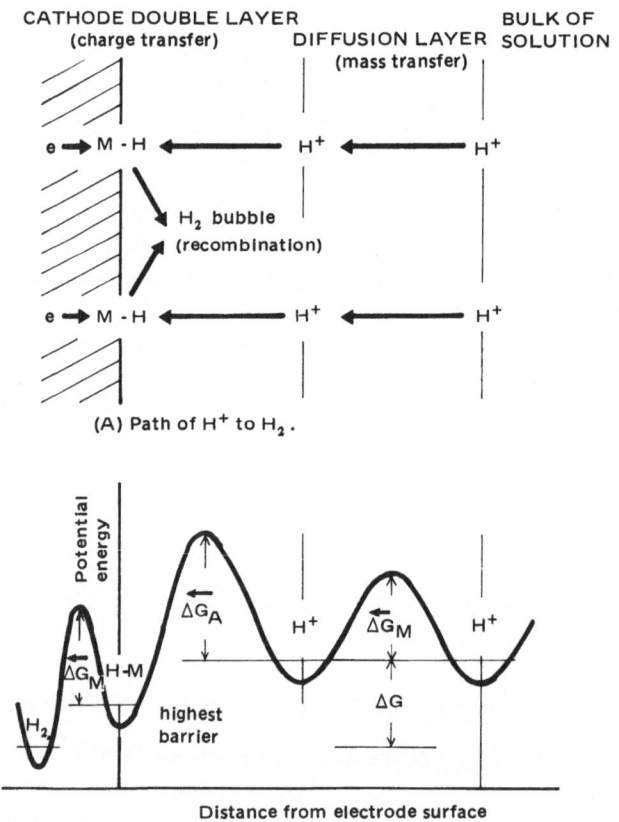

(A) Path of H^+ to H_2.

(B) Schematic profile of potential energy.

FIGURE 3.1. Hydrogen formation and its potential barriers.

(1) Mass transfer: H^+ (double layer) $\leftarrow H^+$ (bulk of solution)   (3.16)

(2) Charge transfer: $H_{ad} \leftarrow H^+ + e$ (3.17)

(3) Recombination: $H_2 \leftarrow H_{ad} + H_{ad}$ (3.18)

Figure 3.1(b) shows the profile of the potential energy near the working electrode. The level of the potential energy for H_2 is sufficiently lower than that of H^+, thus the formation reaction of H_2 will occur preferentially. However, the path is not flat, and the chemical species must climb various mountains and cross several valleys. Of these, the highest peak above the initial energy would be the most difficult barrier to overcome, and is the *rate-determining step* (RDS) for the overall process of hydrogen formation. Generally, the mass transfer step in the solution is considered to be a low hill for hydrogen compared to the charge transfer or the recombination steps.

Now we may assume that the charge transfer step has the highest energy barrier between H^+ and H_2, and is represented by

$$H_{ad} \overset{A}{\leftarrow} H^+ + e \qquad (3.19)$$

where A shows the rate determining step. Figure 3.2 shows a schematic

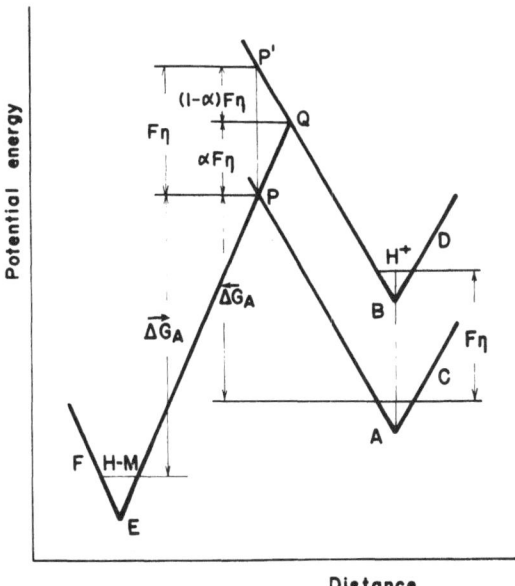

FIGURE 3.2. *Model of potential energy curve.*

profile of the potential barrier vs. distance diagram. The V-shaped line *FEP* shows the potential curve for H_{ad}, and *PAC* for H^+. Assume that only the potential curve for H^+ moves by $F\eta$ when the electrode is polarized cathodically by $-\eta$. The line *QBD* shows the new location of the potential curve for H^+ after polarization. Now the height of the potential barriers for the forward direction $(H^+ \to H_{ad})$ and the backward direction $(H^+ \leftarrow H_{ad})$ become $\Delta \vec{G}_A - \alpha F\eta$ and $\Delta \vec{G}_A + (1-\alpha)F\eta$, respectively, because the intersection of both curves moves from P to Q but not to P' (note that η is negative because the backward reaction is preferential). The transfer coefficient α is the fraction of the energy change $F\eta$ which assists the backward reaction.

With Eq. (3.12)

$$i = i_0 \left\{ \exp\left(\frac{\alpha F\eta}{RT}\right) - \exp\left[-\frac{(1-\alpha)F\eta}{RT}\right] \right\} \tag{3.20}$$

Since

$$\exp\left(\frac{\alpha F\eta}{RT}\right) \ll \exp\left[-\frac{(1-\alpha)F\eta}{RT}\right] \tag{3.21}$$

at $\eta < -70$ mV, Eq. (3.20) becomes

$$i \simeq -i_0 \exp\left[-\frac{(1-\alpha)F\eta}{RT}\right]$$

or

$$-\eta = \frac{RT}{(1-\alpha)F}\left[\ln(-i) - \ln(i_0)\right] = a + b\log(-i) \tag{3.21}$$

where

$$a = -b\log(i_0) \qquad \text{and} \qquad b = 2.303\,RT/(1-\alpha)F$$

At 25°C with an assumption of $\alpha = 0.5$

$$-\eta = 0.120[\log(-i) - \log(i_0)] \tag{3.22}$$

The negative sign for η and i shows cathodic.

It is clear that η is linear with respect to $\log i$ with a slope of 120 mV/decade in the range $\eta < -70$ mV. Equation (3.22) is called the *Tafel equation* in honor of the first investigator of electrode kinetics (1905); $2.3\,RT/(1-\alpha)F$ is called the *Tafel slope*. The exchange current density i_0 can be obtained by extrapolation of the η vs. $\log i$ curve to $\eta = 0$, and is important as a characteristic value of the hydrogen electrode process. It is a

TABLE 3.1
Parameters of the Hydrogen Electrode Process

Electrode	Solution	Temperature (°C)	Tafel slope b (mV)	$-\log i_0$ (A/cm^2)
Ag	$1N$ HCl	RT^a	130 ± 5	3.7 ± 0.4
	$1N$ H$_2$SO$_4$		110	5.7
	$1N$ NaOH	30	120	6.50
Cd	$0.5N$ H$_2$SO$_4$	RT	135 ± 12	10.77 ± 0.75
Cu	$0.1N$ HCl		114 ± 8	6.84
	$0.01N$ NaOH		107 ± 6	6.09
Fe	$0.01N$ HCl		118 ± 15	6.29
	$0.5N$ HCl		133 ± 4	5.18 ± 0.1
	$0.5M$ H$_2$SO$_4$	25 ± 0.03	118	5.65
	$0.1N$ NaOH		120 ± 2	6.06
Hg	$0.1N$ HCl	21 ± 0.1	116	12.11
	$0.1N$ HClO$_4$	25	118	11.82
Ni	$1N$ HCl	20 ± 0.5	109 ± 9	5.4 ± 0.1
	$0.1N$ NaOH	20 ± 0.5	101 ± 4	6.4 ± 0.2
Pb	$1N$ H$_2$SO$_4$		120	12.7
	$0.5N$ NaOH	RT	129 ± 4	6.47 ± 0.26
Pd	$2N$ H$_2$SO$_4$	20	120	3.2
Pt	$0.5N$ H$_2$SO$_4$	RT	30 ± 1	3.53 ± 0.05
	$0.5N$ NaOH	RT	117 ± 8	4.06 ± 0.09
Zn	$1N$ H$_2$SO$_4$	20	120	10.8

^{a} RT = room temperature.

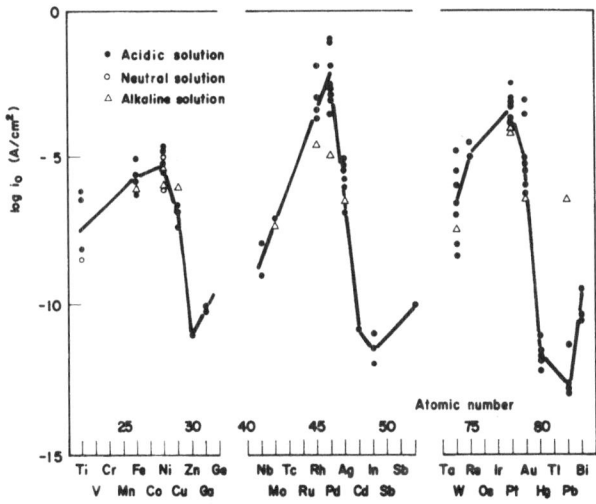

FIGURE 3.3. Exchange current density of the hydrogen electrode process on various metals as a function of the atomic number.

function of the electrode material, the solution composition, the operating conditions such as temperature and pressure, and other factors. Table 3.1 shows i_0 and b for various metals.[1] It is clear that i_0 for Pt, Pd, Ni, Fe, Ag, and Cu is large because those metals are catalytically active for the hydrogen electrode reaction. On the other hand, i_0 for the inactive metals such as Cd, Hg, Pb, and Zn is small.

Kita proposed a significant relationship between the exchange current density of the hydrogen electrode process and the position of metals in the Periodic Table as shown in Fig. 3.3.[1] The metals belonging to the iron–platinum group have large exchange current densities. As is shown in the following section, the hydrogen electrode process in an alkaline solution differs from that in an acidic solution. The Tafel slope of the hydrogen electrode reaction on a Fe and/or Ni cathode in an alkaline solution is about 120 mV/decade, and the current density is almost independent of the solution pH.[2,3] Those results suggest the mechanism of the reaction as follows:

$$H_{ad} + OH^- \leftrightarrow H_2O + e$$

$$\tfrac{1}{2}[H_2 \leftarrow H_{ad} + H_{ad}]$$

$$\overline{\tfrac{1}{2}H_2 + OH^- \leftarrow H_2O + e} \tag{3.23}$$

$$-i = k(a_{H_2O}) \exp[\alpha F(-\eta)/RT] \tag{3.24}$$

3.2.2. Example 3: Oxygen Electrode Process

The oxygen electrode process on a nickel anode has been studied by many authors. Figure 3.4, obtained by Sato and Okamoto, shows the

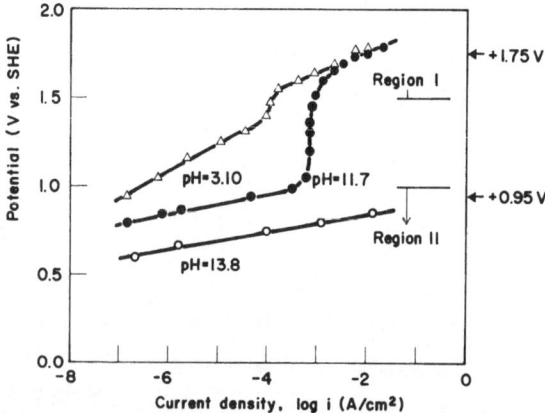

FIGURE 3.4. Polarization curves of the oxygen electrode process on a Ni anode at $25 \pm 0.2°C$.

polarization curves of the oxygen electrode process on a nickel anode in sulfate solution of different pH at 25°C.[4] The Tafel slope in an alkaline solution, pH = 13.8, was 63 mV/decade, but in a solution at pH = 11.7, the anode polarized greatly at high-current densities, and overlapped with the polarization curve in an acid solution having a slope of 127 mV/decade. Figure 3.5 illustrates the relationship between the current density at given potentials and the solution pH. It is clear that the current density at +1.75 V, which is relatively noble, is almost independent of the pH (Region I), whereas the current density at +0.95 V is a function of the OH^- concentration in alkaline solutions (Region II). The slope of the Region II curve is about 2. Consequently, the mechanism of the oxygen electrode process on a nickel anode is considered to be as follows:

Region I (at high-current densities in acidic solution):

$$H_2O \rightleftharpoons OH_{ad} + H^+ + e \qquad (\times 2) \qquad (3.25a)$$

$$OH_{ad} + OH_{ad} \rightarrow \tfrac{1}{2}O_2 + H_2O \qquad (3.25b)$$

$$H_2O \rightarrow 2H^+ + \tfrac{1}{2}O_2 + 2e \qquad (3.25)$$

Since the first step, (3.25a), is considered to be rate-determining for the generation of oxygen, the rate equation is represented by Eq. (3.26) when the overvoltage η is sufficiently large:

$$i \simeq k(a_{H_2O}) \exp(\alpha F\eta/RT) \qquad (3.26)$$

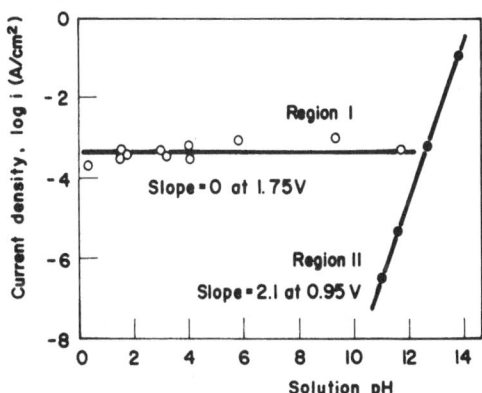

FIGURE 3.5. Current density of the oxygen electrode process as a function of the solution pH.

Region II (at low-current densities in alkaline solution):

$$OH^- \rightarrow OH_{ad} + e \qquad (3.27a)$$

$$OH_{ad} + OH^- \rightarrow O_{ad} + H_2O + e \qquad (3.27b)$$

$$O_{ad} + O_{ad} \rightarrow O_2 \qquad (3.27c)$$

$$\overline{2OH^- \rightarrow H_2O + \tfrac{1}{2}O_2 + 2e} \qquad (3.27)$$

In this case, the first step is in the partial equilibrium state, whereas the second step is rate-determining. Therefore, we apply the Nernst equation for the first step as follows:

$$E' = E^0 + \frac{RT}{F} \ln \left(\frac{a_{OH_{ad}}}{a_{OH^-}} \right)$$

or

$$a_{OH_{ad}} = (a_{OH^-}) \exp \left[\frac{F(E' - E^0)}{RT} \right] \qquad (3.28a)$$

On the other hand, the rate equation for the second step (RDS) is represented by Eq. (3.28b):

$$i \simeq k(a_{OH_{ad}})(a_{OH^-}) \exp \left(\frac{\alpha FE'}{RT} \right) \qquad (3.28b)$$

Substituting Eq. (3.28a) into Eq. (3.28b)

$$i = k'(a_{OH^-})^2 \exp \left[\frac{(1+\alpha)FE'}{RT} \right]$$

$$= k''(a_{OH^-})^2 \exp \left[\frac{(1+\alpha)F\eta}{RT} \right] \qquad (3.28)$$

Reactions (3.25) and (3.27) are the oxygen electrode processes, but the mechanism of both processes differs from each other depending on the solution composition and the electrode potential. The rate equation also differs in both cases as shown by Eqs. (3.26) and (3.28).

3.2.3. Example 3: Chlorine Electrode Process

The chlorine electrode process

$$2Cl^- = Cl_2 + 2e \qquad (3.29)$$

consists of several elementary reactions such as

Mechanism I: $Cl^- = Cl_{ad} + e$ (3.30)

$2Cl_{ad} = Cl_2$ (3.31)

Mechanism II: $Cl^- = Cl_{ad} + e$ (3.32)

$Cl_{ad} + Cl^- = Cl_2 + e$ (3.33)

We assume that one of these reactions is rate-determining and obtain the rate equation and the parameters for the reaction such as the Tafel slope, the stoichiometric number, and the reaction order. The stoichiometric number v is the number of times required for the rate-determining step for an occurrence of the overall reaction under consideration. The reaction order z shows the dependency of the reaction rate on the concentration of species of interest, i.e.,

$$v = \frac{nF}{RT}(i_0)\left(\frac{\partial \eta}{\partial i}\right)_{\eta = 0}$$ (3.34)

and

$$z = \frac{\partial \ln i}{\partial \ln a}$$ (3.35)

The anodic and cathodic polarization curves of a graphite electrode in a saturated NaCl solution containing a maximum concentration of dissolved chlorine at 50°C are shown in Fig. 3.6.[5] The active sites on a graphite

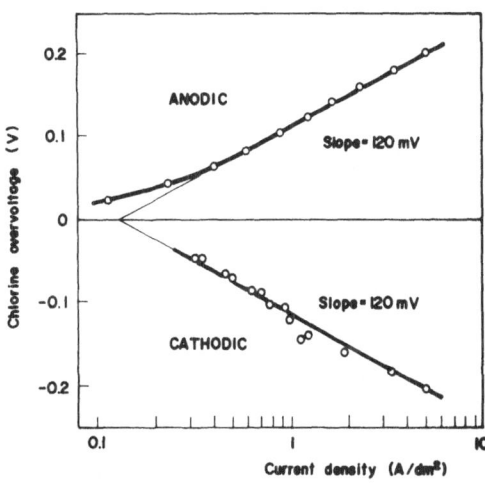

FIGURE 3.6. Polarization curves of graphite electrode in saturated NaCl solution (pH = 0.5 and p_{Cl_2} = 1 atm) at 50°C.

TABLE 3.2
Parameters for the Chlorine Electrode Process

Mechanism		Rate determining step	Stoichiometric number, ν	Tafel slope	Electrochemical reaction order	
					Z_{Cl^-}	Z_{Cl_2}
Anodic	I	$Cl^- = Cl_{ad} + e$	2	$2RT/F$	1	0
		$2Cl_{ad} = Cl_2$	1	∞	0	0
	II	$Cl^- = Cl_{ad} + e$	1	—	—	—
		$Cl_{ad} + Cl^- = Cl_2 + e$	1	$2RT/F$	1	0
Cathodic	I	$Cl_2 = 2Cl_{ad}$	1	$-\infty$	0	1
		$Cl_{ad} + e = Cl^-$	2	$-2RT/F$	0	1/2
	II	$Cl_2 + e = Cl_{ad} + Cl^-$	1	$-2RT/F$	0	1
		$Cl_{ad} + e = Cl^-$	1	$-2RT/3F$	-1	1

				Experimental		
Anodic			1	120 mV	1	0
Cathodic			1	−120 mV	0	1

anode are considered to be occupied almost completely by Cl_{ad}; on the other hand, the electrode surface is bare when it is polarized cathodically.

Figure 3.6 shows that the Tafel slopes of both anodic and cathodic polarization curves in the range of the overvoltage higher than 70 mV is 120 mV/decade or $2RT/F$. These parameters are compared to the theoretical calculations as listed in Table 3.2. It is estimated that mechanism II is possible, and reaction (3.33) is rate-determining, i.e.,

$$Cl^- = Cl_{ad} + e$$

$$Cl_{ad} + Cl^- \rightleftharpoons Cl_2 + e$$

on the graphite electrode in concentrated sodium chloride solutions.

3.2.4. Example 4: Iron Electrode Process

Bockris proposed five routes for an iron electrode process

$$Fe = Fe^{2+} + 2e \tag{3.36}$$

as follows[6]:

Mechanism A: $Fe + OH^- + FeOH = (FeOH)_2 + e$
$(FeOH)_2 \rightleftharpoons 2FeOH$
$FeOH = FeOH^+ + e$
$FeOH^+ = Fe^{2+} + OH^-$

Mechanism B: $Fe + H_2O = FeOH + H^+ + e$
$FeOH = FeOH^+ + e$
$FeOH^+ + Fe \rightleftharpoons Fe_2OH^+$
$Fe_2OH^+ = Fe^{2+} + FeOH + e$
$FeOH + H^+ = Fe^{2+} + H_2O + e$

Mechanism C: $Fe + OH^- \rightleftharpoons Fe(OH)^+ + 2e$
$Fe(OH)^+ = Fe^{2+} + OH^-$

Mechanism D: $Fe + OH^- = FeOH + e$
$FeOH + OH^- \rightleftharpoons FeO + H_2O + e$
$FeO + OH^- = HFeO_2^-$
$HFeO_2^- + H_2O = Fe(OH)_2 + OH^-$
$Fe(OH)_2 = Fe^{2+} + 2OH^-$

Mechanism E: $Fe + H_2O = FeOH + H^+ + e$
$FeOH \rightleftharpoons FeOH^+ + e$
$FeOH^+ + H^+ = Fe^{2+} + H_2O$

Table 3.3 shows the parameters evaluated theoretically compared with experimental results. The table also contains some data obtained by Hine. From the table, mechanism E is estimated to be the most probable route for the iron electrode process.

3.3. ELECTRODE PROCESSES CONTROLLED BY A MASS TRANSFER STEP

3.3.1. Example 1: Copper Electrode Process

The polarization curves of a copper electrode in an acidified 1 M CuSO$_4$ solution are shown in Fig. 3.7. The Tafel slope of the cathodic polarization curve at low current densities is about 110 mV/decade, whereas the slope of the anodic polarization curve is about 65 mV/decade. The potentials shown by the open points in this figure contain the solution *IR* drops between the working electrode and the Luggin–Haber probe. The solution *IR* drop can be eliminated by means of the current interruption technique, as shown by the closed points in the anodic branch. At low-current densities, however, the *IR* drop is insignificant.

The polarization measurement of the cathode is somewhat difficult, especially at high current densities, because rough deposition of copper causes gradual changes of the electrode surface and of the gap between the cathode and the Luggin probe.

TABLE 3.3
Parameters of the Iron Electrode Process

Parameters	Mechanisms					Experimental results (log₁₀ used)	
	A	B	C	D	E	Bockris	Hine
$\dfrac{\partial E_{Fe}}{\partial \ln i_A}$	$\dfrac{RT}{2F}$	$\dfrac{RT}{2F}$	$\dfrac{RT}{F}$	$\dfrac{2}{3}\dfrac{RT}{F}$	$\dfrac{2}{3}\dfrac{RT}{F}$	0.042 ± 0.008	0.042
$\dfrac{\partial E_{Fe}}{\partial \ln i_C}$	$-\dfrac{RT}{2F}$	$-\dfrac{RT}{2F}$	$-\dfrac{RT}{F}$	$-\dfrac{2RT}{F}$	$-\dfrac{2RT}{F}$	-0.116 ± 0.006	-0.124
$\left(\dfrac{\partial \ln i_C}{\partial \ln a_{Fe^{2+}}}\right)_{a_{OH^-}}$	2	2	1	1	1	0.8	1
$\left(\dfrac{\partial \ln i_0}{\partial \ln a_{OH^-}}\right)_{a_{Fe^{2+}}}$	2	1	1	2	1	0.9 ± 0.05	1
$\left(\dfrac{\partial \ln E_{Fe}}{\partial \ln a_{OH^-}}\right)_{a_{Fe^{2+}}}$	$-\dfrac{RT}{F}$	$-\dfrac{RT}{2F}$	$-\dfrac{RT}{F}$	$-\dfrac{4}{3}\dfrac{RT}{F}$	$-\dfrac{2}{3}\dfrac{RT}{F}$		-0.040
$\left(\dfrac{\partial \ln i_0}{\partial \ln a_{Fe^{2+}}}\right)_{a_{OH^-}}$	1	1	$\dfrac{1}{2}$	$\dfrac{3}{4}$	$\dfrac{3}{4}$	0.8 ± 0.1	1
$\left(\dfrac{\partial E_{corr}}{\partial \ln a_{Fe^{2+}}}\right)_{a_{OH^-}}$	$-\dfrac{6}{5}\dfrac{RT}{F}$	$-\dfrac{4}{5}\dfrac{RT}{F}$	$-\dfrac{4}{3}\dfrac{RT}{F}$	$-\dfrac{3}{2}\dfrac{RT}{F}$	$-\dfrac{RT}{F}$	-0.060 ± 0.003	-0.059
$\left(\dfrac{\partial \ln i_{corr}}{\partial \ln a_{OH^-}}\right)_{a_{Fe^{2+}}}$	$-\dfrac{2}{5}$	$-\dfrac{3}{5}$	$-\dfrac{1}{3}$	$-\dfrac{1}{4}$	$-\dfrac{1}{2}$	-0.5 ± 0.01	

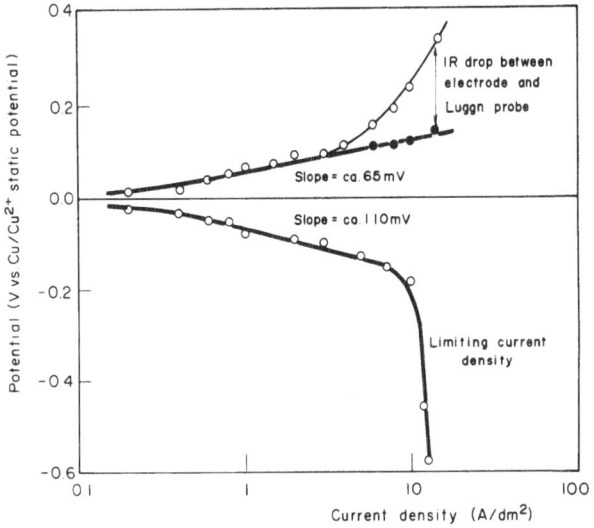

FIGURE 3.7. Polarization curves of a copper electrode in 1 M $CuSO_4$ at 40°C under free convection.

The copper electrode process is considered to be consecutive reactions:

$$Cu(crystal) = Cu_{ad} \tag{3.37}$$

$$Cu_{ad} = Cu^+ + e \tag{3.38}$$

$$Cu^+ \rightleftharpoons Cu^{2+} + e \tag{3.39}$$

$$(Cu^{2+})_{surface} = (Cu^{2+})_{bulk} \tag{3.40}$$

Of these, the third step is rate-controlling at low current densities.

The rate equation of reaction (3.39) is represented by Eq. (3.41):

$$i = i_a - i_c = k \left\{ (a_{Cu^+}) \exp\left(\frac{\alpha F \eta}{RT}\right) - (a_{Cu^{2+}}) \exp\left[-\frac{(1-\alpha) F \eta}{RT} \right] \right\} \tag{3.41}$$

where k is the rate constant. When $i_a \gg i_c$, we have

$$i \simeq k(a_{Cu^+}) \exp\left(\frac{\alpha F \eta}{RT}\right) \tag{3.42}$$

Since the first and the second steps, (3.37) and (3.38), are in partial equilibrium, the Nernst equation can be applied [see Eqs. (3.55)–(3.57)]. Consequently, the overvoltage η can be represented as follows:

$$\eta = E' - E = \frac{RT}{F} \ln \frac{a_{Cu^+}}{a_{Cu^+}^0} \tag{3.43}$$

where E' and E are the electrode potential with and without electrolytic current, respectively, and $a^0_{Cu^+}$ is the activity of Cu^+ at zero current and is considered to be a constant under given conditions. Substituting Eq. (3.43) into Eq. (3.42)

$$i \simeq k(a^0_{Cu^+}) \exp \left[\frac{(1+\alpha)F\eta}{RT} \right] \qquad (3.44)$$

When $i_a \ll i_c$, on the other hand, the rate equation (3.41) becomes Eq. (3.45):

$$-i \simeq k(a_{Cu^{2+}}) \exp \left[-\frac{(1-\alpha)F\eta}{RT} \right] \qquad (3.45)$$

where the negative sign of i and η refers to the cathodic.

If so, the Tafel slopes for the both directions are obtained theoretically:

$$\frac{\partial \eta}{\partial \ln(i)} = \frac{RT}{(1+\alpha)F} \qquad (3.46)$$

and

$$\frac{\partial(-\eta)}{\partial \ln(-i)} = \frac{RT}{(1-\alpha)F} \qquad (3.47)$$

	Tafel slopes at 25°C (mV)	
	Anodic	Cathodic
α	$\dfrac{2.3RT}{(1+\alpha)F}$	$\dfrac{2.3RT}{(1-\alpha)F}$
0.4	42	98
0.5	39	118
0.6	37	148

That is, the charge transfer step involving Cu^{2+} and Cu^+ is rate-determining, and the subsequent charge transfer and crystallization steps are considered to be in the state of partial equilibrium at low current densities. Of course, physicochemical studies on crystallization of deposited metal are important and have been investigated by many authors, because crystallization of metals is closely related to the morphology of the deposited layer.[7]

The cathodic overvoltage becomes very high at high current densities as shown in Fig. 3.7 due to the delay of the transfer of Cu^{2+} from the bulk of the solution to the electrode surface. This step, (3.40), is considered to be the rate-determining step for the overall reaction of copper deposition at high current densities.

The mass transfer of Cu^{2+} from the bulk of the solution to the electrode surface is conducted by two factors: migration and diffusion. That is,

$$v = v_m + v_d \qquad (3.48)$$

$$v_m = \frac{\lambda C}{\kappa} \frac{i}{F} = t \frac{i}{F} \qquad (3.49)$$

$$v_d = -D \frac{dC}{dx} \qquad (3.50)$$

where v is the flux or reaction rate (mol/cm² sec), t is the transport number, C is the concentration (mol/cm³), D is the diffusion coefficient (cm²/sec), λ is the equivalent conductance (mho cm²/g-eq.), κ is the conductivity of solution (mho/cm), and the subscripts m and d refer to migration and diffusion, respectively.

Now we assume that the solution contains sufficient sulfuric acid, so that the migration term of Cu^{2+} in the rate equation is negligible compared to the diffusion term. Such an electrolyte is called the *supporting electrolyte* or the *independent electrolyte*. Addition of the independent electrolyte is useful in order to increase the electrical conductivity of the solution.

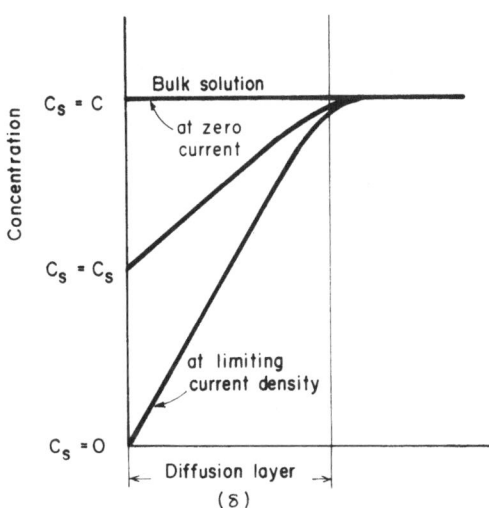

FIGURE 3.8. Concentration of species near the diffusion controlled cathode.

Cupric ions may be reduced to Cu on the cathode surface immediately after arrival from the bulk of the solution since the charge-transfer step is considered to be fast. Therefore, the Cu^{2+} concentration at the cathode decreases as shown in Fig. 3.8, where C_s is the Cu^{2+} concentration on the cathode surface. According to Fick's law, under a steady state,

$$v = -D \left(\frac{dC}{dx} \right)_{x=0} \tag{3.50'}$$

For simplicity, a linear distribution of concentration in the diffusion layer of thickness of δ is assumed. Thus we have

$$v = D \left(\frac{C - C_s}{\delta} \right) \tag{3.51}$$

The reaction rate v becomes maximum when C_s tends to zero. Because $i = nFv$, we have

$$i_d = nFD \left(\frac{C - C_s}{\delta} \right) \tag{3.52}$$

$$i_L = nFD \left(\frac{C}{\delta} \right) \tag{3.53}$$

$$\frac{i_d}{i_L} = 1 - \frac{C_s}{C} \tag{3.54}$$

where i_d is the current density controlled by diffusion. It is clear that i_L is the maximum current density at $C_s = 0$, and is called the *limiting current density*.

Since the charge transfer step is assumed to be nearly reversible as described above, the Nernst equation can be applied:

$$E = E^0 + \frac{RT}{nF} \ln C \qquad \text{at} \quad i = 0 \tag{3.55}$$

$$E' = E^0 + \frac{RT}{nF} \ln C_s \qquad \text{at} \quad i = i_d \tag{3.56}$$

By convention

$$\eta = E' - E = \frac{RT}{nF} \ln \left(\frac{C_s}{C} \right) \tag{3.57}$$

Substituting Eq. (3.54) into Eq. (3.57)

$$\eta = \frac{RT}{nF} \ln \left(1 - \frac{i_d}{i_L} \right) \tag{3.58}$$

This equation shows that η becomes infinite when i_d tends to i_L (see Fig. 3.7). Substituting the data:

$$-i_L = 12 \text{ A/dm}^2 = 0.12 \text{ A/cm}^2$$

$$C = 1 \, M/1 = 10^{-3} \, M/\text{cm}^3$$

$$D \simeq 10^{-5} \text{ cm}^2/\text{sec}$$

$$nF = 2 \times 96,500 \text{ A sec}/M$$

into Eq. (3.53), we have

$$\delta = \frac{2 \times 96,500 \times 10^{-5} \times 10^{-3}}{0.12} \simeq 1.6 \times 10^{-2} \text{ cm}$$

The thickness of the diffusion layer is in the range 10^{-4}–10^{-2} cm, depending on the hydrodynamic conditions at the electrode–electrolyte interface. It will be described later in detail.

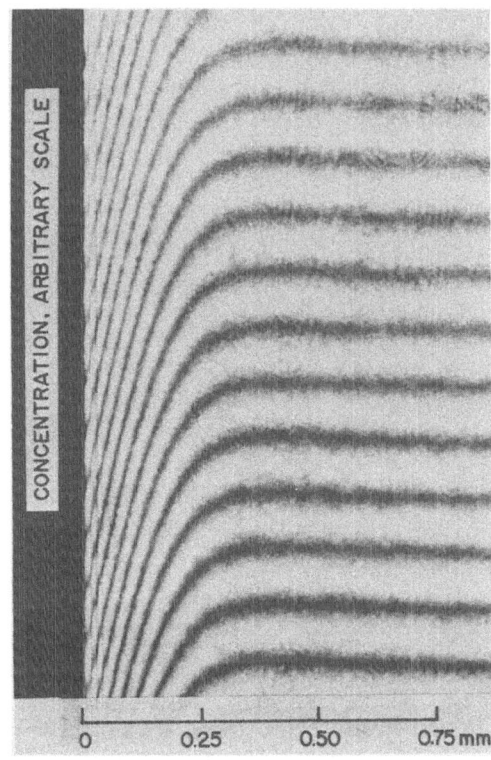

FIGURE 3.9. An example of enlarged holographic interferogram near the copper cathode (Kondo et al., Kyoto University). Photograph was taken soon after current-off. Conditions: Electrolyte = 0.6M CuSO₄; temperature = 23°C; current density = 2 mA/cm².

According to Fig. 3.7, up to about 10 A/dm^2, charge transfer is the only rate-controlling mechanism; however, it is also true that the polarization is small compared with the mass transfer induced polarization that occurs at $i \gtrsim 10$ A/dm^2.

Early electrochemists such as Nernst assumed a concentration distribution at the working electrode and a diffusion layer for the mass transfer controlled reaction as shown in Fig. 3.8. Recently the concentration distribution near the cathode surface has been observed by means of optical methods by Ibl and Muller,[8] and by subsequent researchers. Figure 3.9 is an example of the holographic interferogram near the copper cathode taken by Kondo and co-workers at Kyoto University.

3.3.2. Example 2: Oxygen Cathode Reaction

The overall reactions for the oxygen electrode in acidic and alkaline solutions are as follows:

$$2H_2O + O_2 = 4H^+ + 4e \qquad E^0 = 1.227 \text{ V vs. SHE} \qquad (3.59)$$

$$4OH^- = O_2 + 2H_2O + 4e \qquad E^0 = 0.400 \text{ V vs. SHE} \qquad (3.60)$$

Formation of H_2O_2 or HO_2^- as intermediate for the reaction is well known,[9] that is, in acid solution

$$H_2O_2 = O_2 + 2H^+ + 2e \qquad E_0 = +0.68 \text{ V vs. SHE} \qquad (3.61)$$

$$2H_2O = H_2O_2 + 2H^+ + 2e \qquad E^0 = +1.77 \text{ V vs. SHE} \qquad (3.62)$$

and in alkaline solution

$$HO_2^- + OH^- = O_2 + H_2O + 2e \qquad E^0 = -0.08 \text{ V vs. SHE} \qquad (3.63)$$

$$3OH^- = HO_2^- + H_2O + 2e \qquad E^0 = +0.88 \text{ V vs. SHE} \qquad (3.64)$$

The reduction process of oxygen on a rotating Pt cathode is controlled by diffusion of dissolved oxygen in solution as shown by Fig. 3.10(a). The limiting current density i_L is proportional to the square root of the rate of revolution ω [Fig. 3.10(b)]. It has been confirmed that the limiting current density is also proportional to the concentration of dissolved oxygen, or the partial pressure of oxygen in the gas phase at equilibrium, p_{O_2}. Therefore, Eq. (3.65) can be obtained as

$$i_L = k p_{O_2} \sqrt{\omega} \qquad (3.65)$$

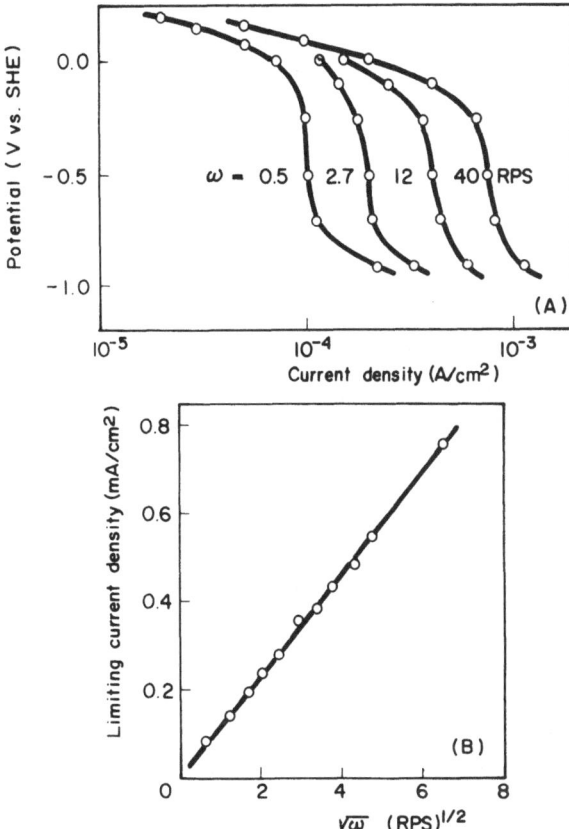

FIGURE 3.10. (a) Polarization curves of the oxygen cathodic reduction process, and (b) the dependency of the limiting current density on the revolution rate of a Pt rotating cathode. $C_{O_2} \times 3 \times 10^{-4}M$.

3.3.3. Example 3: Fe^{2+}/Fe^{3+} Redox System

The redox-type electrode process such as Fe^{2+}/Fe^{3+} is also controlled by diffusion. Figure 3.11 shows the polarization curves of a rotating Pt electrode in a mixed solution of $FeSO_4$ and $Fe_2(SO_4)_3$ at 1800 rpm.[10] The limiting current density for the cathode reaction is proportional to the Fe^{3+} concentration and the square root of the revolution rate.

For the redox-type electrode process, we need to consider the back diffusion of the product from the working electrode to the bulk of solution. In the case of the cathodic reduction of the Fe^{2+}/Fe^{3+} system, for example, Fe^{2+} is the product: $Fe^{3+} + e \rightarrow Fe^{2+}$, and hence the Fe^{2+} concentration on the electrode surface is large compared with that in the bulk of solution, resulting in transport of Fe^{2+} to the solution. Thus, the cathodic process can be affected by mass transport limitation of product Fe^{2+}. Since the solution

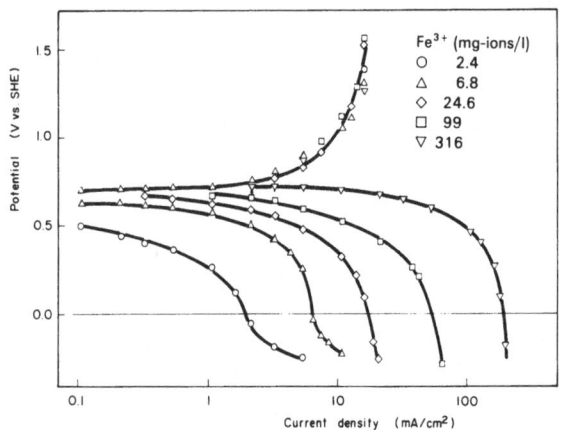

FIGURE 3.11. Polarization curves of a rotating Pt electrode in a mixed solution of $FeSO_4$ and $Fe_2(SO_4)_3$.

contained some amount of Fe^{2+} (24.6 mg ions/l in the case of the experiment shown in Fig. 3.11), the effect of Fe^{2+} on the cathodic polarization curves was negligible. The anodic polarization curves obtained in the solutions containing a small amount of Fe^{3+}, say 25 mg ions/l, were also independent of the Fe^{3+} concentration as shown in Fig. 3.11 due to the same reason as that of the cathodic process.

It is important that we can evaluate the concentration of the ionic species in the solution by means of the limiting current density [see Eq. (3.53). The electrode process takes place in a given potential range depending mostly on the reversible potential, so that we also may estimate the electrode process occurring. Even in a mixed solution containing various kinds of ionic species, those species can be analyzed qualitatively and quantitatively by polarization curves which can resolve respective potentials and limiting current densities. This is the basic concept of polarography.

3.3.4. Example 4: Sodium Amalgam Electrode

The sodium amalgam electrode process in concentrated NaCl solution

$$Na\text{-}Hg = Na^+ + (Hg) + e \qquad (3.66)$$

is fast, and at very high current densities, the limiting current densities appear on both the anodic and cathodic branches of the polarization curves. A schematic profile of the concentrations of Na in the amalgam and of Na^+ in the sodium chloride solution near the amalgam–electrolyte interface is shown in Fig. 3.12. When the amalgam electrode is polarized anodically, the

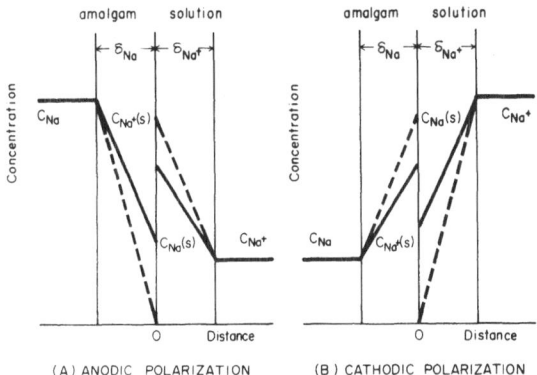

FIGURE 3.12. *Concentration profile of Na and Na$^+$ near the interface between sodium amalgam and NaCl solution.*

Na concentration at the interface, $C_{Na}(s)$, decreases with increase of the current density, and it becomes zero at a maximum current density, $i_L(A)$, i.e.,

$$i_L(A) = \frac{F D_{Na} C_{Na}}{\delta_{Na}} \qquad (3.67)$$

On the other hand, for cathodic reduction of Na$^+$, the Na$^+$ concentration at the interface, $C_{Na^+}(s)$, diminishes, and tends to zero, at which the current density $i_L(C)$ is represented as follows:

$$i_L(C) = \frac{F D_{Na^+} C_{Na^+}}{\delta_{Na^+}} \qquad (3.68)$$

where D is the diffusion coefficient and δ is the thickness of the diffusion layer. The subscripts Na and Na$^+$ represent the amalgam side and the solution side, respectively.

It is clear from Eqs. (3.67) and (3.68) that $i_L(A)$ and $i_L(C)$ are related to C_{Na} and C_{Na^+}, respectively, and hence the anodic reaction is controlled by diffusion of Na in the amalgam, whereas the cathodic process is controlled by Na$^+$ in the solution.

3.4. MORE COMPLICATED ELECTRODE PROCESSES

3.4.1. Electrode Process Controlled by Chemical and Mass Transfer Steps

In the previous section, we assumed that the electrolytic solution contained a sufficient amount of independent salt, so that the migration term in the rate equation could be neglected.

Here we may consider the effect of migration on the current density of the cathodic process. The current density is represented as follows[11]:

$$i = i_m + i_d = nFD \left(\frac{nFC\phi}{RT} - \frac{dC}{dx} \right) \tag{3.69}$$

where i_m and i_d are the terms of migration and diffusion, respectively, and ϕ is the intensity of electric field in V/cm. The current density i is also shown by Eq. (3.20) with some modifications:

$$i = \frac{i_d}{1-t} = i_0 \left[\exp(f\eta) - \frac{C_s}{C} \exp(-f'\eta) \right] \tag{3.70}$$

where $t \; (= i_m/i)$ is the transport number, and C_s is the concentration of ionic species of interest on the electrode surface. Also, $f = \alpha nF/RT$, and $f' = (1 - \alpha)nF/RT$.

Substituting Eq. (3.54) into Eq. (3.70), we have

$$i = \frac{i_0 [\exp(f\eta) - \exp(-f'\eta)]}{1 - (1 - t)(i_0/i_L) \exp(-f'\eta)} \tag{3.71}$$

If diffusion is fast, and the reaction step is slow, i.e.,

$$(1 - t) \frac{i_0}{i_L} \exp(-f'\eta) \ll 1$$

Eq. (3.71) becomes

$$i = i_0 [\exp(f\eta) - \exp(-f'\eta)] \tag{3.72}$$

which agrees with Eq. (3.20). On the contrary, for a diffusion-controlled process

$$(1 - t) \left(\frac{i_0}{i_L} \right) \exp(-f'\eta) \gg 1$$

therefore, since $f + f' = nF/RT$,

$$i = \frac{-i_L}{1 - t} \left[\exp \left(\frac{nF\eta}{RT} \right) - 1 \right]$$

$$\eta = \frac{RT}{nF} \ln \left[1 - (1 - t) \left(\frac{i}{i_L} \right) \right] \tag{3.73}$$

If $t = 0$, Eq. (3.73) agrees with Eq. (3.58).

Now we may discuss the electrode process controlled by both diffusion and reaction steps occurring on a rotating disk electrode. The relationship between the current density i and the rotating speed ω in this case is as follows

$$i = A(C - C_s)\sqrt{\omega} \qquad (3.74)$$

$$i_L = AC\sqrt{\omega} \qquad (3.75)$$

where A is a coefficient, and C and C_s are the concentration of ionic species under discussion in the bulk of solution and at the electrode surface, respectively.

It is required that the current density is also a function of the concentration C_s

$$i = k(C_s)^p \qquad (3.76)$$

where k is the specific reaction rate and p is the reaction order. If ω tends to infinity, C_s becomes C, and hence,

$$i = i_r = k(C)^p \qquad (3.77)$$

where i_r is the limiting current density due to slow surface reaction. For simplicity, assume that $p = 1$, i.e., a first-order reaction, then substituting Eqs. (3.76) and (3.77) into Eq. (3.74), we obtain

$$\frac{1}{i} = \frac{1}{i_r} + \frac{1}{AC\sqrt{\omega}} \qquad (3.78)$$

Thus we can evaluate i_r with the intersection on the abscissa of the $1/i$ vs. $1/\sqrt{\omega}$ plots.

Here we may consider the cathodic reaction of ClO^- in alkaline solution:

$$Cl^- + 2OH^- = ClO^- + H_2O + 2e$$

$$E^0 = +0.88 \text{ V vs. SHE at } 25°C \qquad (3.79)$$

The polarization curves of a rotating Pt disk cathode in mixed solutions containing ClO^- of different concentrations are shown in Fig. 3.13.[12] The limiting current density is proportional to the ClO^- concentration and the square root of the rotation speed, as shown in Figs. 3.14 and 3.15, respectively. The $1/i$ vs. $1/\sqrt{\omega}$ plots in Fig. 3.16 show straight lines intersecting the origin. Therefore, it is concluded that there is no reaction limited current density i_r in this case.

According to Vetter,[13] the rate equation for the successive charge transfer reaction

$$S_r = S_m + e \qquad (3.80)$$

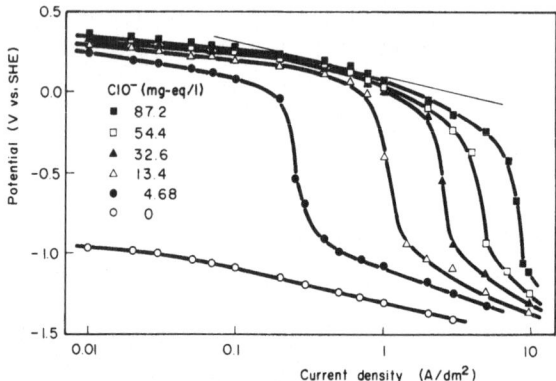

FIGURE 3.13. Polarization curves of a rotating Pt cathode in the mixed solutions of 2 M NaOH and 3 M NaCl containing ClO⁻ at 1660 rpm.

and

$$S_m = S_0 + e \qquad (3.81)$$

is shown by Eq. (3.82)

$$i = -2k'a(1 - \theta) \exp\left[-\frac{(1 - \alpha)F\eta}{RT}\right] \qquad (3.82)$$

where k' is the specific reaction rate, a is the activity of S_0, θ is the coverage, and α is the transfer coefficient.

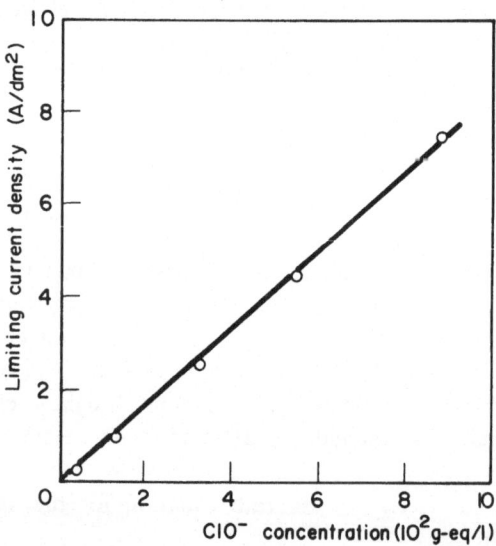

FIGURE 3.14. Limiting current density as a function of the ClO⁻ concentration.

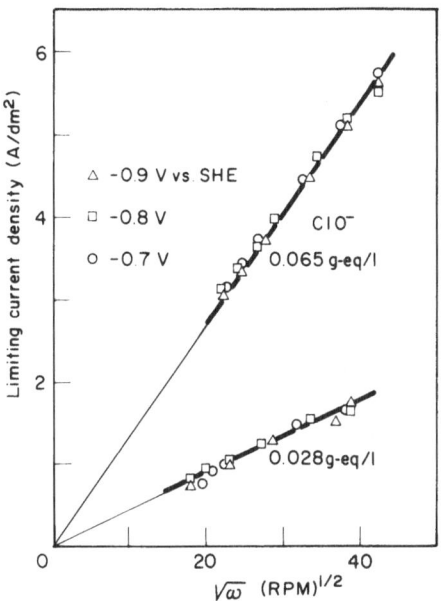

FIGURE 3.15. Limiting current density as a function of the rotating rate.

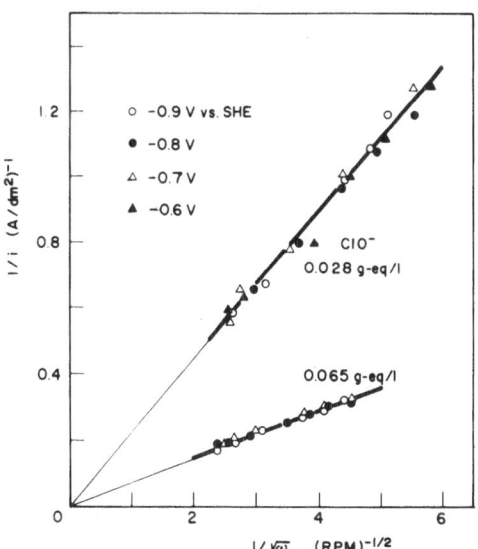

FIGURE 3.16. An example of $1/i$ vs. $1/\sqrt{\omega}$.

If the activity of the species on the electrode surface a_s is affected by slow diffusion [see Eq. (3.54)]

$$\frac{a_s}{a} = 1 - \frac{i}{i_L} \tag{3.83}$$

Substituting this equation into Eq. (3.82)

$$\frac{i}{1 - (i/i_L)} = -2k'a(1 - \theta)\exp\left[-\frac{(1-\alpha)F\eta}{RT}\right] \tag{3.84}$$

The thin line in Fig. 3.13 shows the results of calculation with Eq. (3.84), and the line has a slope of about 120–140 mV/decade. This figure also shows that the current density i is proportional to the ClO^- concentration at the potential of $+0.25$ V. Consequently, the reaction under discussion is assumed to be first-order with respect to ClO^-.

In conclusion, the mechanism of the overall reaction of the cathodic reduction of ClO^-, Eq. (3.79), can be represented as follows:

$$ClO^- + H_2O + e = Cl_{ad} + 2OH^- \tag{3.85}$$

$$Cl_{ad} + e = Cl^- \tag{3.86}$$

$$\overline{\hspace{6cm}}$$

$$ClO^- + H_2O + 2e = Cl^- + 2OH^- \tag{3.79}$$

3.4.2. Dimensional Analysis of Mass Transfer on an Electrode

We have considered in the previous section that the diffusion layer forms on the working electrode controlled by mass transfer, mostly diffusion. The concentration of the ionic species varies with the distance from the electrode surface to the bulk of the solution because of insufficient mixing due to the viscosity of the solution.

According to hydrodynamics, liquid flow is retarded in the immediate vicinity of the plate, as shown in Fig. 3.17. The affected region is called the *velocity boundary layer*. The solution is partially mixed even in the boundary layer, so that the diffusion layer is located within the boundary layer. In other words, the diffusion layer is thinner than the velocity boundary layer:

$$\delta = k\delta_f \qquad k < 1$$

where δ and δ_f are the thickness of the diffusion layer and the velocity boundary layer, respectively, and k is the ratio of δ to δ_f.

The velocity boundary layer and the diffusion layer have been studied by many authors in detail.[14–18] Also the profile of concentration

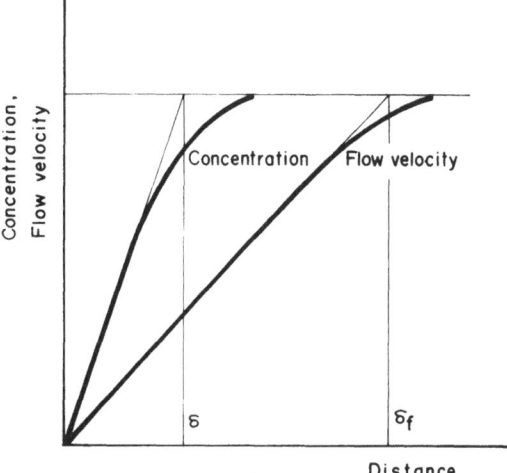

FIGURE 3.17. Distributions of concentration and flow velocity near the electrode–electrolyte interface.

distribution in the diffusion layer has been established experimentally by optical methods as shown in Fig. 3.9.

The concept and the mathematical analysis of the velocity boundary layer are already described in several sources.[19–20] In general, the flow in the boundary layer is obtained by simplification of the Navier–Stokes and the continuity equations, namely;

$$\frac{\partial u_x}{\partial t} + u_x \frac{\partial u_x}{\partial x} + u_y \frac{\partial u_x}{\partial y} = -\frac{1}{\rho}\frac{\partial p}{\partial x} + v\left(\frac{\partial^2 u_x}{\partial x^2} + \frac{\partial^2 u_x}{\partial y^2}\right)$$

$$\frac{\partial u_y}{\partial t} + u_x \frac{\partial u_y}{\partial x} + u_y \frac{\partial u_y}{\partial y} = -\frac{1}{\rho}\frac{\partial p}{\partial x} + v\left(\frac{\partial^2 u_y}{\partial x^2} + \frac{\partial^2 u_y}{\partial y^2}\right)$$

(3.87)

and

$$\frac{\partial u_x}{\partial x} + \frac{\partial u_y}{\partial y} = 0 \qquad \text{(incompressible fluid)}$$

(3.88)

where x is the axis along the flat plate and y is perpendicular to the plate, u_x and u_y are the velocity components in the x- and y-directions, respectively, t is time, p is pressure, ρ is the density of solution, and v is the kinematic viscosity. For a flat plate having a leading edge in a laminar flow as shown in Fig. 3.18, for example, the boundary conditions are as follows:

$$u_x = u_y = 0 \qquad \text{at } y = 0$$

$$u_x = u_0 \qquad \text{at } y \geqslant \delta_f$$

$$\frac{\partial u_x}{\partial y} = 0 \qquad \text{at } y = \delta_f$$

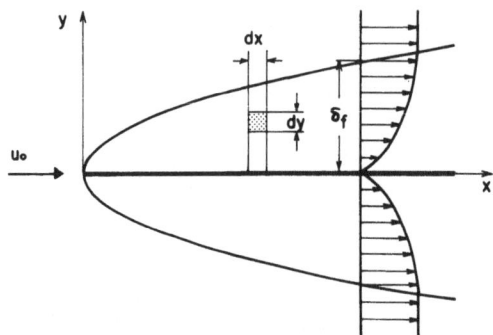

FIGURE 3.18. Boundary layer near a flat plate and velocity profile.

where u_0 is the flow velocity in the bulk of solution. With several assumptions for simplicity, the next equation is obtained and is well known as Kàrmàn's integral equation for momentum.

$$\frac{d}{dx}\int_0^{\delta_f} u_x^2 \, dy - u_0 \frac{d}{dx}\int_0^{\delta_f} u_x \, dy = -\frac{1}{\rho}\frac{dp}{dx} - v\left(\frac{\partial u_x}{\partial y}\right)_{y=0} \qquad (3.89)$$

The thickness δ_f is a function of x. Schlichting introduces the sine function to conform to the requirement of zero velocity gradient in the direction perpendicular to the plate.[22] Assume that

$$u_x = u_0 \sin\left(\frac{\pi y}{2\delta_f}\right) \qquad (3.90)$$

then the next equations can be obtained, namely,

$$\frac{d\delta_f}{dx} = \left(\frac{\pi^2}{4-\pi}\right)\left(\frac{v}{u_0 \delta_f}\right)$$

and

$$\delta_f = \left(\frac{2\pi^2}{4-\pi}\right)^{1/2}\left(\frac{vx}{u_0}\right)^{1/2} = 4.80\left(\frac{vx}{u_0}\right)^{1/2} = 4.80\frac{x}{(Re)^{1/2}}$$

$$(3.91)$$

where Re is the Reynolds number.

The material balance in a volume element of $dx \cdot dy$ shown in Fig. 3.18 for unit time is given by

$$\left(u_x + \frac{du_x}{dx}dx\right)C\,dy - u_x C\,dy = \frac{du_x}{dx}C\,dx\,dy \qquad (3.92)$$

where C is the concentration and the width of the volume element is considered to be unity. Since the chemical species entering into the diffusion layer will be consumed by cathodic reaction, we have

$$-D \left(\frac{\partial C}{\partial x} \right)_{y=0} = \frac{d}{dx} \int_0^\delta u_x C \, dy \qquad (3.93)$$

where D is the diffusion coefficient. This equation can be evaluated if u_x and C are given as the functions of x and y.

According to Bird et al.,[23] the ratio δ/δ_f is expected to be constant, independent of position along the plate, which is an important underlying feature for the mathematical treatment.

Vielstich obtained a simple form[14]:

$$\frac{\delta}{\delta_f} = (\text{Sc})^{-1/3} \qquad (3.94)$$

$$\text{Sc} = \frac{v}{D} \qquad \text{(Schmidt number)} \qquad (3.95)$$

The rate of mass transfer through the boundary layer, v in kg moles/m^2 hr is

$$v = \frac{D}{\delta} (\Delta C) = k(\Delta C) \qquad (3.96)$$

where D is the diffusion coefficient in m^2/hr, k is the mass transfer coefficient in m/hr, and ΔC is the difference of concentration in kg moles/m^3. It is clear that k is a function of δ and/or Re. It appears in nondimensional form in the Sherwood number (Sh) as follows:

$$\text{Sh} = \frac{kL}{D} \qquad (3.97)$$

where L is called the characteristic length and/or the equivalent diameter. Since Sc contains both v and D [see Eq. (3.95)], Sh is a function of Re and Sc:

$$\text{Sh} = f_1(\text{Sc}, \text{Re}) \qquad (3.98)$$

k in Eq. (3.96) can be evaluated by Eq. (3.98), and hence the reaction rate is obtained by Eq. (3.96). If

$$v = -D(\Delta C/\Delta y)_{y=0}$$

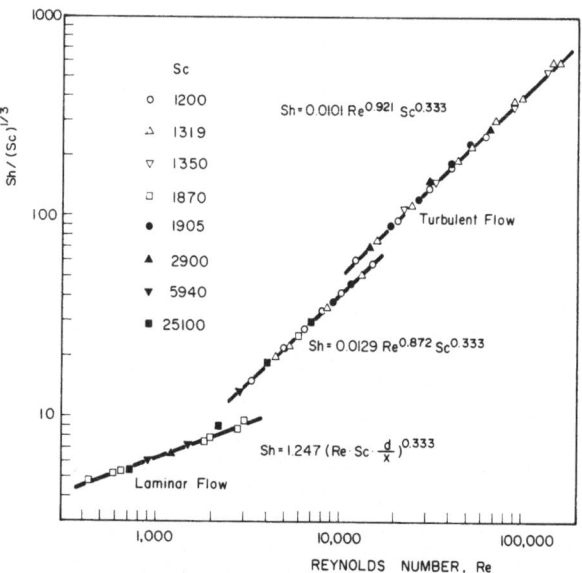

FIGURE 3.19. Relationship between the mass transfer and the flow velocity in an electrolytic cell.

for example, v is represented as follows:

$$v = \frac{0.327D(C_0 - C_s)}{x}(\text{Re})^{1/2}(\text{Sc})^{1/3} \tag{3.99}$$

Figure 3.19 shows the experimental results obtained in the wide ranges of Re and Sc by Landau *et al.*,[15] where d is the equivalent diameter and x is the distance from the leading edge of the flat-plate electrode. On the other hand, the rate of heat transfer through the boundary layer, q in kcal/m² hr, is as follows:

$$q = \frac{\lambda}{\delta_f}(\Delta t) = h(\Delta t) \tag{3.100}$$

where λ is the specific conductance for heat transfer in kcal/m hr °C, h is the coefficient for heat transfer in kcal/m² hr °C, and Δt is the temperature difference in °C. The heat transfer coefficient is also represented by a nondimensional Nusselt number (Nu):

$$\text{Nu} = \frac{hL}{v} \tag{3.101}$$

Nu is a function of the Prandtl number:

$$\text{Pr} = \frac{v}{D_h} \tag{3.102}$$

TABLE 3.4

Comparison between Mass Transfer and Heat Transfer
through the Boundary Layer

	Mass transfer	Heat transfer
Velocity	$v = \dfrac{D}{\delta}(\Delta C) = k(\Delta C)$	$q = \dfrac{\lambda}{\delta_f}(\Delta t) = h'(\Delta t)$
Transfer coefficient	$\text{Sh} = \dfrac{kL}{D}$	$\text{Nu} = \dfrac{hL}{\lambda}$
Diffusion coefficient	$\text{Sc} = \dfrac{v}{D}$	$\text{Pr} = \dfrac{v}{D_h}$
Reynolds number	$\text{Re} = \dfrac{u_x L}{v}$	$\text{Re} = \dfrac{u_x L}{v}$
Function	$\text{Sh} = f_1(\text{Sc}, \text{Re})$	$\text{Nu} = f_2(\text{Pr}, \text{Re})$

where v is the kinematic viscosity in m^2/hr, and D_h is the diffusion coefficient of heat in m^2/hr. Consequently,

$$\text{Nu} = f_2(\text{Pr}, \text{Re}) \qquad (3.103)$$

Equations (3.96)–(3.99) are analogous with Eqs. (3.100)–(3.103), because the basic concepts for heat transfer are the same as for mass transfer (see Table 3.4).

When a copper plate positioned vertically in an unstirred solution of sulfuric acid dissolves anodically, the concentration of copper ions near the anode increases, and downward flow of the electrolytic solution due to natural convection takes place. The Navier–Stokes equation in this case is

$$\frac{\partial u_x}{\partial t} + u_x \frac{\partial u_x}{\partial x} + u_y \frac{\partial u_x}{\partial y} = v \frac{\partial^2 u_x}{\partial y^2} + g\beta \, \Delta C \qquad (3.104)$$

and

$$\beta = \frac{1}{\rho_0}\frac{\partial \rho}{\partial C} = \frac{1}{\rho_0}\frac{(\rho_0 - \rho)}{\Delta C} \qquad (3.105)$$

where ρ_0 and ρ are the density of solution on the electrode surface and in the bulk of solution, respectively. The result shown in Eq. (3.108) is obtained by integration of Eq. (3.104) with the boundary conditions:

$$u_x = 0 \quad \text{and} \quad C = C_s \qquad \text{at } y = 0$$

$$u_x = u_{\max} \qquad\qquad\qquad \text{at } y = \tau$$

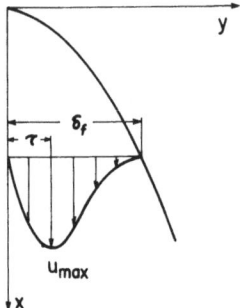

FIGURE 3.20. Boundary layer under free convection.

and

$$u_x = 0 \quad \text{and} \quad C = C_0 \quad \text{at } y = \delta_f$$

as shown in Fig. 3.20.

$$\text{Sh} = \alpha(\text{Gr})^{1/4}(\text{Sc})^{1/4} \tag{3.106}$$

where the Grashof number (Gr) is defined as follows:

$$\text{Gr} = \frac{gL^3}{\nu^2}\beta(C_0 - C_s) \tag{3.107}$$

In Eq. (3.106), α is close to unity.

It is well known that the rotating disk electrode is a useful tool for electrochemical studies. However, the Navier–Stokes and continuity equations present a formidable array of terms for velocity components. Here only the vector notation is given as follows:

$$\frac{\partial u}{\partial t} + \nabla \cdot uu = -\frac{1}{\rho}\nabla p + \nu\nabla^2 u \tag{3.108}$$

and

$$\nabla \cdot u = 0 \qquad \text{(incompressible flow)} \tag{3.109}$$

The boundary conditions in this case are

$$u_r = 0, \quad u_z = 0, \quad \text{and} \quad u_\theta = r\omega \qquad \text{at } z = 0$$

$$u_r = 0 \quad \text{and} \quad u_\theta = 0 \qquad\qquad \text{at } z = \infty$$

where r is the radius, z is the axis perpendicular to the disk, θ is the angle, and ω is the rotating speed in revolutions per second (RPS). The thickness of the velocity boundary layer is given by the equation[21]

$$\delta_f = 6.194 \left(\frac{\nu}{\omega}\right)^{1/2} \tag{3.110}$$

where

$$\text{Re} = \frac{\omega r_0^2}{\nu} \tag{3.111}$$

If a disk electrode is operated at very high speed,

$$\delta_f = 0.522 \, r \left(\frac{\nu}{\omega r^2}\right)^{1/5} \tag{3.112}$$

The critical Reynolds number is said to be about 5×10^4.

The limiting current density, i_L, under laminar flow conditions ($\text{Re} < 5 \times 10^4$) has been investigated in detail, yielding the result:

$$i_L = nFk(\Delta C)\sqrt{\omega} \tag{3.113}$$

That is, i_L is proportional to $\sqrt{\omega}$ (see Section 3.3.2).

REFERENCES

1. H. Kita, *Electrode Processes* (E. Yeager and E. Eisenmann, ed.), p. 79, Electrochemical Society (1970).
2. F. Hine, M. Yasuda, S. Bamba, K. Yamakawa, and T. Nakagawa, *Denki Kagaku* **38**, 893 (1970).
3. A. C. Makrides, *J. Electrochem. Soc.* **109**, 977 (1962).
4. N. Sato and G. Okamoto, *Electrochim. Acta* **10**, 495 (1965).
5. F. Hine and M. Yasuda, *J. Electrochem. Soc.* **121**, 1289 (1974).
6. J. O'M. Bockris and A. K. N. Reddy, *Modern Electrochemistry*, Vol. 2, p. 1091, Plenum, New York (1970).
7. J. O'M. Bockris and G. A. Razumney, *Fundamental Aspects of Electrocrystallization*, Plenum, New York (1967).
8. N. Ibl and R. Muller, *Z. Elekrochem.* **59**, 671 (1955).
9. K. J. Vetter, *Electrochemical Kinetics*, p. 632, Academic, New York (1967).
10. F. Hine, M. Yasuda, and F. Matsui, *Zairyo* (J. Soc. Materials Sci. Japan) **22**, 956 (1973).
11. S. Glasstone, K. J. Laidler, and H. Eyring, *The Theory of Rate Processes*, p. 552, McGraw-Hill, New York (1941).
12. F. Hine and M. Yasuda, *J. Electrochem. Soc.* **118**, 170 (1971).
13. K. J. Vetter, *Electrochemical Kinetics*, p. 149, Academic, New York (1967).
14. W. Vielstich, *Z. Elektrochem.* **57**, 646 (1953).

15. U. Landau and C. Tobias, Abstract No. 266, Electrochemical Society meeting, Washington, D.C., May 1976.
16. Y. Awakura, Y. Takenaka, and Y. Kondo, *Electrochim. Acta* **21**, 789 (1976).
17. H. Blasius, *Z. Math. Phys.* **56**, 1 (1908).
18. K. Asada, F. Hine, S. Yoshizawa, and S. Okada, *J. Electrochem. Soc.* **197**, 242 (1960).
19. R. Bird, W. E. Stewart, and E. N. Lightfoot, *Transport Phenomena*, p. 142, Wiley, New York (1960).
20. V. G. Levich, *Physicochemical Hydrodynamics*, p. 9, Prentice-Hall, Englewood Cliffs (1962).
21. D. Hershey, *Transport Analysis*, p. 162, Plenum/Rosetta, New York (1974).
22. H. Schlichting, *Boundary Layer Theory*, pp. 108 and 242, McGraw-Hill, New York (1960).
23. R. Bird, W. E. Stewart, and E. N. Lightfoot, *Transport Phenomena*, pp. 368 and 605, Wiley, New York (1960).

VOLTAGE BALANCE AND ENERGY BALANCE IN AN ELECTROLYTIC CELL

4.1. CONDUCTIVITY OF ELECTROLYTIC SOLUTIONS

4.1.1. Specific Conductance, Molar and Equivalent Conductivities

Electrolytic current will flow from the anode to the cathode in electrolytic solution in a cell by means of transport of both cations and anions to the respective directions of the electric field. The mechanism of conduction in electrolytic solution differs from that in a metallic conductor in which free electrons move in the inverse direction of flow of electric current. However, the voltage drop through the electrolytic solution is governed by Ohm's law

$$IR = Il/A\kappa = il/\kappa \qquad (4.1)$$

where I is the current in amperes, i the current density in A/cm^2, l the electrode spacing in cm, A the uniform cross-sectional area in cm^2, and κ the conductivity in mhos/cm.

The conductivity of the HCl solution (see Fig. 4.1) increases with an increasing concentration and temperature, whereas the conductivity decreases in solutions more concentrated than 6 N HCl.[1] Sodium hydroxide and potassium hydroxide solutions show similar properties, and have maximum conductivity at concentrations of 20–40% by weight, as shown in Figs. 4.2 and 4.3, respectively.[2]

The molar conductivity, Λ_m in mho cm^2/mol, and the equivalent conductivity, Λ in mho cm^2/g-eq., are shown by Eq. (4.2) and (4.3) as

$$\Lambda_m = 1000\,\kappa/C \qquad (4.2)$$

and

$$\Lambda = 1000\,\kappa/c \qquad (4.3)$$

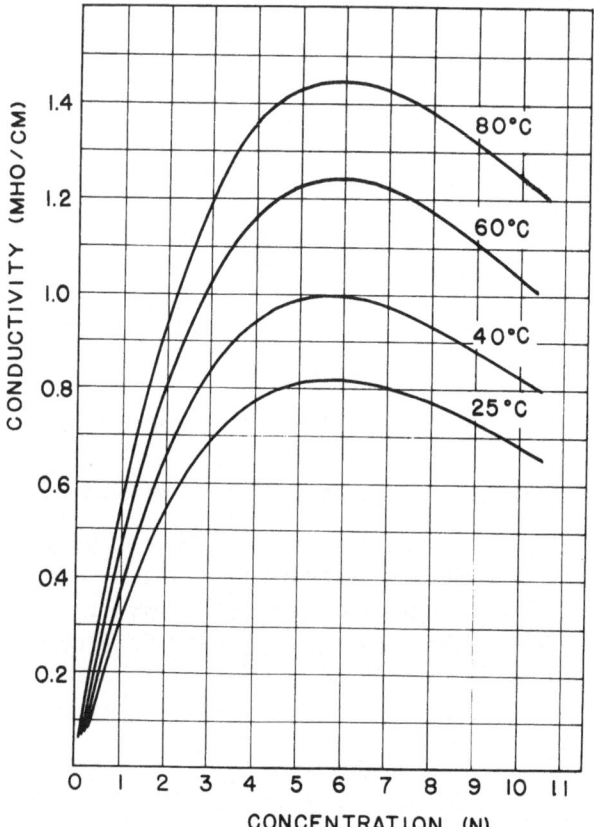

FIGURE 4.1. *Conductivity of HCl solutions.*

where C and c are the molar concentration in mol/liter and the equivalent concentration in g-eq./liter, respectively.

The number of ions which will form from 1 g-mol of electrolyte is $\alpha v_i N$, where α is the degree of dissociation, v_i is the number of ions of interest in a molecule, and N is the Avogadro number. The subscript i refers to the ionic species under consideration. The charge transfer for an ionic species is $n_i e u_i$ in an electric field of intensity 1 V/cm, where u_i is the ion mobility, and n_i is equivalents per mole. Therefore, with $Ne = F$, we have

$$\Lambda_m = \sum_i \alpha v_i N n_i e u_i = \alpha F \sum_i v_i n_i u_i \qquad (4.4)$$

In the case of a binary electrolyte,

$$v_+ n_+ = v_- n_- = vn$$

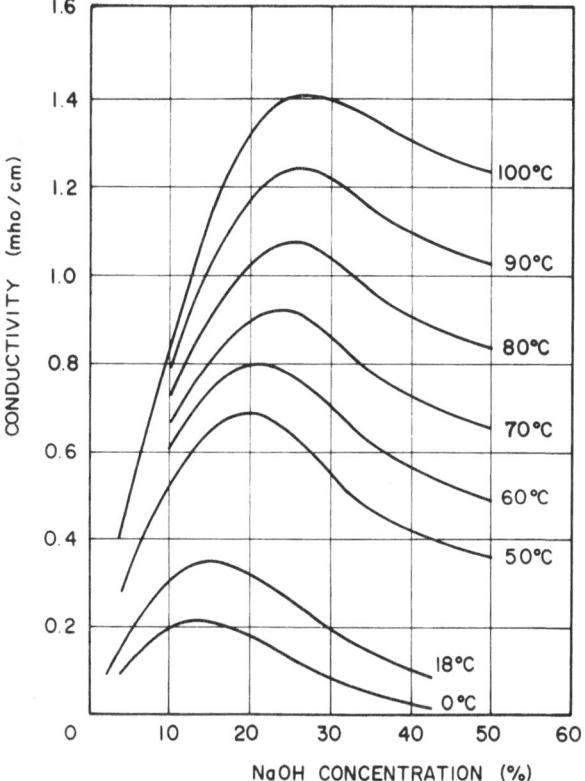

FIGURE 4.2. Conductivity of NaOH solutions.

Thus

$$\Lambda_m = \alpha F v n (u_+ + u_-) \tag{4.5}$$

and

$$\Lambda = \alpha F (u_+ + u_-) \tag{4.6}$$

where the plus and minus subscripts show the cation and anion, respectively. That is, ionic species having the respective velocity of u_i are considered to move independently from one electrode to another depending on the direction of the unit electric field in the cell. Movement of ionic species might be retarded in concentrated solutions, whereas it is easy to move in dilute solutions. That is, the equivalent conductance at the infinite dilution Λ_∞ is represented by the sum of the equivalent conductance $\lambda_i = F u_i$ of all the ions existing in the solution.

It is clear in Eq. (4.6) that the cation and anion each contribute to a

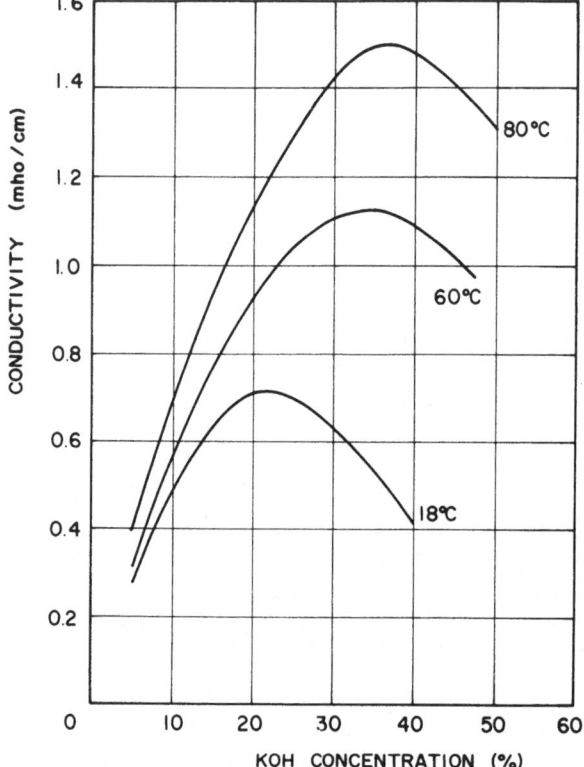

FIGURE 4.3. *Conductivity of KOH solutions.*

part of electric conduction, i.e., Fu_+ and Fu_-, respectively. The fractional contributions of both ions to electric conduction are

$$t_+ = \frac{u_+}{u_+ + u_-} \qquad \text{and} \qquad t_- = \frac{u_-}{u_+ + u_-} \qquad (4.7)$$

in which t is named the transport number or the transference number.

The transport numbers of H^+ and OH^- are particularly large in comparison with other ions as shown in Table 4.1, attributed to a different mechanism of conduction for these ions.[3]

4.1.2. Example 1: Conductivity of Concentrated KCl Solutions

Electrolysis of NaCl solutions has a large share in the chlor-alkali industry, but the capacity of KCl electrolysis is also increasing. While both processes are essentially the same, details such as voltage balance are

TABLE 4.1
Equivalent Conductance of Ions in Solutions of Infinite Dilution

Cations	0°C	25°C	100°C	Anions	0°C	25°C	100°C
$H^+(H_2O)$	225	349.7	637	OH^-	105	200	446
Li^+	19.1	38.68	120	F^-		55.4	
Na^+	25.85	50.10	150	Cl^-	41.4	76.32	207
K^+	40.3	73.50	200	ClO_3^-	36	64	172
Rb^+	43.5	76.4		ClO_4^-	37.3	68	179
Cs^+	44	76.8	200	Br^-	43.1	78.3	
NH_4^+	40.3	73.7	184.3	BrO_3^-	31	56	155
$\frac{1}{2}Mg^{2+}$	28.5	53.06	170	I^-	42.0	76.8	
$\frac{1}{2}Ca^{2+}$	30.8	59.50	187	IO_3^-	21	41.0	127
$\frac{1}{2}Ba^{2+}$	33.6	63.7	200	IO_4^-		55.6	
$\frac{1}{3}Ce^{3+}$		67		$\frac{1}{2}SO_4^{2-}$	41	79.8	256
$\frac{1}{3}Cr^{3+}$		67		NO_3^-	40.2	71.42	189
$\frac{1}{2}Mn^{2+}$	27	53.5		$H_2PO_4^-$		36	
$\frac{1}{2}Fe^{2+}$	28	53.5		$\frac{1}{2}HPO_4^{2-}$		57	
$\frac{1}{3}Fe^{3+}$		68		HCO_3^-		44.5	
$\frac{1}{2}Co^{2+}$	28	54		$\frac{1}{2}CO_3^{2-}$	36	72	
$\frac{1}{2}Ni^{2+}$	28	54		CN^-		78	
$\frac{1}{2}Cu^{2+}$	28	56		$\frac{1}{2}CrO_4^{2-}$	42	85	
Ag^+	33	61.9	180	$\frac{1}{2}MoO_4^{2-}$		74.5	
$\frac{1}{2}Zn^{2+}$	28	53.5		MnO_4^-	36	62.8	
$\frac{1}{2}Cd^{2+}$	28	54		CH_3COO^-	20	41	130
Tl^+	43.3	74.9		$HCOO^-$		47 at 18°C	
$\frac{1}{2}Pb^{2+}$	37.5	70		$\frac{1}{2}(C_2O_4)^{2-}$		63 at 18°C	

different due to the difference in physicochemical properties between the two salts.[4-12]

For the performance of electrolytic cells, a general formula for calculating the conductivity of the aqueous solutions of KCl electrolysis is required.

An example of the conductivity versus temperature curve is shown in Fig. 4.4. However, representation of specific conductance as a function of the KCl concentration at various temperatures, such as shown in Fig. 4.5, is more feasible for use. Thus, we may put

$$\kappa = a + bC + cC^2 \tag{4.8}$$

where κ is the conductivity in mho/cm and C is the concentration in mol/liter. Three coefficients, a, b, and c, are functions of temperature as follows:

$$a = a_0 + a_1 \left(\frac{t}{100} \right) + a_2 \left(\frac{t}{100} \right)^2 + a_3 \left(\frac{t}{100} \right)^3 \tag{4.9}$$

$$b = \beta_0 + \beta_1 \left(\frac{t}{100} \right) + \beta_2 \left(\frac{t}{100} \right)^2 \qquad (4.10)$$

$$c = \gamma_0 + \gamma_1 \left(\frac{t}{100} \right) + \gamma_2 \left(\frac{t}{100} \right)^2 \qquad (4.11)$$

$$\alpha_0 = +0.06714771 \qquad \beta_0 = -0.02683625 \qquad \gamma_0 = +0.02169544$$
$$\alpha_1 = -0.27901096 \qquad \beta_1 = +0.59002202 \qquad \gamma_1 = -0.11169075$$
$$\alpha_2 = +0.21209696 \qquad \beta_2 = -0.28912264 \qquad \gamma_2 = +0.06978237$$
$$\alpha_3 = +0.00089866$$

Minimum deviation of this formula is 1.5%, and is enough for engineering use.

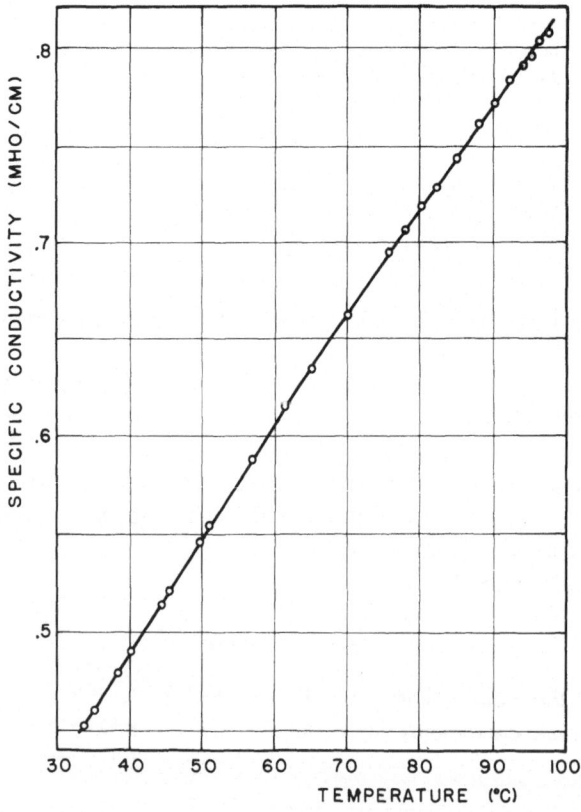

FIGURE 4.4. *Specific conductance of 5 mol/1 KCl solution at various temperatures.*

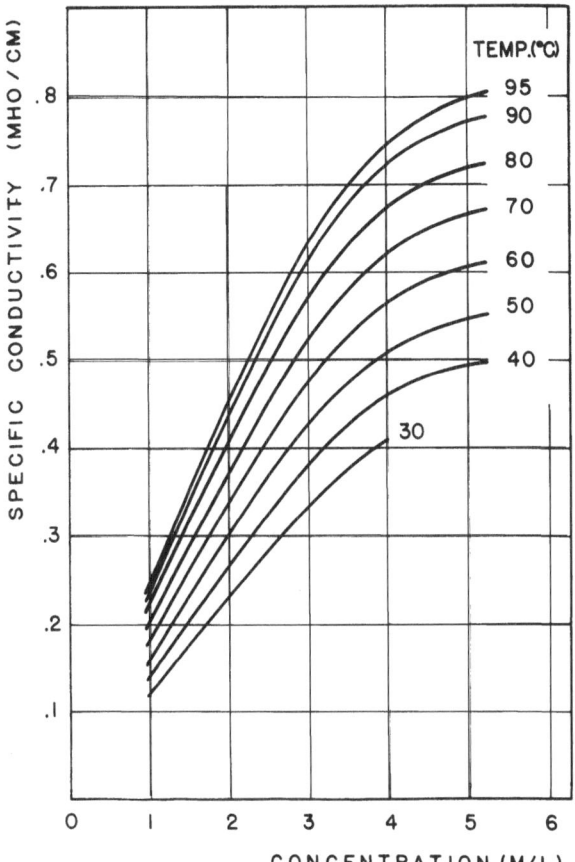

FIGURE 4.5. *Specific conductance of concentrated KCl solutions at high temperature ranges.*

The slope of the potential versus distance curve near the working electrode depends on the conductivity of solution

$$\Delta E = i/\kappa$$

where ΔE is the potential gradient in V/cm. However, it is well known that the conductivity evaluated in this manner with the potential versus distance curve near the graphite anode in KCl solution, at which chlorine evolution takes place, is about 50% larger than one of pure solution calculated by Eq. (4.8). The pH of the solution near the graphite anode decreases considerably during electrolysis due to dissolution of chlorine, and excess H^+ may assist the conduction. Chlorine bubbles disturb passage of electric current at very high current density, and the apparent conductivity of

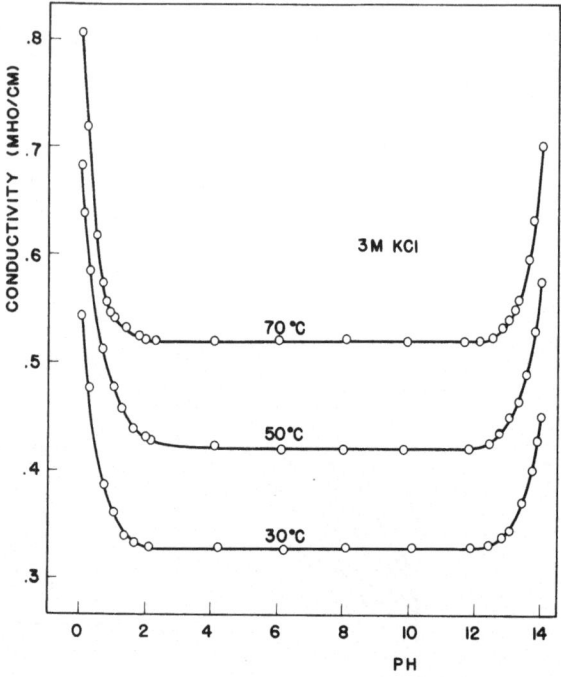

FIGURE 4.6. Conductivity vs. pH of 3 M KCl at constant temperature.

TABLE 4.2

Equivalent Conductance of NaCl
Solution at 25°C[a]

c (g-eq./liter)	Λ (mho · cm^2/g-eq.)
0.0001	125.59
0.0002	125.24
0.0005	124.50
0.001	123.74
0.002	122.67
0.005	120.65
0.01	118.51
0.02	115.76
0.04	112.34
0.06	110.01
0.08	108.23
0.10	106.74
0.12	105.48
0.14	104.42
0.16	103.43
0.18	102.52
0.20	101.70
0.22	100.96

[a] From Ref. 13.

solution near the working electrode decreases in a practical cell (see Sec. 4.1.4).

Figure 4.6 is the conductivity of KCl at various conditions of concentration, temperature, and pH. The conductivity of both acidified and alkaline solutions is large in comparison with neutral solution due to the large mobilities of H^+ and OH^-.

The equivalent conductance Λ of dilute strong electrolytes is shown by the well-known Onsager's equation

$$\Lambda = \Lambda_0 - K\sqrt{c} \tag{4.12}$$

where Λ_0 is the equivalent conductance at the infinite dilution and K is a constant. This equation was obtained theoretically with several assumptions for simplicity. Therefore, Eq. (4.12) (Onsager's equation) is applicable to a limited range of the electrolyte concentration. MacInnes has described the theory and the concepts in detail.[13] Table 4.2 shows the equivalent conduc-

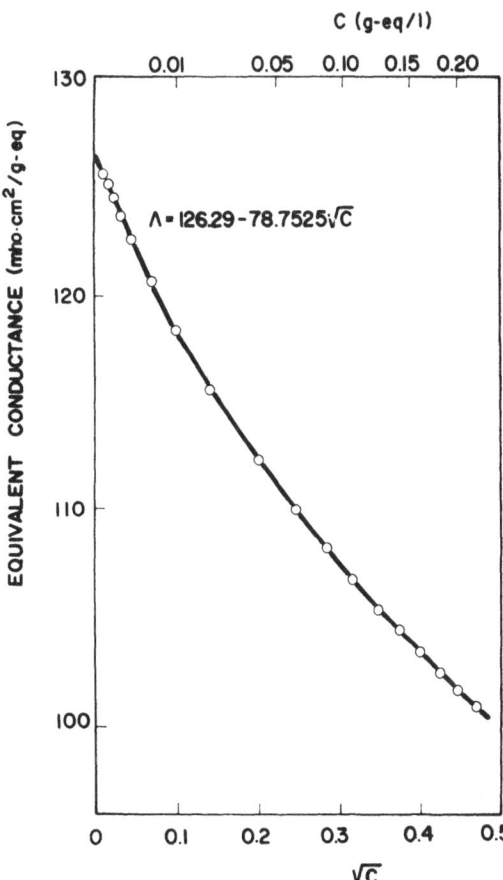

FIGURE 4.7. Onsager's equation for NaCl solutions at 25°C.

tance of dilute NaCl solutions at 25°C. As is shown in Fig. 4.7, the Λ vs. $\sqrt{c}$ curve is straight only in the concentration range up to 0.010 g-eq./liter. A number of modification of Onsager's equation have been proposed to obtain appropriate relationship between Λ and c for more concentrated solutions such as

$$\Lambda = \Lambda_0 - k_1\sqrt{c} + k_2 c \tag{4.13}$$

and

$$\frac{\Lambda + \sigma\sqrt{c}}{1 - \theta\sqrt{c}} = \Lambda_0 + B_1 c + B_2 c^2 \qquad \text{(Shedlovsky's equation)} \tag{4.14}$$

where k_1, k_2, B_1, and B_2 are coefficients, and would be obtained experimentally. On the other hand, σ and θ in Eq. (4.14) are functions of solvent and temperature. For aqueous solutions, $\sigma = 59.78$ and $\theta = 0.2273$ at 25°C.[14]

Equation (4.14) fits well the experimental results as shown in Fig. 4.8, and the coefficients B_1 and B_2 are also given in the figure. Of course, this equation is valid only for more dilute solutions than 0.2 g-eq./liter at 25°C.

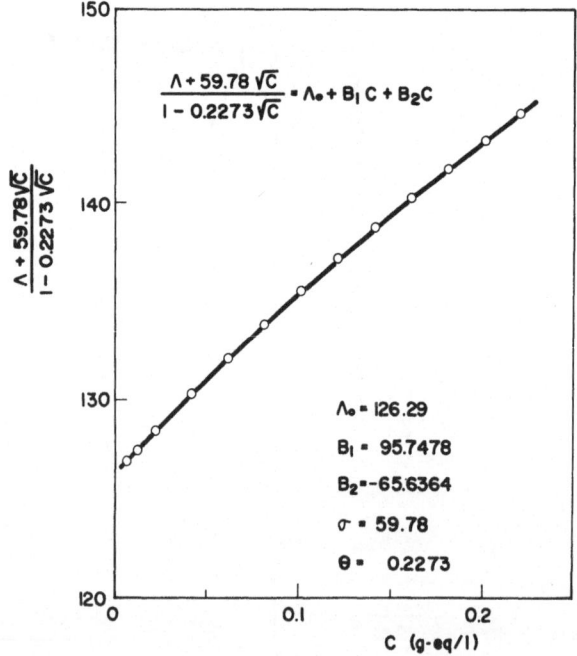

FIGURE 4.8. Shedlovsky's equation (4.14) for NaCl solution at 25°C.

4.1.3. Example 2: Conductivity of the Mixed Solution of HCl and $CuCl_2$

For the direct electrolysis of hydrochloric acid, the electrolyte is a single solution of HCl. A single and saturated solution of NaCl is used as electrolyte in chlor-alkali cells. However, mixed solutions are also used in some cases of electrochemical processes.

Very few papers on the conductivity of mixed solutions have been published. Therefore, measurements are performed as needed. Procedures for the measurement of conductivity of electrolytic solution are straightforward except for cases of high-temperature and high-pressure conditions. Many textbooks on experimental work have been published (see Ref. 13, Chap. 3).

Here we may discuss HCl solutions containing $CuCl_2$, which have been used as electrolyte for an indirect electrolysis of hydrochloric acid.[1,15]

The conductivity of HCl solution decreases when $CuCl_2$ is added, as shown in Fig. 4.9. Decrease of the solution conductivity affects the terminal voltage, especially at high-current densities. For example, the conductivity of a mixed solution consisting of 5.6 M HCl and 1.5 M $CuCl_2$ is 0.87 mho/cm at 50°C compared to 1.13 mho/cm for a single solution of 5.6 M HCl.

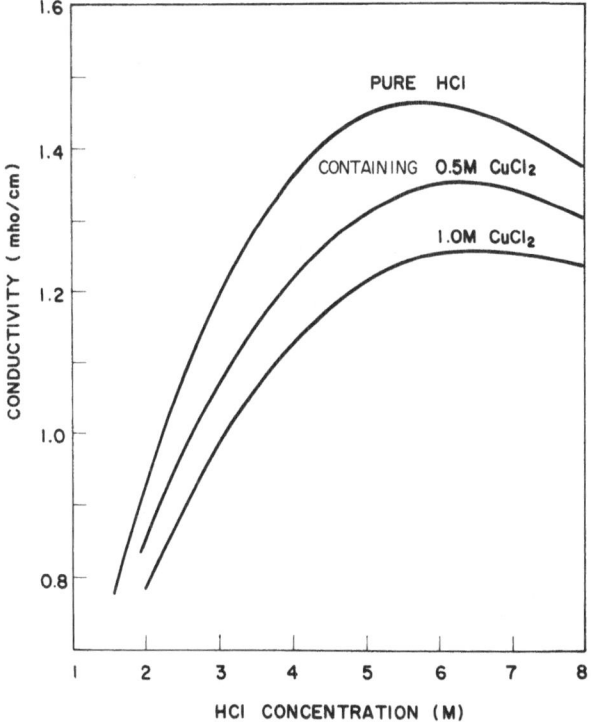

FIGURE 4.9. Conductivity of HCl–$CuCl_2$ mixed solutions at 80°C.

Therefore, the solution IR drops in both solutions for 1 cm of the interelectrode gap at 1 A/cm^2 are:

In single solution: $\dfrac{1 \times 1}{1.13} = 0.88$ V

In mixed solution: $\dfrac{1 \times 1}{0.87} = 1.15$ V

Difference: 0.27 V

Even though it is not large compared to the overall IR drop in the electrolyte solution, it is still a factor to be considered for reducing cell voltage to a minimum.

The conductivity of molten salt electrolyte also decreases when a foreign salt is added. For example, NaF melt has a large conductivity, but the conductivity of the NaF–AlF$_3$ mixture decreases with increase of the AlF$_3$ content, whereas it increases with addition of LiF. These electrolyte melts are used in the well-known Hall–Heroult type of aluminum cell (see Chap. 9).

Physicochemical properties of concentrated electrolyte solutions such as electric conductivity, activity, and viscosity are still veiled in mystery in comparison with that of dilute solutions, while most industrial electrochemical processes are using concentrated solutions or saturated solutions in some cases.

4.1.4. Conductivity of the Electrolyte Solutions Containing Gas Bubbles

An important item in the voltage balance of an industrial electrolytic cell is the IR drop in the electrolyte across the gap between the electrodes. When the electrode reaction evolves gas, the presence of gas bubbles dispersed in the electrolyte may greatly increase the IR drop. This is particularly true if the electrode is not perforated, in which case all the evolved gas must pass through the electrolysis gap.

There are several ways of minimizing the unwanted "bubble effects":

1. *Use of perforated electrodes* permits some, if not all, of the evolved gas to be removed from the electrolysis gap. Thus, in amalgam-type chlorine cells equipped with horizontal block anodes of graphite, these blocks may be machined and perforated in various ways to minimize accumulation of chlorine gas bubbles on the underside of these blocks.[16-19] Also, modern amalgam cells equipped with "spaghetti"-type metal anodes almost eliminate the bubble effect.

In modern diaphragm-type chlorine cells equipped with vertical dimensionally stable anodes (DSA), which are usually perforated, a major part of the evolved chlorine is removed from the electrolysis gap by passing through the perforations. This also permits closer spacing between anode and diaphragm. Thus, the *IR* drop can be minimized.

2. *Recirculation of electrolyte* decreases the gas void fraction in the electrolysis zone, and thus improves the conductivity of the two-phase mixture of electrolyte and gas in the gap.

In older diaphragm-type chlorine cells equipped with solid graphite blade anodes, a return path is provided which allows the anolyte brine to recirculate freely. The recirculation is driven by the gas lift effect in the anode–diaphragm gap.[5,9]

In newer diaphragm-type chlorine cells equipped with box-type DSAs, the inside of the DSA can be utilized as a downcomer for recirculating the anolyte. Box-type electrodes can also be used in cells for water electrolysis to improve the circulation of caustic potash electrolyte, thus decreasing the void fraction of hydrogen on one side of the diaphragm, and of oxygen on the other.[20]

3. *Miscellaneous* methods, such as electrolysis under high pressure, decrease the volume of the gas bubbles and improve conductivity. Increase in the resistivity of electrolyte due to suspensions of nonconducting particles has been studied by De La Rue and Tobias,[21] and Tobias has also described a theoretical treatment of the effect of gas on the current distribution in a vertical cell.[22] Hine *et al.* studied the *IR* drops in the chlor-alkali cells, and discussed preferable geometry and arrangement of the graphite anode for the horizontal cell.[16–19,23]

Gardiner stated that the apparent resistivity of the brine between the graphite anode and the amalgam cathode of the Olin E-11 cell was practically free of chlorine bubbles when the anode-to-cathode gap was more than 0.3 in., whereas the apparent resistivity of brine was affected by bubbles if the gap was narrow.[24] MacMullin pointed out that the vertical-type diaphragm cell such as the Hooker cell is a "good pump," because gas lift near the anode enhances circulation of the brine and causes significant reduction of the *IR* drop between two electrodes.[5,9] The cell voltage of the amalgam-type chlorine cells operated at high flow rate of mercury cathode is relatively low even at high-current densities, because chlorine bubbles under the graphite anodes are removed quickly.[25–27]

Other electrolytic cells with gas evolution include water electrolyzers. In some cases, water electrolyzers are operated with forced flow of solution along the electrode, and the *IR* drop depends on the flow rate.

Additional useful papers with respect to the bubble effects on the current distribution and the voltage balance in electrolytic cells can be found.[28–30] Funk and Thorpe studied the void fraction and the current distribution in a water electrolysis cell.[28] They concluded that the slip ratio was of great importance in defining the void fraction and in the analysis of hydrodynamics or pressure drop in the electrolytic cell.

Although there is no doubt that the *IR* drop and the overvoltage near the electrode are affected by bubbles, enhancement of mass- and heat-transfer is also of importance in electrolytic gas evolution. The thickness of the hydrodynamic film or the boundary layer along the electrode is reduced, and hence the heat transfer coefficient in the cell is increased by circulation of the electrolytic solution.[31-39] MacMullin *et al.* studied the enhancement of heat transfer in an electrolyzer where hydrogen evolution took place at the cathode.[34] The gas evolution promotes rapid circulation of the electrolyte resulting in a significant increase of the fluid film heat transfer coefficient by a factor of 2 to 3.

Flow of the gas–electrolyte mixture along the vertical electrode during cell operation is a "two-phase flow" or a "backmix flow" depending on the liquid flow velocity and the void fraction of bubbles.[40-44] In some cases, gas bubbles grow in size, and cover the electrode surface, thus the overvoltage becomes high. On the other hand, some gas bubbles leaving the electrode disperse in the solution, and the *IR* drop between two electrodes increases. That is, "bubble effects" are classified into (i) the surface area reduction on the working electrode, and (ii) increase in the resistivity of electrolytic solution.

Tobias and his collaborators propose Bruggeman's equation

$$\rho/\rho_0 = (1 - \varepsilon)^{-3/2} \tag{4.15}$$

to correlate the resistivity versus gas void fraction in electrolytic cells.[21,22,43] The gas void fraction is closely related to the bubble size, i.e., the larger the bubble size the smaller is the void fraction in general. This relationship is valid for electrolytic gas bubbles over a wide range of operation, as shown in Fig. 4.10.[45,46]

Figure 4.11 shows examples of the gas void fraction ε as a function of the distance from cell bottom. Tobias made a mathematical treatment on the gas void fraction ε_x and the local current density i_x, where subscript x indicates local position.[21] Some assumptions proposed are as follows:

● The electrode is equipotential.
● Size distribution of bubbles at all levels is the same.
● The rise velocity of bubbles is given by Stokes' law.
● The rise velocity is independent of depth.
● Rise velocity is independent of volume concentration of gas.
● No coalescence of gas bubbles occurs.

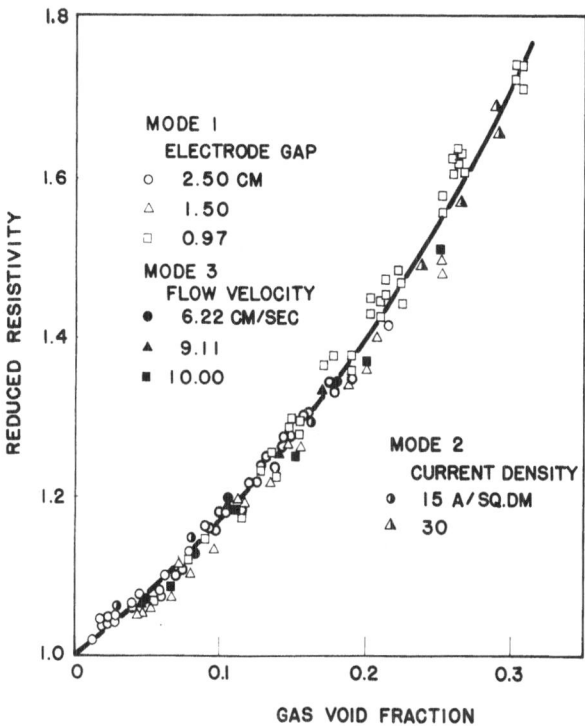

FIGURE 4.10. Relationship between the reduced resistivity and the gas void fraction under blocked convection (Mode 1), natural circulation (Mode 2), and forced circulation (Mode 3). See Fig. 4.13.

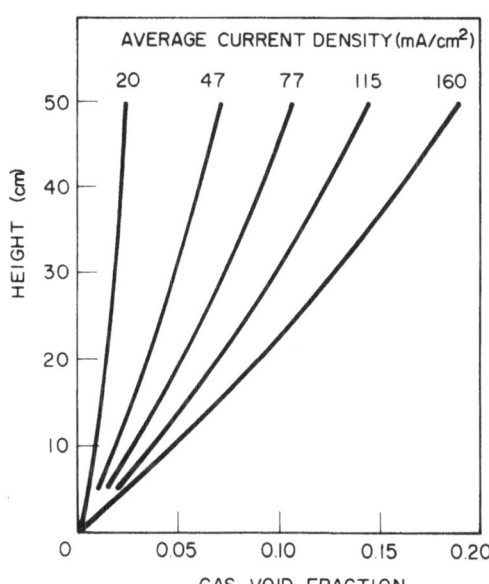

FIGURE 4.11. Gas void fraction as a function of the distance from cell bottom.[12]

- The velocity of liberation of gases depends on pressure, temperature, and electrode material.
- Polarization is linear over the range of current density under discussion.
- The flow of current is unidirectional.

With those assumptions, he obtained simple equations for ε and i as follows

$$\varepsilon_x = [(KX)^2 + 4KX]/(KX + 2)^2 \tag{4.16}$$

and

$$i_x = 8K/hC(KX + 2)^3 \tag{4.17}$$

where K is the gas effect parameter, X is the dimensionless height ($=x/h$), C a coefficient, and h is the total height.

Since the average current density i_a is given by the equation

$$i_a = \frac{1}{h} \int_0^h i_x \, dx = K(K + 4)/Ch(K + 2)^2 \tag{4.18}$$

the dimensionless current density and resistivity ρ are shown as follows:

$$i_x/i_a = 8(K + 2)^2/(KX + 2)^2(K + 4) \tag{4.19}$$

and

$$\rho/\rho_0 = (K + 2)^2/(K + 4) \tag{4.20}$$

where ρ_0 is the resistivity of solution freed from bubble. The average gas void fraction ε_u is

$$\varepsilon_a = K/(K + 2) \tag{4.21}$$

Figure 4.12 shows the current distribution along the vertical electrodes of different height.[46] It is important that the i_x/i_a vs. x/h is almost independent of the electrode height. The thick line shows the empirical equation obtained by the method of least squares with the data plotted. The thin line is the curve calculated by Tobias' equation (4.19) with $K = 0.6$.

Tobias also showed the effect of polarization on the local current density along the electrode at which electrolytic gas bubbles evolve.

$$i_x/i_a = \tfrac{3}{2}K\{(\mu + 1)^2 + 3KX\}^{-1/2}[\{(\mu + 1)^2 + 3K\}^{1/2} - (\mu + 1)]^{-1} \tag{4.22}$$

$$KX = \mu\varepsilon_x + 2(1 - \varepsilon_x)^{-1/2} - 2 \tag{4.23}$$

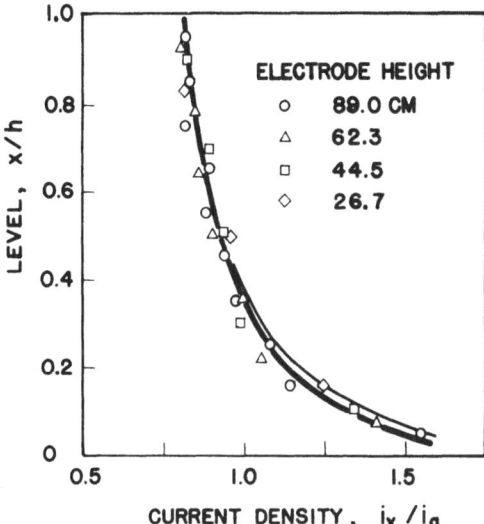

FIGURE 4.12. Current distribution curve with electrodes of different heights. Electrode gap = 3.90 cm. Average current density = 37.4 A/dm². Thin line calculated by Eq. (4.19) with K = 0.6.

where

$$\mu = b/\rho\delta$$

b is the sum of the Tafel slopes at the anode and the cathode, and δ is the interelectrode gap. Tobias states that the current distribution becomes uniform when the polarization increases.

Hine studied the effect of electrolyte circulation on the performance of gas-evolving cells, particularly with reference to current distribution and voltage drop in the electrolysis zone.[46] Figure 4.13 illustrates three types of

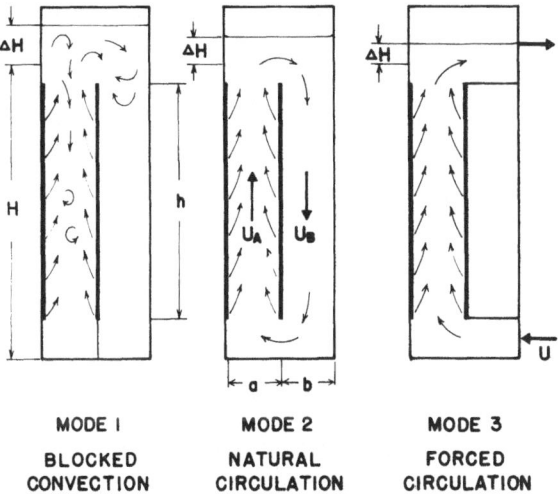

FIGURE 4.13. Flow pattern in vertical cells.

flow pattern in a vertical cell: blocked convection (Mode 1), natural circulation (Mode 2), and forced circulation (mode 3). In Mode 1, the gas–liquid mixture rises to the cell top, and the solution freed of gas bubbles comes down through a part of the electrode gap, thus complicating the flow pattern. In Mode 2, the gas–solution mixture rises through the electrode gap by gas lift action and the solution flows down through the downcomer when the circulation path is provided. Separation of gas bubbles from the solution is stimulated by rapid flow, and hence the gas void fraction decreases in this case. On the other hand, when the solution containing gas bubbles is forced into the reservoir from the cell top by pumping, the gas void fraction may be still further reduced, as in Mode 3. In Fig. 4.13, H represents the level of electrolyte before the current is turned on. With current on, the level increases by ΔH, due to hold up of gas bubbles in the electrolyte.

For Mode 1, the spacing $b = 0$, and the average gas void fraction, ε_a is

$$\varepsilon_a = \frac{H}{H + \Delta H} \qquad \text{(approximately)} \qquad (4.24)$$

For Mode 2

$$\varepsilon_a = \frac{a + b}{a} \frac{\Delta H}{H + \Delta H} \qquad \text{(approximately)} \qquad (4.25)$$

In this case, the solution in the downcomer is assumed to be bubble-free.

Suppose that the gas void fraction at the center level is ε_a. The current density at the center on the vertical electrode is almost average. The data shown in Fig. 4.10 were obtained by this manner.

In a cell operated under natural circulation (Mode 2), the flow velocity reaches a maximum at an electrode gap of 0.75 cm, as shown in Fig. 4.14. Also shown is the effect of solution flow on reduction of the IR drop or the reduced resistivity of solution (center).

The terminal voltage was minimum at 5–7 mm of the electrode gap. In a narrow channel, less than 5 mm, the flow pattern becomes slug, and ρ/ρ_0 is large. On the other hand, ρ/ρ_0 is almost independent of the electrode gap when the gap is larger than 10 mm. Consequently, the solution IR drop and the terminal voltage at constant current density increase with increase of the electrode gap (top, Fig. 4.14).

The reduced resistivity in a cell without circulation, Mode 1, is shown by the closed points. Since convection occurs only in the space between two electrodes, the larger the electrode gap, the smaller is the resistivity. Also, ρ/ρ_0 is affected greatly by the current density, which indicates difficulty of separation of gas bubbles from the electrolytic solution.

In this connection, the size of the downcomer seems to be an important factor for natural circulation. According to experiments, a gap of 12–15 mm was sufficient for solution circulation.

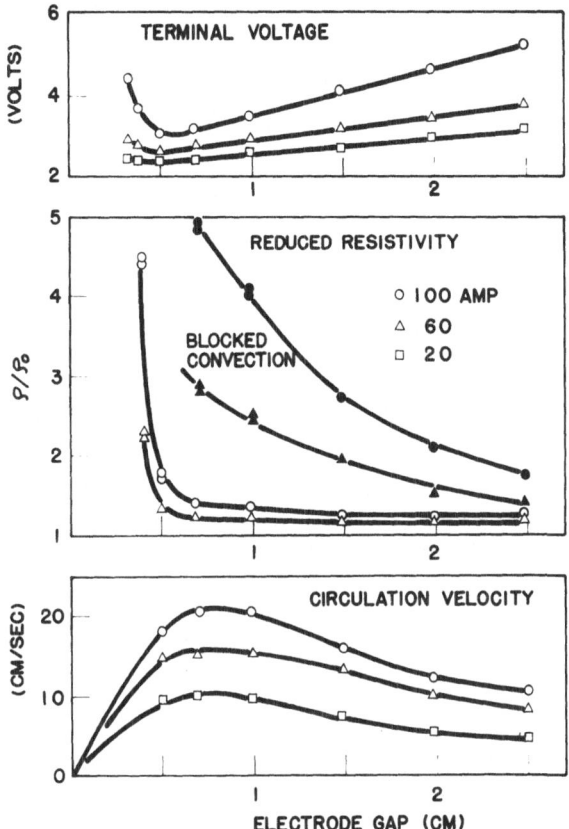

FIGURE 4.14. Terminal voltage, reduced resistivity, and circulation velocity as functions of the electrode gap and the current. Average current density: 37.2 A/dm² at 100 A, 22.7 A/dm² at 60 A, and 7.5 A/dm² at 20 A. Closed points = blocked convection (Mode 1).

The experiments described in the paper in Ref. 46 were conducted with solid electrodes, and so the electrolysis zone was crowded with evolved gases. This was the case for the old-style diaphragm-type chlorine cells using graphite blade anodes. In the new-type cells using DSAs, the anodes are perforated, so that the evolved chlorine passes through the electrodes to the outer space, and hence the gas void fraction in the electrode gap is reduced considerably, resulting in low voltage drops.

The bubble effect in a horizontal cell under pumped circulation has also been studied.[25,46] The reduced resistivity of electrolytic solution containing gas bubbles was a function of the ratio of gas volume to liquid volume (V_g/V_1), the electrode gap, and the Reynolds number as follows

$$\frac{\rho}{\rho_0} = k \left(\frac{V_g}{V_1} \right)^\alpha \left(\frac{\delta}{L} \right)^\beta (\mathrm{Re})^\gamma \qquad (4.26)$$

where V is the flow rate, L is the cell width, δ is the electrode gap, and k, α, β, and γ are coefficients. The Reynolds number, Re, in Eq. (4.26) is a weak factor in this case since V_g/V_1 contains some influences of Re.

4.2. CURRENT EFFICIENCY AND ENERGY EFFICIENCY

4.2.1. Current Efficiency

It is well known that electrochemical conversion, w in kg, is proportional to the product of current, I in kA, and time of electrolysis, θ in hr:

$$w = \left(\frac{M}{nF}\right) I\theta \qquad \text{(Faraday's law)} \qquad (4.27)$$

where M is the molecular weight in kg-mol, n is the number of electrons involved in charge transfer in eq./mol, and F is the Faraday's constant ($=26.8$ kAh/kg-eq.). In practice, however, the amount of production, w' kg, for $I\theta$ kAh is smaller than w kg in many cases due to several losses. Thus, ξ represented by

$$\xi = \frac{w'}{w} \times 100 \ (\%) \qquad (4.28)$$

is called the *current efficiency*. In some cases, such as dissolution of the soluble anodes in metal refining electrolyzers, the current efficiency sometimes exceeds 100% because of several reasons such as corrosion by dissolved oxygen in solutions and mechanical loss of crystal grains of metals due to intergranular corrosion.

Deviation of the current efficiency from 100% is caused by several factors:

(i) In the electrolytic cell:

- Physical factors such as mixing, convection, diffusion, entrainments, etc. For example, a part of product at the cathode is sent to the anode by mixing, and may be reoxidized. Such a loss in current efficiency is presumed to be different from chemical factors.

- Chemical factors such as side reaction, decomposition, recombination, etc.

(ii) For the overall process including electrolytic cells:

- Physical and chemical losses during preparation of feed stock, polishing of products, and other processing.

Figure 4.15 illustrates the mass transfer in the diaphragm-type chlor-alkali cell as an example. The concentration of active chlorine dissolved in the

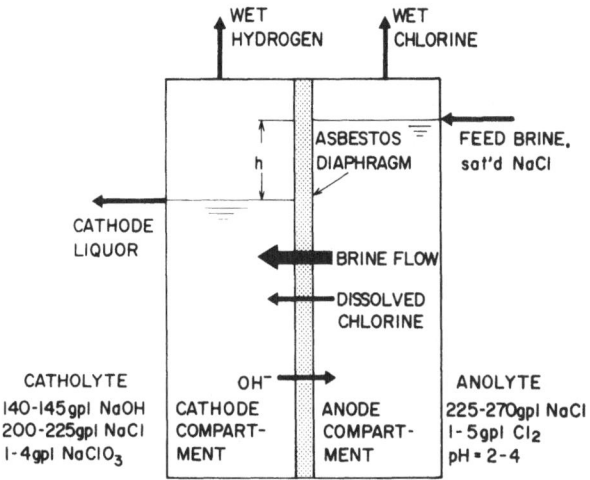

FIGURE 4.15. Mass transfer through the asbestos diaphragm in a chlor-alkali cell.

anolyte is as high as about 0.4–0.5 g/liter because the solubility of chlorine in aqueous solution of NaCl is relatively large. Dissolved chlorine with the brine may permeate through the diaphragm into the cathode compartment, and the current efficiency with respect to chlorine produced at the anode decreases by about 0.5% on this account.

Dissolved chlorine passing through the asbestos diaphragm hydrolyzes and reacts with alkali to form HClO and/or ClO$^-$, and a part of those species may convert further into NaClO$_3$ and/or ClO$_3^-$ via both chemical and electrochemical routes:

$$Cl_2 + H_2O = HClO + H^+ + Cl^-$$

$$HClO = H^+ + ClO^-$$

$$2HClO + ClO^- = ClO_3^- + 2H^+ + 2Cl^-$$

$$6HClO + 3H_2O = \tfrac{3}{2}O_2 + 4Cl^- + 2ClO_3^- + 12H^+ + 6e$$

These chemical reactions would also be factors in the loss of chlorine. Incidentally, chlorate is hardly reduced at the cathode, whereas it is a strong oxidizer. Thus, NaClO$_3$ may contaminate the catholyte liquor, and hence the caustic soda product of diaphragm-type chlor-alkali cells.

A part of OH$^-$ formed at the cathode back-migrates through the asbestos diaphragm to the anode compartment, and causes a current loss of 0.1–1.0%, depending on the characteristics of the diaphragm and the operating conditions.

With this example, it is evident that the current efficiency with respect to the anode product is not equal to that for the cathode product because the mechanism and the route of losses for both products differ from each other. Thus, it is important to indicate clearly the chemical species of interest when the current efficiency is discussed.

From the viewpoint of chemical economy, it is convenient and useful to calculate losses of materials in the overall process, which involves the electrolytic cells, as the total current inefficiency. Figure 4.16 shows a flowsheet for chlorine in the amalgam-type chlor-alkali cell plant as an example. Wet chlorine, called the cell gas, leaving the amalgam cells, is dried and sent to liquefaction. The waste gas containing an amount of oxygen and nitrogen as well as chlorine is purged and thus there seems to be a loss of chlorine. However, the weak chlorine is washed out completely with alkali to produce hypochlorite. Since the return brine leaving the cell is saturated with chlorine, concentrated chlorine gas is recovered under vacuum, and the gas is processed with the cell gas. The brine containing a small amount of chlorine is dechlorinated further by a suitable method such as aeration, and the waste gas is treated by caustic solution to recover chlorine as hypochlorite. In some cases, dechlorination of the brine is carried out by addition of a reducing agent such as sulfite or sulfide. In recent practice, a small amount of chlorine has been allowed to remain in the brine so as to prevent precipitation of mercury compounds. These processes may consume a part of the chlorine produced, and hence the chlorine current efficiency may decrease somewhat.

The catholyte liquor leaving the diaphragm-type chlorine cell is evaporated, and the slurry of salt deposited is separated from the caustic liquor. In this process, some caustic leaving with recrystallized salt is lost. Another factor which contributes to caustic loss is washing of the caustic

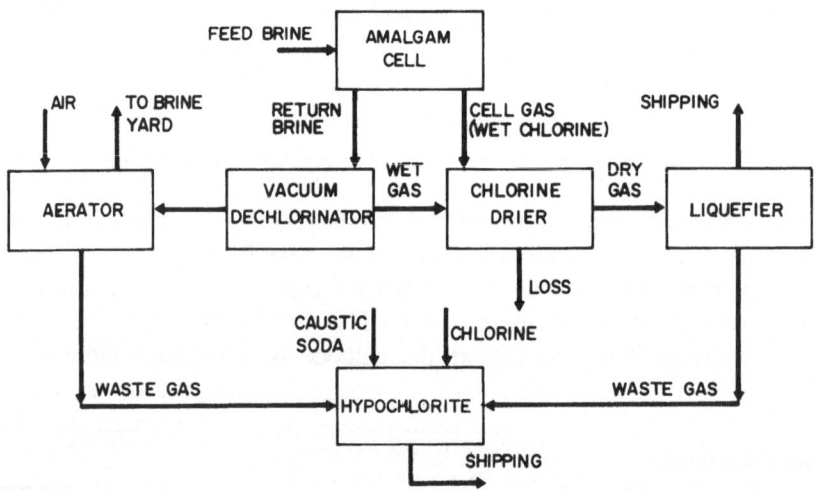

FIGURE 4.16. An example of flow of chlorine in the amalgam-type chlor-alkali cell plant.

evaporator, which is sometimes carried out but which is used less frequently in order to improve energy consumption for evaporation. Other product losses depend on the system used.

4.2.2. Energy Balance

The terminal voltage V_T of the cell operated at the current I is

$$V_T = E_d + \Delta V \tag{4.29}$$

$$\Delta V = \eta_A + (-\eta_C) + \Sigma IR \tag{4.30}$$

where E_d is the decomposition voltage, η_A and η_C are the overvoltage at the anode and cathode, respectively, and ΣIR is the sum of the ohmic voltage drops. ΔV is a part of the terminal voltage, and $I \cdot \Delta V$ converts to Joule heat.

The electric energy W for unit production is

$$W = IV_T \tag{4.31}$$

On the other hand, the minimum requirement W_0 based on thermodynamics is

$$W_0 = I_0 \cdot E_d \tag{4.32}$$

where I_0 is the current based on the Faraday law. Thus, the current efficiency ξ in percent is

$$\xi = \frac{I_0}{I} \times 100 \ (\%) \tag{4.33}$$

From Eqs. (4.29)–(4.33), the energy efficiency ζ in percent is as follows:

$$\zeta = \frac{W_0}{W} \times 100 \ (\%) = \frac{E_d}{V_T} \times \xi \ (\%) \tag{4.34}$$

In the amalgam-type chlorine cell, for example,

$$E_d = 3.05 \ \text{V}, \qquad V_T = 4.20 \ \text{V}, \qquad \text{and} \qquad \xi = 95\%$$

Therefore,

$$\zeta = \frac{3.05 \times 95}{4.20} = 69.0\%$$

It is clear with Eq. (4.34) that there are two ways to improve the energy efficiency for electrolysis: the one is to minimize the useless voltage drop ΔV, and the other is to maximize the current efficiency ξ. For minimizing ΔV, it is necessary to develop suitable electrode materials having low overvoltage characteristics and to improve the cell design, which could reduce the *IR* drops in the electrolysis gap and of the metal hardware. The chlorine overvoltage on the graphite anode in chlor-alkali cells, both amalgam-type and diaphragm-type, has been large. In modern chlorine cells equipped with DSA consisting of the mixed oxides of RuO_x and TiO_x, the chlorine overvoltage is very small compared to other factors of the terminal voltage.[47-50]

The cathode overvoltage in the amalgam cell is very low (see Fig. 4.17), and hence it is a small factor of the terminal voltage. On the other hand, the hydrogen overvoltage of steel cathode in the diaphragm cell equipped with the DSAs is a major factor as well as the brine gap, so that it should be investigated to minimize the terminal voltage.[51]

The *IR* drop between two electrodes is considered to be nearly optimum in almost all practical electrochemical cells, and it is difficult to do further reduction even by improved design technology for the structure of electrolyzers. It is well known that the adjustment of the brine gap between the anode and the cathode, which is called the *anode adjustment* in practical plants, is an important subject of operation in amalgam cell plants. A number of anode plates are positioned only 3–5 mm above the flowing amalgam cathode 1–3 m wide and 10–15 m long. In the chlorine cell using graphite anodes, the anode adjustment is carried out frequently, say once or

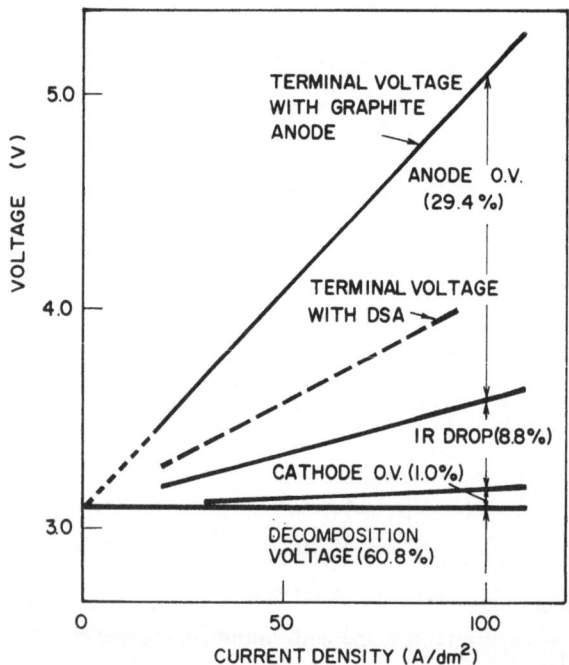

FIGURE 4.17. An example of the voltage balance of the amalgam-type chlorine cell.

twice a day, to minimize the solution IR drop because the graphite anodes are consumed gradually by electrolysis. The metal anodes are dimensionally stable, but they must be controlled exactly because active mass on the metal substrate is degraded severely if the anode surface touches ripples of the amalgam flow. For the purpose of anode adjustment, process computers are being widely used today.[52,53]

Electrolytic gas bubbles displace electrolyte, resulting in a large IR drop between two electrodes. Some bubbles cover the electrode surface, and hence the overvoltage becomes high. Such bubble effects have been investigated by several authors so as to minimize the cell voltage (see Sec. 4.1.4).

In old-type diaphragm cells, graphite blades about 2.5 cm thick were employed as anodes. Evolved chlorine bubbles rise along the electrode, and hence the gas void fraction increases from the bottom to the top of the cell (see Fig. 4.11). There is a space between the cathode fingers at the center of cell, and it works as the brine downcomer. The solution IR drop in the electrolysis zone is reduced in such a manner by natural circulation of the electrolytic solution. In modern chlorine cells equipped with the box-type DSA, natural circulation of solution takes place through the anode box. Since the DSAs are made of expanded titanium sheet, chlorine bubbles escape quickly into the anode box through the mesh, and further reduction of the solution IR drop is achieved.

It might be possible to reduce the solution IR drop be use of supporting electrolyte, but this method is limited because the solution composition is related greatly to the mechanism of the electrode process under consideration.

Change of the cell voltage, and thus the energy consumption, with time of operation, is another problem in practice. It is caused by several reasons such as rise of the overvoltage due to degradation of the active mass and increase of the IR drop through the diaphragm due to plugging. Of course, it is necessary to conduct periodic maintenance to keep the energy consumption low.

4.3. VOLTAGE BALANCE

As described in Sec. 4.2.2, the cell voltage consists of the decomposition voltage, the overvoltages at the anode and the cathode, the solution IR drop, and the IR drop in the metal hardware, that is,

$$V_T = E_d + \eta_A + (-\eta_C) + IR(\text{solution}) + IR(\text{metal}) \qquad (4.35)$$

Those components of the terminal voltage except E_d are functions of the current density: the lower the current density the lower the voltage drops. The overvoltage versus current density is related to the reaction mechanism, and has a complicated relationship. However, in practice, it is frequently represented by a linear function for practical convenience. Thus we have

$$V_T = A + Bi \qquad (4.36)$$

instead of a complicated relationship for Eq. (4.35). Figure 4.17 shows an example of the voltage balance of the amalgam-type chlorine cell and applicability of Eq. (4.36) over a range of current density. The intersection on the ordinate at $i = 0$ is the factor A in Eq. (4.36), and is close to E_d. Thus, it is called the *superficial decomposition voltage*. Factor B is the slope of the voltage versus current density curve having a unit in ohm $\cdot$ dm^2 when i is in A/dm^2 and V_T is in volts. It is called the *voltage slope*, and is also an important factor for evaluating the electrochemical system. Of course, B should be minimized to save energy.

In other words, the purpose of analyzing the voltage balance, sometimes called *breakdown*, is to clarify the target for the project of cell voltage reduction. In Fig. 4.17, for example, the chlorine overvoltage on the graphite anode is very large compared to other factors, whereas the cathode overvoltage and the voltage drop in the metal hardware are small. The anode overvoltage in modern cells equipped with the DSAs has become quite low as shown by the dotted line.

As described in Sec. 4.1.4, electrolytic gas bubbles raise the cell voltage due to dispersion in the solution and covering of the electrode surface.

The IR drop in the metal hardware of each cell is low in general, but it is not negligible when many electrolytic cells are operated in series. Heavy copper busbars must be used to keep the IR drops low, adding to the investment cost. The metallic conductor between adjacent cells in the bipolar type of electrolyzer is quite short and hence the voltage drop is low. This is a clear advantage of the bipolar construction.

Although the decomposition voltage is, of course, fixed by thermodynamics of the reaction under consideration, we could still have a small possibility to reduce it by means of electrolysis at high temperatures under high pressure. There would, of course, be attendant difficulty of design and operation, and increased cost.

The decomposition voltages of the electrolytic processes through different routes differ from each other even if they obtain the same products. It is of interest to discuss and compare those processes from the viewpoint of industrial economy. Comparison between the amalgam cell process and the diaphragm cell process for making chlorine and caustic is an example.

4.3.1. Example 1: Electrolysis of NaCl Solutions—Comparison between the Amalgam Process and the Diaphragm Process

Chlorine is liberated at the anode in both cells, but the cathode reaction in the diaphragm cell differs from that in the amalgam cell:

Amalgam Process:

At the anode: $Cl^- = \frac{1}{2}Cl_2 + e$

$$\Delta G_A = 31 \text{ kcal/mol} \tag{4.37}$$

At the cathode: $Na\text{-}Hg = Na^+ + (Hg) + e$

$$\Delta G_{Na} = -42 \text{ kcal/mol} \tag{4.38}$$

Overall reaction: $NaCl + (Hg) = \frac{1}{2}Cl_2 + Na\text{-}Hg$

$$\Delta G(M) = 73 \text{ kcal/mol} \tag{4.39}$$

Decomposition voltage: $E_d(M) = 3.16 \text{ V}$

Diaphragm Process:

At the anode: $Cl^- = \frac{1}{2}Cl_2 + e$

$$\Delta G_A = 31 \text{ kcal/mol} \tag{4.37}$$

At the cathode: $OH^- + \frac{1}{2}H_2 = H_2O + e$

$$\Delta G_H = -19 \text{ kcal/mol} \tag{4.40}$$

Overall reaction: $Cl^- + H_2O = \frac{1}{2}Cl_2 + \frac{1}{2}H_2 + OH^-$

$$\Delta G(D) = 50 \text{ kcal/mol} \tag{4.41}$$

Decomposition voltage: $E_d(D) = 2.17 \text{ V}$

Calculation shown above indicates a difference of the decomposition voltage by 1 V between two processes. The amalgam cell is operated at high-current densities, say 100 A/dm^2 or more, but the cathode overvoltage is small. On the other hand, the hydrogen overvoltage on the steel mesh cathode and the *IR* drop through the asbestos diaphragm of the diaphragm cell are relatively large even at low-current densities such as 30 A/dm^2. Consequently, the terminal voltage of the amalgam cell is only about 0.3 V higher than that of the diaphragm cell as shown in Table 4.3.

Caustic soda is produced directly by the diaphragm cell [reaction (4.41)]; on the other hand, in the amalgam process, sodium amalgam formed at the amalgam cathode must be decomposed in a separate cell, named the amalgam decomposer, to convert into caustic soda. Amalgam decomposition is an electrochemical process, but exothermic, and hence it takes place without an external power supply. That is, the sodium amalgam anode and the hydrogen cathode on graphite blades or lumps in the decomposer form a

TABLE 4.3

*Voltage Balances of the Amalgam-type and the
Diaphragm-type Chlorine Cells*

	Amalgam cell at 100 A/dm²	Diaphragm cell at 25 A/dm²
Decomposition voltage	3.16 V	2.17 V
Anode overvoltage, DSA	0.20 V	0.03 V
Cathode overvoltage	0.05 V	0.30 V
Solution IR, including bubble effects	0.44 V	0.35 V
Diaphragm		0.60 V
Metal hardware	0.05 V	0.20 V
(Sum) Terminal voltage	3.91 V	3.65 V

short-circuited local cell. (Details will be described in Sec. 7.3.) Thus, the free energy change for this process must be considered:

Amalgam Decomposition:

At the anode:
$$Na\text{-}Hg = Na^+ + (Hg) + e$$
$$\Delta G_{Na} = -42 \text{ kcal/mol} \tag{4.38}$$

At the cathode:
$$OH^- + \tfrac{1}{2}H_2 = H_2O + e$$
$$\Delta G_H = -19 \text{ kcal/mol} \tag{4.40}$$

Overall reaction:
$$Na\text{-}Hg + H_2O = Na^+ + OH^- + \tfrac{1}{2}H_2 + (Hg)$$
$$\Delta G(A) = -23 \text{ kcal/mol} \tag{4.42}$$

Electromotive force: $E_f = 0.99$ V

The overall process consisting of the amalgam formation reaction (electrolysis) and the amalgam decomposition reaction is, of course, equal to that of the diaphragm cell. Therefore, the Gibbs free energy changes for the amalgamation $\Delta G(M)$ and for the amalgam decomposition $\Delta G(A)$ are related to the free energy for the diaphragm cell $\Delta G(D)$ as follows:

$$\Delta G(M) + \Delta G(A) = \Delta G(D) \tag{4.43}$$

Of course, $\Delta G(A)$ is released from the system as heat.

The decomposition voltage of the amalgam cell system will become equal to that of the diaphragm cell if the free energy change for amalgam decomposition is recovered as dc electric power by an adequate procedure such as fuel cell technology, that is,

$$E_d(M) - E_f = E_d(D) = 2.17 \text{ V}$$

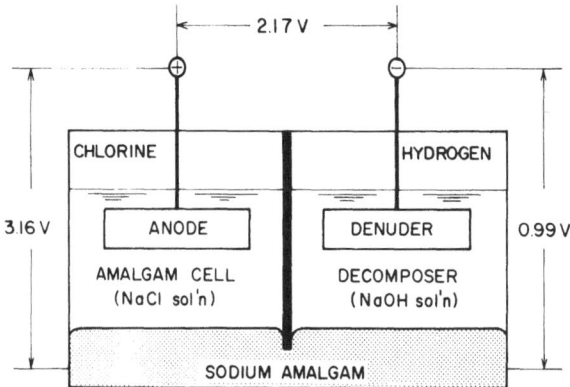

FIGURE 4.18. Combination of the amalgam cell and the amalgam decomposition cell.

Figure 4.18 shows an idea on this concept, where the electric energy equivalent to about 1 V could be recovered by the amalgam decomposition fuel cell connected with the amalgam cell.

The anode overvoltage of sodium amalgam electrode is relatively low, as shown by curve A in Fig. 4.19, but the hydrogen overvoltage on graphite cathode is high even at low-current densities as shown by curve B, which is troublesome for the fuel cell.[54] The terminal voltage can be kept high, 0.6 V at 10 A/dm² for example, if an activated cathode consisting of the Ni-S electroplated material obtained from $NiSO_4$ solution containing KCNS is

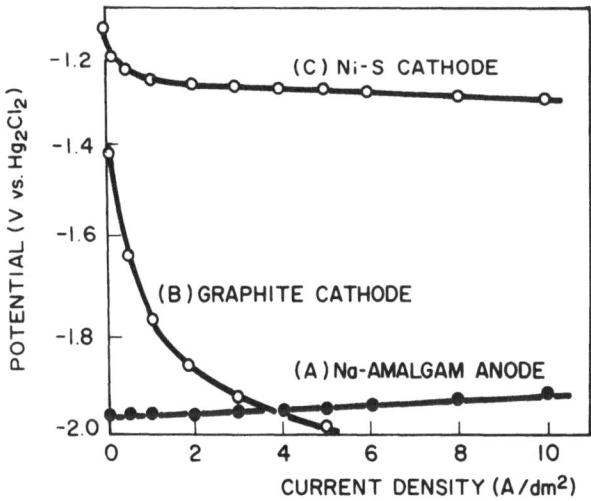

FIGURE 4.19. Anodic polarization curve of a 0.2% Na-amalgam (A), and the cathodic polarization curves of graphite (B), and Ni-S electroplated cathode (C) in 30% NaOH at 50°C.

employed (curve C). A problem is amalgamation and deactivation of the cathode in this case.

A large emf, say 2.2 V, can be obtained if the oxygen reduction cathode is applied in the amalgam decomposition fuel cell instead of the hydrogen evolution cathode:

At the positive electrode: $OH^- = \frac{1}{4}O_2 + \frac{1}{2}H_2O + e$
$$E_C^0 = +0.40 \text{ V} \tag{4.44}$$

At the negative electrode: $Na\text{-}Hg = Na^+ + (Hg) + e$
$$E_A = -1.8 \text{ V} \tag{4.38}$$

Overall reaction: $Na\text{-}Hg + \frac{1}{4}O_2 + \frac{1}{2}H_2O = Na^+ + OH^- + (Hg)$
$$E_f = E_C - E_A = 2.2 \text{ V} \tag{4.45}$$

Yeager has published a paper on a high-duty fuel cell consisting of the sodium amalgam anode and the oxygen cathode, which is of interest from the viewpoint of energy conservation, although there might be many difficulties for practical application.[55] The system shown in Fig. 4.18 has a serious problem. The current efficiency of the amalgam cell is 92–95%. Therefore, the anodic oxidation of mercury may take place due to imbalance of $4 - 8\%$ of current efficiency in the amalgam decomposition cell.

4.3.2. Example 2: Electrorefining and Electrowinning of Copper

In the copper electrorefining cell, an acidified solution of $CuSO_4$ is electrolyzed with a crude copper anode and a pure copper sheet cathode. The anode dissolves into Cu^{2+}, and copper deposits at the cathode. The electrode processes in this case are as follows:

At the anode: $Cu(\text{crude}) = Cu^{2+} + 2e$ (4.46)
At the cathode: $Cu(\text{pure}) = Cu^{2+} + 2e$ (4.47)

Overall reaction: $Cu(\text{crude}) = Cu(\text{pure})$
$$\Delta G \simeq 0 \qquad E_d \simeq 0 \tag{4.48}$$

Since the anode reaction is similar to the cathode reaction, ΔG's for both reactions are about equal to each other. Therefore, the free energy change for the overall reaction, and thus the decomposition voltage, are almost equal to zero. It is a main reason for low terminal voltage of the copper electrorefining cell. The power consumption is, of course, low.

On the other hand, an adequate insoluble anode is employed in the electrowinning cell because Cu^{2+} dissolved in the electrolytic solution must be recovered at the cathode in metallic form. In this case, oxygen formation takes place at the anode, whereas the cathode reaction is electrodeposition of copper:

At the anode: $H_2O = 2H^+ + \frac{1}{2}O_2 + 2e$ $E_A = 1.23$ V   (4.49)

At the cathode: $Cu(\text{pure}) = Cu^{2+} + 2e$ $E_C = 0.33$ V   (4.50)

Overall reaction: $Cu^{2+} + H_2O = Cu(\text{pure}) + 2H^+ + \frac{1}{2}O_2$
$$E_d = E_A - E_C = 0.90 \text{ V} \qquad (4.51)$$

Thus, the decomposition voltage is large, about 0.9 V, as is also the oxygen overvoltage on the Pb alloy anode, resulting in a very large consumption of electric power. The terminal voltage of the electrowinning cell is as high as 2.0–2.2 V in comparison with 0.16–0.17 V of the electrorefining cell.

4.3.3. Example 3: Application of the Oxygen Cathode to HCl Electrolysis

Although HCl electrolysis cells are being commercialized at present, reduction of the energy consumption and thus of the terminal voltage is an important task.[56-62] One approach is the application of the oxygen cathode instead of the conventional hydrogen evolution cathode.[1,63-65] This process can be considered to be a combination of electrolysis with the Deacon-type oxidation process,[67] or an application of fuel cell technology.

The anodic and the cathodic processes of the oxidation electrolysis of HCl are as follows:

$$2Cl^- = Cl_2 + 2e \qquad\qquad E_A^0 = +1.36 \text{ V} \qquad (4.52)$$
$$H_2O = 2H^+ + \tfrac{1}{2}O_2 + 2e \qquad E_C^0 = +1.23 \text{ V} \qquad (4.53)$$

$$2HCl + \tfrac{1}{2}O_2 = Cl_2 + H_2O \qquad E_d = \;\; 0.13 \text{ V} \qquad (4.54)$$

The reversible potential for the cathode reaction, Eq. (4.53), is quite noble compared to that for the hydrogen electrode in the direct HCl electrolysis cell:

$$H_2 = 2H^+ + 2e, \qquad E_C^0 = 0.00 \text{ V} \qquad (4.55)$$

Therefore, it is preferable if reaction (4.53) proceeds with an adequate rate.

Materials for use as cathode must resist attack by the electrolytic solution of concentrated HCl containing active chlorine at high temperatures. Although graphite is corrosion resistant, the electrode process for oxygen reduction on the graphite cathode is slow, as shown by curve A in Fig. 4.20, obtained in oxygen saturated 6.2 M HCl at 18°C. However, the diffusion current density increased greatly when $CuCl_2$ was added (curve B). The

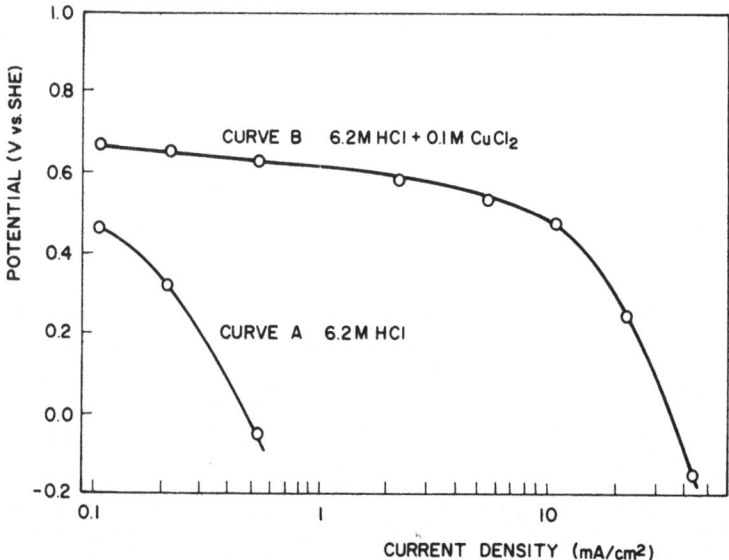

FIGURE 4.20. Polarization curves of a graphite cathode in oxygen saturated solutions at 18°C.

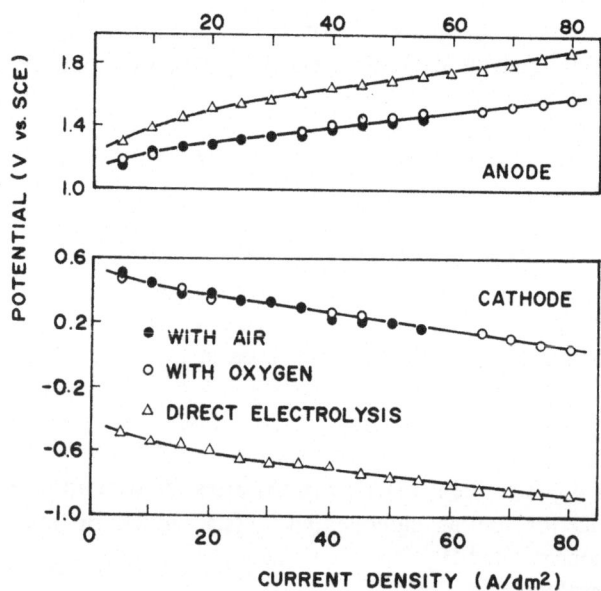

FIGURE 4.21. Anodic and cathodic polarization curves in 6 N HCl. The electrolytic solution in the Kyoto process cell with air and oxygen also contains 1 M CuCl_2. Temperature = 80°C.

limiting current density in the mixed solutions of HCl and $CuCl_2$ depends on the $CuCl_2$ concentration. Also, the reversible potential for the cathode reaction in the mixed solution differs from that in a single solution of HCl.

The mechanism of the electrode process in the mixed solution has been determined experimentally.[63] Cuprous and cupric ions may exist in the mixed solution as the complex ions: $CuCl_m^{-(m-1)}$ and $CuCl_n^{-(n-2)}$, respectively. The cupric chloro-complex ions are reduced cathodically:

$$CuCl_n^{-(n-2)} + e = CuCl_m^{-(m-1)} + (n - m)\, Cl^- \qquad (4.56)$$

followed by oxidation with oxygen dissolved in the bulk of solution:

$$CuCl_m^{-(m-1)} + \tfrac{1}{4}O_2 + H^+ + (n - m)\, Cl^- = CuCl_n^{-(n-2)} + \tfrac{1}{2}H_2O \quad (4.57)$$

The cathode process, Eq. (4.56), is fast compared to reaction (4.53). Chemical oxidation of cuprous ions is also fast.

Figure 4.21 illustrates the anodic and the cathodic polarization curves of graphite electrode in 6 M HCl at 80°C. The cathode potential in the mixed solution of HCl and $CuCl_2$ containing dissolved oxygen is about 1 V more noble than the potential for the hydrogen evolution process in a single HCl solution. Therefore, the terminal voltage of the oxidation electrolysis cell is considerably less in comparison with that of the direct electrolysis cell as shown in Fig. 4.22.

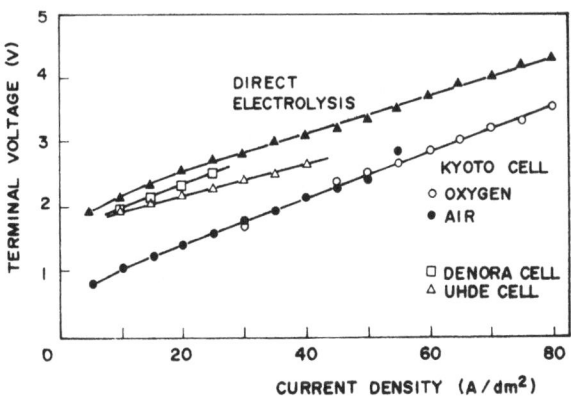

FIGURE 4.22. Terminal voltage of HCl cells. Data for the DeNora cell[66] and the Uhde cell[57] were taken from references indicated.

REFERENCES

1. F. Hine, S. Yoshizawa, K. Yamakawa, and Y. Nakane, *Electrochem. Technol.* **4**, 555 (1966).
2. F. Hine, *Electrochem. Technol.* **2**, 79 (1964).
3. *Kagaku Binran* (Handbook of Chemistry), p. 862, Maruzen, Tokyo (1964).
4. F. Hine, M. Yasuda, and S. Inuta, *Denki Kagaku* (J. Electrochem. Soc. Japan) **39**, 934 (1971).
5. R. B. MacMullin, *Electrochem. Technol.* **1**, 5 (1963).
6. R. B. MacMullin, *Electrochem. Technol.* **2**, 106 (1964).
7. F. Hine, *Denki Kagaku* **35**, 838 (1968).
8. R. B. MacMullin, *J. Electrochem. Soc.* **116**, 416 (1969).
9. R. B. MacMullin, *Denki Kagaku* **38**, 570 (1970).
10. F. Hine, *Denki Kagaku* **39**, 60 (1971).
11. *International Critical Tables* **6**, 231 (1929).
12. F. Hine, *Denki Kagaku* **39**, 438 (1971).
13. D. A. MacInnes, *The Principles of Electrochemistry*, p. 322, Reinhold, New York (1939).
14. T. Shedlovsky and D. A. MacInnes, *J. Am. Chem. Soc.* **58**, 1970 (1936).
15. F. Hine and K. Yamakawa, Abstract No. 184, Electrochemical Society meeting, Cleveland, OH, May 1966.
16. F. Hine, S. Yoshizawa, S. Okada, and T. Uesugi, *Kogyo Kagaku Zasshi* (J. Chem. Soc. Japan, Industrial Section) **58**, 554 (1955).
17. F. Hine, S. Yoshizawa, and S. Okada, *Denki Kagaku* **24**, 370 (1956).
18. S. Okada, S. Yoshizawa, F. Hine, and Z. Takehara, *Denki Kagaku* **26**, 165, 211 (1958).
19. S. Yoshizawa, F. Hine, Z. Takehara, and M. Yamashita, *Denki Kagaku* **28**, 205 (1960).
20. C. L. Mantell, *Electrochemical Engineering*, p. 308, McGraw-Hill, New York (1960).
21. R. E. De La Rue and C. Tobias, *J. Electrochem. Soc.* **106**, 827 (1959).
22. C. W. Tobias, *J. Electrochem. Soc.* **106**, 833 (1959).
23. R. B. MacMullin, Electrolysis of brines in mercury cells, in *Chlorine*, edited by J. S. Sconce, ACS Monograph 154, Reinhold, New York (1962).
24. W. C. Gardiner, *Electrochem. Technol.* **1**, 71 (1963).
25. F. Hine, M. Yasuda, R. Nakamura, and T. Noda, *J. Electrochem. Soc.* **122**, 1185 (1975).
26. M. Murozumi, *Electrochem. Technol.* **5**, 236 (1967).
27. H. Shibata and Y. Yamasaki, *Electrochem. Technol.* **5**, 239 (1967).
28. J. E. Funk and J. F. Thorpe, *J. Electrochem. Soc.* **116**, 48 (1969).
29. N. D. Koshel' and O. S. Ksenzhek, *Soviet Electrochem.* **8**, 436 (1972).
30. P. A. Danna, *J. Electrochem. Soc.* **121**, 1286 (1974).
31. N. Ibl, *Chem. Ing. Tech.* **35**, 353 (1963).
32. L. J. J. Janssen and J. G. Hoogland, *Electrochim. Acta* **15**, 1013 (1970); **18**, 543 (1973).
33. N. Ibl, J. Venczel, E. Schalch, and E. Adam, *Chem. Ing. Tech.* **43**, 202 (1971).
34. R. B. MacMullin, K. L. Mills, and F. N. Ruehlen, *J. Eletrochem. Soc.* **118**, 1582 (1971).
35. M. G. Fouad, G. H. Sedahmed, and H. A. El-Abd, *Electrochim. Acta* **18**, 279 (1973).
36. M. G. Fouad and G. H. Sedahmed, *Electrochim. Acta* **20**, 615 (1975).
37. I. Rousar and V. Cezner, *Electrochim. Acta* **20**, 289 (1975).
38. I. Rousar, J. Kacin, E. Lipper, F. Smirous, and V. Cezner, *Electrochim. Acta* **20**, 295 (1975).
39. H. Vogt, *Electrochim. Acta* **23**, 1019 (1978).
40. D. S. Scott, Properties of cocurrent gas–liquid flow, in *Advances in Chemical Engineering*, Vol. 4, edited by T. B. Brew *et al.*, p. 199, Academic, New York (1963).
41. *Chemical Engineers' Handbook*, edited by J. H. Perry, pp. 4-21, and 5-38, McGraw Hill, New York (1963).

42. T. Z. Fahidy, *Can. J. Chem. Eng.* 174 (June, 1966).
43. R. E. Meredith and C. W. Tobias, *J. Electrochem. Soc.* **110**, 1257 (1963).
44. R. R. Lessard and S. A. Zieminski, *Ind. Eng. Chem. Fundam.* **10**, 260 (1971).
45. F. Hine and T. Sugimoto, Abstract No. 459, Electrochemical Society meeting, Seattle, WA, May 1978.
46. F. Hine and K. Murakami, *J. Electrochem. Soc.* **127**, 292 (1980).
47. O. DeNora, *Chem. Ing. Tech.* **42**, 222 (1970).
48. O. DeNora, *Chem. Ing. Tech.* **43**, 182 (1971).
49. J. Horacek and S. Puechaver, *Chem. Eng., Prog.* **67**(3), 71 (1971).
50. G. Faita and G. Fiori, *J. Appl. Electrochem.* **2**, 31 (1972).
51. J. E. Currey, A. T. Emery, and C. S. McLarly, *Chlorine Bicentennial Symposium*, p. 187, Electrochemical Society (1974).
52. R. W. Ralston, *Chlorine Bicentennial Symposium*, p. 145, Electrochemical Society (1974).
53. J. P. Lombard, Abstract No. 463, Electrochemical Society meeting, Seattle, WA, May 1978.
54. K. Aoki and K. Nagaike, *Denki Kagaku* **30**, 652 (1962).
55. E. Yeager, ONR Contract, No. 2391(00) (1958 and 1960).
56. H. Hölmann, *Chem. Eng.* p. 63 (July 25, 1960).
57. *Uhde HCl Cell* (catalogue), Achema (1964).
58. F. S. Low, U. S. Patents 2,468,766 and 2,470,073.
59. C. P. Roberts, *Chem. Eng. Prog.* **46**, 436 (1950).
60. Anonymous, *Chem. Ind.* **66**, 501 (1950).
61. D. W. Schroeder, *Ind. Eng. Chem.* **1**, 141 (1962).
62. W. Teske und H. Hölmann, *Z. Elektrochem.* **66**, 787 (1962).
63. F. Hine and K. Yamakawa, *Electrochim. Acta* **13**, 2119 (1968).
64. F. Hine and K. Yamakawa, *Electrochim. Acta* **15**, 769 (1970).
65. F. Hine and K. Yamakawa, Abstract No. 258, Electrochemical Society meeting, Los Angeles, CA, May 1970.
66. P. Gallone and G. Messner, *Electrochem. Tech.* **3**, 321 (1965).
67. A. Redniss, HCl oxidation processes, in *Chlorine*, edited by J. S. Sconce, p. 250, ACS Monograph 154, Reinhold, New York (1962).

II   ELECTROCHEMICAL INDUSTRIES

Chapter 5

WATER ELECTROLYSIS

Production of ammonium sulfate fertilizer via synthetic ammonia was a national project in Japan just after World War II, and water electrolysis as the source of hydrogen was active. For producting 1 ton of ammonia, 2,100 Nm^3 of hydrogen and 700 Nm^3 of nitrogen are required, as shown in Table 5.1.[1] Of these, relatively cheap nitrogen can be supplied by means of air liquefaction, and hence the manufacturing cost of ammonia depends largely on the price of hydrogen.

Assume that the electric consumption of water electrolysis is 5 kWh/Nm^3 of hydrogen, and that the price of electricity is 15 yen/kWh (6 ¢/kWh), which is a current low estimate. The electric power cost per ton of ammonia is computed as 16,000 yen or $640, and is much higher than the production cost. It means that hydrogen by electrolysis is no longer acceptable as raw material. The preferred major source of hydrogen has become cracking and reforming of hydrocarbons (see Table 5.2). Byproduct carbon dioxide from the process is used for synthesizing urea:

$$2NH_3 + CO_2 = NH_2COONH_4 \text{ (carbamate)}$$

$$NH_2COONH_4 = NH_2CONH_2 + H_2O$$

It is also a strong point of the hydrocarbon process for hydrogen production in comparison with the electrolytic process.

Urea is desirable as a nitrogen-type fertilizer. Since ammonia and urea

TABLE 5.1.

Consumption for Synthesis of Ammonia as 1000 kg NH_3

Hydrogen	2,100 Nm^3
Nitrogen	700 Nm^3
Electric power	1,000 kWh
Water	160 m^3
Steam	15 kg
Catalyst	0.1 kg

TABLE 5.2

Reactions for Producing Hydrogen

$C + \frac{1}{2}O_2 = CO$	Partial oxidation
$C + H_2O = H_2 + CO$	Water gas reaction
$C_nH_{(2n+2)} + \dfrac{n}{2}O_2 = nCO + (n+1)H_2$	Partial oxidation
$C_nH_{(2n+2)} + nH_2O = nCO + (2n+1)H_2$	Reforming
$CO + H_2O = H_2 + CO_2$	Conversion

TABLE 5.3

Production and Demand of Ammonia and Urea in Japan (10^6 tons/year)[a]

	1965	1970	1974	1975
Ammonia				
Capacity	2.31	3.38	4.43	4.43
Production	2.25	3.32	3.99	(3.03)
Industrial use	0.82	1.27	1.44	1.53
Urea				
Capacity	1.46	2.99	3.81	3.87
Production	1.35	2.37	3.39	(2.16)
Export	0.81	2.09	2.45	(1.25)
Domestic demand	0.41	0.58	0.65	
Ammonium sulfate				
Capacity	2.68	2.10	1.85	2.21
Export	1.28	1.23	0.61	(0.67)
Domestic demand	1.18	0.94	0.91	
Source of hydrogen (%)				
Coal	5.3	0.0	0.0	0.0
Electrolysis	5.0	1.9	0.0	0.0
Coke oven gas	16.3	11.6	1.2	1.2
Oil cracking	39.2	21.9	5.4	5.4
Natural gas	18.0	12.2	8.8	8.8
Petroleum waste gas	6.8	2.7	16.4	16.4
Butane	0.0	16.5	22.9	22.9
Naphtha	9.4	33.2	45.3	45.3
Total	100	100	100	100

[a] Figures in parentheses are estimates.

are important basic chemicals today, demand for industrial use is increasing continuously as shown in Table 5.3.[2] The table also shows complete shut down of water electrolysis in Japan in the 1970s with very few exceptions, although the electrolyzer is still being operated in European countries where the electric power cost from hydropower plants is cheap.

There have been recent feasibility studies of water electrolysis as a result of the emphasis on the development of new energy sources in many countries including Japan and the United States, particularly for the so-called "hydrogen economy." It has been estimated that transportation of pressurized hydrogen by pipe line over 600 miles is cheaper than the price of electric power for the same distance, and hence there would be an appreciable advantage for a large consolidated clean power plant located at a site far from residential and/or industrial areas. Load leveling is said to be important for base load generating equipment such as a nuclear power plant. The water electrolyzer could be operated at night to use available off-peak power. Hydrogen produced would be stored and burned at the steam power plant in the peak demand hours. The hydrogen–air fuel cell could also be used in the electricity generation system. Additional studies of water electrolysis as a source of "clean energy" are needed because competitive processes for hydrogen production are also being investigated.[3-9]

5.1. WATER ELECTROLYZER

Figure 5.1 shows an example of a water electrolyzer, the Pintsch Bamag SC 90, which is a filter press-type cell. There are also monopolar cells in use such as the Hitachi cell.

As described in Sec. 2.1.2, the decomposition voltage of water is independent of the solution pH, and therefore either alkaline or acid solution could be used. In practice, however, only a pure solution of alkali hydroxide, mostly caustic potash, is used as an electrolyte because there are some difficulties of design and material selection for cells using acid solution.

The electrical conductivity of caustic alkali solutions depends on the concentration and increases with increase in temperature, as shown in Figs. 4.2. and 4.3. Because both metallic and nonmetallic components used in the electrolyzer are corroded seriously by hot concentrated caustic alkalis, practical cells are generally operated in the temperature range 50°C–80°C.

The vessel of the electrolyzer is made of carbon-steel sheet, and gaskets consist of nonmetallic corrosion-resistant materials such as asbestos and rubber, both natural and synthetic. Asbestos fiber fabric is used as a diaphragm. It is coarse, and hence the electric resistance is small since it is only required to separate hydrogen and oxygen bubbles from each other.

The cathode is made of steel sheet, which is corrosion resistant to

FIGURE 5.1. Pintsch Bamag water electrolysis cell SC 90. Capacity = 300 Nm^3 H_2 and 150 Nm^3 O_2 per hour. (Quoted from catalogue by permission of Pintsch Bamag Aktiengesellschaft.)

caustic solution. Also, the hydrogen overvoltage of steel is relatively low, probably the best level among practical materials (see Sec. 3.2.1). Since iron and steel dissolve slightly in alkali solutions when they are polarized anodically, heavily nickel-electroplated steel is used as an anode. Nickel is resistant even at noble potentials, and its oxygen overvoltage is suitably low as described in Sec. 3.2.2.

Gas bubbles of hydrogen and oxygen cover a certain amount of the surface of the respective electrodes, and the electrolytic solution is displaced by the gases. A great effort in cell design has been devoted to eliminating these unwanted bubble effects.

The conductivity of caustic alkali solution depends on the concentration and the temperature, as shown in Figs. 4.2 and 4.3. The electrolytic solution in the water electrolyzer consists of 20% NaOH or 30% KOH for maximum conductivity at the operating temperatures of 50°C–80°C. As mentioned, at high temperatures, corrosion problems will occur. Here we may assume the operating conditions for further discussion to be as follows:

Electrolyte: 20% NaOH
Temperature: 80°C
Current density: 5 A/dm^2
Anode-to-cathode gap: 23 mm
Diaphragm thickness: 3 mm

The terminal voltage V_T is represented as follows:

$$V_T = E_d + \eta_A + (-\eta_C) + \sum IR \qquad (5.1)$$

where $\sum IR$ is the sum of the ohmic voltage drops. The decomposition voltage E_d is 1.146 V under the given conditions shown above [see Eq. (2.21) in Sec. 2.1.2]. Figure 5.2 shows some examples of the hydrogen overvoltage, η_H, and the oxygen overvoltage, η_O, diagrams of a nickel-electroplated iron electrode in 2.5N NaOH (about 11 wt%) at 80°C. At 5 A/dm^2, η_H and η_O are −0.47 V and +0.35 V, respectively.

Since the conductivity of 20% NaOH solution at 80°C is 1.02 mho/cm (see Fig. 4.2), the ohmic voltage drop IR(solution) is

$$IR(\text{solution}) = ipl = il/\kappa = \frac{0.05 \times 2}{1.02} = 0.098 \text{ (V)}$$

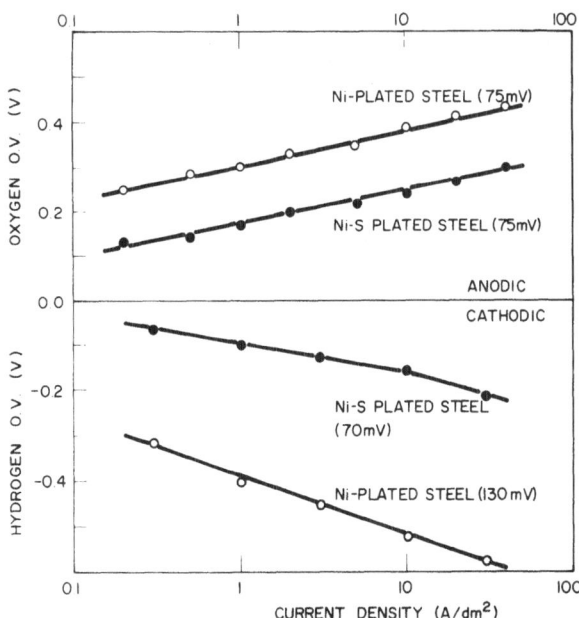

FIGURE 5.2. Anodic and cathodic polarization curves of nickel-electroplated electrode and a black nickel (Ni-S) electroplated electrode in 2.5 N NaOH at 80°C. ( ) indicates the slope of the curve in mV/decade.

The ohmic drop through the asbestos cloth diaphragm is a small factor, i.e., the conductivity in 20% NaOH at 80°C was measured to be 3 mho/cm^2 or 0.017 V (=0.05/A cm^2/3 mho/cm^2).

Furthermore, we must consider the bubble effect, which is about 0.1 V, and also the ohmic drops in the hardware. We tabulate them as follow:

Decomposition voltage:	E_d	1.146 V
Oxygen overvoltage:	η_O	0.35 V
Hydrogen overvoltage:	$-\eta_H$	0.47 V
Solution IR drop:	IR(solution)	0.098 V
Diaphragm IR drop:	IR(diaphragm)	0.017 V
Bubble effect:	IR(bubble)	0.1 V
Hardware IR drop:	IR(metal)	0.03 V
Terminal voltage:	V_T	2.21 V

It is clear that both hydrogen and oxygen overvoltages are the major factors of the cell voltage. For reducing the voltage or the energy consumption, many trials for development of low overvoltage materials have been made. Electrodeposition of a black nickel containing sulfur, which can be obtained from a solution containing $Ni(CNS)_2$, is an example.[10] This material shows relatively low overvoltages for both hydrogen and oxygen as shown in Fig. 5.2: -0.14 V and $+0.23$ V, respectively, at 5 A/dm^2. Therefore, the cell voltage could be reduced to 1.8 V when black nickel is used.

Reduction of cell voltage offers a further advantage. A large part of cell

TABLE 5.4
Energy Balance of Water Electrolyzer

	kcal/mol	%
Heat in		
Electric power at 2.1 V	96.86	99.20
Sensible heat of feed water	0.78	0.80
	97.64	100.00
Heat out		
Reaction heat ($-\Delta H$)	68.52	70.18
Sensible heat of H_2, O_2, and water vapor	1.01	1.03
Latent heat of vaporization	4.93	5.05
Heat loss	23.18	23.74
	97.64	100.00

electricity, say 0.8 V, which is $V_T - E_d$, converts into thermal energy and heats the cell. Table 5.4 shows an example of the heat balance of a water electrolyzer.[11] About 24% of the output is wasted. Reduction of the cell voltage could yield a decrease in waste heat since the voltage balance of the electrolyzer is related closely to the heat balance. The operating temperature could be kept high by means of thermal insulation.

The high-pressure water cell was studied during World War II.[12,13] The decomposition voltage increases slightly under high pressure conditions, e.g., about 10% at 200 atm compared to the cell operated under normal pressure. The overvoltages, both anode and cathode, are almost independent of the pressure. The bubble effect is reduced due to decrease in the gas volume under high pressure. It is of interest to note that the electrolyzer could supply high-pressure hydrogen and oxygen directly without any mechanical compressor. These advantages suggest that water electrolyzer systems should be fitted for high pressure operation.

Water electrolyzers, especially high-pressure cells, could have an opportunity for revival as a part of the "hydrogen economy" or "energy conservation" projects.

5.2. PRODUCTION OF HEAVY WATER

Heavy water is used as a moderator of neutrons in a nuclear reactor. Protium (H), deuterium (D), and tritium (T), are the isotopes of hydrogen with mass number 1, 2, and 3, respectively. Of these, T is a radioisotope, and exists only ca. $10^{-17}\%$ in natural water.

Since natural water contains about 0.014% of heavy water, D_2O or HDO, it should be extracted into a concentrated form to more than 99% for reactor use. Table 5.5 shows the physical properties of heavy water.[14]

Electrochemical technique plays two parts in the production of heavy water: the first stage for concentrating heavy water from natural water to 1–10%, called *electrolytic concentration*, and the second stage for making the final product of 99.7–99.8% D_2O from an intermediate containing 20–30% heavy waier, which is the electrolytic recovery and distillation.

The cathodic process for electrolysis of water consisting of H_2O and D_2O (or HDO) is as follows[15]

$$H_2O + HDO + 2e = H_2 + OH^- + OD^- \tag{5.2}$$

and

$$H_2O + HDO + 2e = HD + 2OH^- \tag{5.3}$$

Since the rate constant for HD (or D_2) formation differs from that for H_2

TABLE 5.5

Physical Properties of Light Water and Heavy Water

Property	Unit	Light water	Heavy water
Molecular weight		18.0160	20.0284
Specific gravity	at 25°C	1.00000	1.10775
Maximum density	g/ml	1.000 (at 3.98°C)	1.1060 (at 11.2°C)
Melting point	°C	0.00	3.80
Boiling point	°C	100.0	101.43
Critical temperature	°C	374.2	371.5
Critical pressure	atm	218.53	218.6
Critical density	g/ml	0.325	0.363
Molecular heat of vaporization	kcal/mol	9.719	9.927
Molecular heat of melting	kcal/mol	1.436	1.515
Molecular heat of sublimation	kcal/mol	12.170	12.631
Specific heat	at 20°C	18.00	20.15
Viscosity	cp at 25°C	0.893	1.100
Refractive index	n_D^{20}	1.3330	1.3283
Dissociation constant	at 25°C	1×10^{-14}	1.98×10^{-15}

formation, the content of D in the electrolytic gas differs from that in the electrolytic solution. The separation factor α is thus defined as follows:

$$\alpha = \frac{[1 - D(g)]/D(g)}{[1 - D(l)]/D(l)} \tag{5.4}$$

where $D(g)$ and $D(l)$ are the atomic fraction of deuterium in the electrolytic gas and in the electrolyte, respectively. The numerator of the right side of Eq. (5.4), $[1 - D(g)]/D(g)$, shows the atomic ratio of H to D in the electrolytic gas; on the other hand, the denominator, $[1 - D(l)]/D(l)$, is that in

TABLE 5.6

Separation Factor of Deuterium on Metal Cathodes

Cathode metal	Separation factor
Fe	6–9
C	7–8
Cu	5.5–6.5
Ag	5–6
Ni	4–6
Smooth Pt	5–7
Platinized Pt	3.5–4.5
Pb	3–4
Hg	3–4.5

the electrolytic solution. The cathode materials are divided into two groups as shown in Table 5.6, one having a separation factor of $\alpha \simeq 6$ and the other $\alpha \simeq 3$, suggesting different mechanisms for the cathode process of each group.[16] The separation factor thus depends on the electrode material, but is almost independent of the content of heavy water, the operating temperature, and the current density. Now it is clear that heavy water in the electrolytic solution is concentrated somewhat by electrolysis because H^+ is more easy to discharge to form H or H_2 than D^+ to D or D_2.

As described in Sec. 5.1, the water electrolyzer for producing hydrogen and oxygen is generally operated at 60°C–80°C. Since water is consumed by electrolysis itself as well as entrainment with hydrogen and oxygen, purified water must be supplied continuously into the cell. Both hydrogen and oxygen leaving the cell are cooled to remove water and alkali mists. The condensate contains heavy water at some extent in comparison with the feed electrolyte or natural water because of the slow rate of the cathodic reduction of D^+ as described. The condensate is sent to the next electrolyzer to obtain hydrogen and oxygen, and the condensate, containing more heavy water than before, is collected.

Since HD and D_2 are released with H_2 from the electrolyzer, they must be recovered efficiently by means of the exchange reaction between steam and water. Figure 5.3 shows the catalyst column for the exchange reaction of

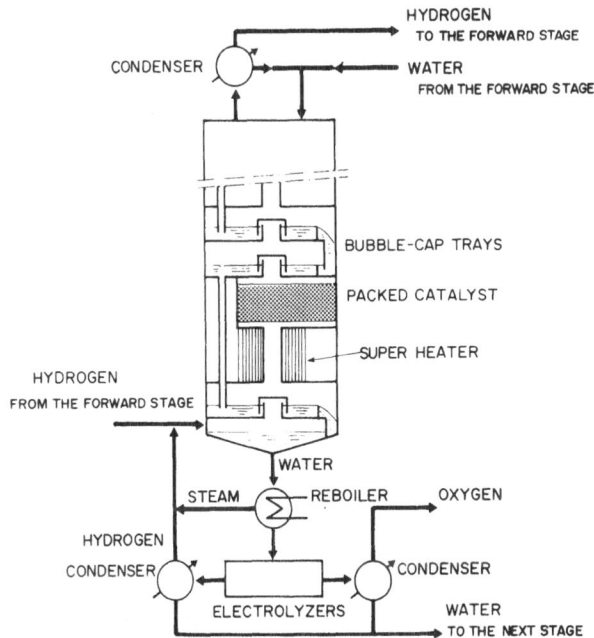

FIGURE 5.3. Catalyst column for exchange reaction of deuterium.

deuterium. Water at the bottom of the column is boiled, and steam generated is mixed with electrolytic hydrogen. The steam–hydrogen mixture which contains deuterium to be recovered is sent to the catalyst-packed column to proceed in the exchange reaction:

$$HD(gas) + H_2O(steam) \rightarrow H_2 + HDO \qquad (5.5)$$

The mixture of H_2 and HDO is washed with water coming from the forward stage by the bubble-cap trays to recover HDO. The gas–steam mixture is superheated before the exchange reaction, since otherwise the catalyst may wet and become inactive.

On the other hand, the equilibrium constant of reaction (5.5)

$$K = \frac{(HDO)_l \cdot (H_2)_g}{(H_2O)_l \cdot (HD)_g} \qquad (5.6)$$

is 3 at 20°C and 2 at 100°C.[13] Since K decreases with increasing temperature, the operating temperature is kept at 75°C, at which K is 2.88.[15] Platinum on carbon substrate and nickel-chromium are used as a catalyst.[1]

A part of the solution at the bottom of the tower is sent back to the electrolyzer. These procedures are repeated to concentrate heavy water to 1–4% by weight depending on the process flowsheet and economic conditions.

The process has been first commercialized at the Trail, B.C., plant of

TABLE 5.7

Dimensions of the Four-stage Electrochemical Recovery of Heavy Water

		Stage			
	Unit	1st	2nd	3rd	4th
Number of cells		2693	378	96	30
Number of catalyst columns		3	1	1	1
Column diameter	ft	8.5	5	2.5	1.5
Column height	ft	112	108	97	96.5
Number of plates		35	32	32	32
Number of catalyst chambers		13	13	13	13
Plate-to-plate gap	in.	8	8	8	8
Thickness of catalyst	in.	4.25	6	6	4.25
Volume of catalyst	ft³	480	111	27	5.4
Catalyst		Pt	Pt	Pt	Ni-Cr
Capacity of reboiler	Btu/hr	2,250,000	580,000	187,000	18,700
Capacity of condenser	Btu/hr	2,240,000	1,160,000	216,000	58,300

TABLE 5.8

An Example of the Cell Dimensions and the Operating Conditions of the
Electrolytic Concentration Process to Produce 99.8% Heavy Water

	Unit	Stage				
		1st	2nd	3rd	4th	5th
Rated amperage	A	1000	100	50	50	15
Number of cells		2	6	2	1	1
Electrode area	dm^2	94	10	4.14	4.14	0.26
Current density	A/dm^2	10.6	10	12.1	12.1	11.9
Electrode spacing	mm	50	35	35	35	35
Feed water	l/month	379.1	127.7	44.0	15.6	5.1
Product	l/month	96.3	33.6	12.7	5.1	2.2

the Consolidated Mining and Smelting Company of Canada. The capacity
was 2000 lb/day of 3.5% HDO.[13] Table 5.7 shows the dimensions of the
plant.[1]

Heavy water must be concentrated further to 99.8% to use it as a
moderator of neutrons in a nuclear reactor. The electrolytic concentration is
an adequate method for the purpose. The process consists of batchwise electrolysis and hydrogen burning. The process principle resembles electrolytic
concentration. Since hydrogen and oxygen from the electrolyzer are wet,
these gases are cooled to recover HDO containing water. The solution is sent
to the next cell for further electrolytic concentration. Electrolytic hydrogen
containing deuterium is burned with oxygen to recover HDO, and the water
is sent back to the forward electrolyzer. These combinations of electrolysis
and burning are repeated several times to yield HDO of an adequate concentration, i.e., 99.8% or more. Table 5.8 shows an example of the cell
dimensions and the operating conditions.[1]

The electric energy consumption in the Trail plant is said to be
4 kWh/lb or 8.8 kWh/kg HDO at 99.8% in concentration.[13]

Other processes such as rectification are being studied for producing
heavy water[17] as a part of energy conservation efforts.

5.3. PROSPECTS OF WATER ELECTROLYSIS

Water electrolysis cells are being operated in European countries,
whereas there is no installation at all in Japan. In the United States, research
and development for water electrolysis based on new concepts are
progressing in connection with load leveling for nuclear power plants and
also as part of hydrogen economy efforts.

Subjects such as

● improvement of hardware of electrolyzers,
● R & D for electrode materials of low overvoltages,
● development of large-scale cells and analysis of economy,
● maintenance and safety problems, and
● new concepts and technologies such as SPE and acid electrolysis

need investigation to minimize electric energy consumption and to scale-up electrolysis plants for reducing the cost of hydrogen.

Hydrogen is a main product of water electrolysis. On the other hand, there is only a limited market for byproduct oxygen and it should be enlarged to reduce the manufacturing cost of hydrogen. Production of ammonia is a major market for electrolytic hydrogen at present, but demand for hydrogen as a raw material for chemical process industries and as a clean fuel feedstock could become more important in the near future. Transportation of large amounts of hydrogen over long distances from the electrolytic cell plant located close to the atomic power station to customers in urban areas is an important task from the viewpoints of both economy and safety.

Table 5.9 shows some examples of large-scale water electrolysis plants now in operation. Of these, the capacity of the Aswan, Egypt, plant would be doubled to 32,000 m^3/hr.[18,19]

Almost all the electrolyzers for these projects are the bipolar filter press-type design, and the unit capacity is 250–350 Nm^3 H_2/hr cell (see Table 5.10).[20] It is reported that giant cells with capacities of 1000 N · m^3/hr or more are now being designed at several places. The capacity of the water electrolysis plant to be sited at a nuclear power station would be as large as 400,000 m^3/hr. For this purpose, more than 100 thousand 10 kA-cells would be required, which is impractical. The DeNora

TABLE 5.9
Large-scale Water Electrolysis Plants

| Location | Plant size | | Fabricator |
	MW	Nm^3 H_2/hr	
Rjukan (Norway)	165	27,900	Norsk Hydro
Glomfjord (Norway)	160	27,100	Norsk Hydro
Nangal (India)	125	21,100	DeNora
Aswan (Egypt)	100	16,900	BBC
Reykjavik (Iceland)	20	3,000	Norsk Hydro

TABLE 5.10

Dimensions of Modern Water Electrolyzers

		BBC	Norsk Hydro
Capacity	$Nm^3 H_2/hr$ stack	260	340
Temperature	°C	80	80
Pressure	mmH_2O		500
Current density	A/dm^2	20	17
Cell voltage	V/cell	2.04	1.7–1.8
Gas quality	Hydrogen (%)	>99.8	99.80–99.90
	Oxygen (%)	>99.6	99.30–99.70

Co. is working to design a large-scale stack of 120 kA consisting of 11 elements. Some 720 stacks could meet the requirement.[20]

Another target for new developments in water electrolyzers is how to reduce the cell voltage or the energy consumption as well as scaling-up of the cell geometry. The current density must be increased to some $100 A/dm^2$ from the present value of $20 A/dm^2$ in the near future. Consequently, among problems being investigated are

- how to operate the cell under high pressures at high temperatures;
- research and development of new materials of electrodes, both anode and cathode, having low overvoltages; and
- reduction of the solution *IR* drop, mostly elimination of bubble effects.

It is clear that the energy consumption of the cell operated at high pressures and temperatures is small, whereas the investment cost of such electrolyzers would be high. The terminal voltage of the cell operated at 1 atm and at 300°C is estimated to be 1.48 V,* and 1.60 V at 40 atm and at 230°C, compared to about 1.8 V for the best performance of present cells. These high-temperature cell voltages are close to the decomposition voltages for the respective conditions, so that the energy efficiency for electrolysis is almost 100%.[21] Selection of corrosion-resistant materials for use in hot concentrated caustic alkali under such conditions is difficult, and so studies of new materials for electrodes and diaphragms must be conducted.

Easy operation and maintenance and safety are the important points for design and construction of new electrolyzers and auxiliary equipment and apparatus. There is a growing awareness by workers of industrial safety and health hazard concerns. On the other hand, there is increased use of unskilled labor with the attendant difficulties for management and main-

---

* Practically impossible because of a higher temperature than the boiling point.

tenance of safety in almost all chemical process industries, including the electrochemical industries. In the case of a water electrolysis plant, it is important that safety be emphasized to prevent any accidents and operating problems because installation of water electrolyzers is increasing in underdeveloped countries where there is a shortage of well-trained labor.

Demand of hydrogen in North America is around 10^{11} Nm^3/year and the world needs are about three times greater. Use of hydrogen is estimated to increase by some 50% in 1985 and 300% in the year 2000, and the market value of hydrogen will rise. Incentive is thus provided for new development of large scale water cells which could be operated even by unskilled labor without difficulty.

A major part of hydrogen cost is the cost of electric power, and hence the cell voltage or the power consumption should be minimized to reduce the product cost. The investment cost is also a large factor. The bipolar filter press-type cell is common at present, but the unipolar cell must be reconsidered because its design is simple and the investment cost is low.[22]

The National Fertilizer Development Center of the Tennessee Valley Authority, Muscle Shoals, Alabama, is continuing to investigate applications of hydrogen, especially production of ammonia. They have compared technologies of hydrogen production now available, not only the electrolytic process but also the chemical route, and have stated that production cost of hydrogen with hydrocarbons would not be economical when the price of crude oil increases in the near future.

The price of electrolytic hydrogen is a function of the investment cost, electric energy efficiency, percent operating time, and the price of electricity. Figure 5.4 shows the relationship between the hydrogen cost and the cell cost.[23] It is clear that the larger the operating time percentage, the lower is the hydrogen price even at high fixed charges for cell installation.

Production of heavy water and enlargement of market and efficient use of byproduct oxygen are ways for reducing the price of electrolytic hydrogen. The economic effect of such efforts is estimated to be large although detailed evaluation is difficult.[22-25]

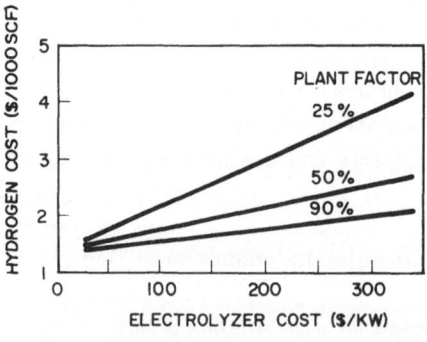

FIGURE 5.4. Relationship between hydrogen cost and electrolyzer cost for various plant factors. Cell efficiency = 70%. Power cost = 1 ¢/kWh.

REFERENCES

1. *Seizo Kotei Zu Zenshu* (Comprehensive Treatise of Chemical Processes), Vol. 1, p. 164, Kagaku Kogyo Sha, Tokyo (1968)
2. Makoto Kuwahara, *Kagaku Keizai* (Chemical Economy), p. 12 (July 1976).
3. L. W. Jones, *Science* **174**, 367 (1971).
4. G. Weismantel, *Chem. Eng.*, p. 46 (November 27, 1972).
5. N. P. Chopey, *Chem. Eng.*, p. 24 (December 25, 1972).
6. J. O'M. Bockris, *Science* **177**, 875 (1972).
7. C. Marchetti, *Chem. Econ. Eng. Rev.*, p. 7 (January 1973).
8. A. M. Bueche, *J. Electrochem. Soc.* **120**, 295C (1973).
9. C. A. McAuliffe, *Chem. Br.* **9**, 559 (1973).
10. K. Kanzaki and K. Fukatsu, *Denki Kagaku* (J. Electrochem. Soc. Japan) **19**, 255 (1951); **23**, 169 (1955).
11. N. Kawashima, *Ryuan Gijutsu* (Ammonium Sulfate Technology) **7**(1), 9 (1954).
12. Y. Hinonishi, *Kogyo Butsuri Kagaku* (Industrial Physical Chemistry) **4**, 73, Corona Co., Tokyo (1949).
13. C. L. Mantell, *Electrochemical Engineering*, p. 315, McGraw-Hill, New York (1960).
14. *Kagaku Purosesu Shusei* (Collection of Chemical Processes), p. 384, Tokyo Kagaku Dozin, Tokyo (1970).
15. P. Gallone, *The Encyclopedia of Electrochemistry*, edited by C. A. Hampel, p. 696, Reinhold, New York (1964).
16. J. Horiuchi, *Suiso Denkyoku Hanno* (Hydrogen Electrode Processes), p. 110, Kawade Publishers, Tokyo (1950).
17. N. P. Chopey, *Chem. Eng.*, p. 118 (February 20, 1961).
18. M. J. Braun, Abstract No. 473, Electrochemical Society meeting, Seattle, WA, May 1978.
19. K. Christiansen and T. Grundt, Abstract No. 474, Electrochemical Society Meeting, Seattle, WA, May 1978.
20. P. M. Spaziante, L. Giuffre, and A. Nidola, Abstract No. 478, Electrochemical Society Meeting, Seattle, WA, May 1978.
21. A. J. Appleby and J. Crepy, Abstract No. 349, Electrochemical Society meeting, Philadelphia, PA, May 1977.
22. R. L. LeRoy and A. K. Stuart, Abstract No. 482, Electrochemical Society meeting, Seattle, WA, May 1978.
23. D. P. Gregory, K. F. Blurton, and N. P. Biedermen, Abstract No. 476, Electrochemical Society Meeting, Seattle, WA, May 1978.
24. A. J. Smith and J. D. Hatfield, Abstract No. 484, Electrochemical Society Meeting, Seattle, WA, May 1978.
25. S. D. Gupta, H. P. Dhar, J. Jacobs, S. Mohanta, and N. White, Abstract No. 495, Electrochemical Society Meeting, Seattle, WA, May 1978.

Chapter 6

ELECTROLYSIS OF HYDROCHLORIC ACID SOLUTION

Chlorine is generally produced by electrolysis of sodium chloride solution. Table 6.1 shows the end-use pattern of chlorine in Japan in 1979 and 1980.[1] More than 70% of chlorine is consumed by its producers. Of these, more than half is sent to the synthesis of vinyl chloride monomer and vinylidene chloride monomer. About 10% of chlorine is used for trichloroethylene and propylene oxide. Those figures are almost the same as in more advanced countries. In the United States, for example, about 20% of chlorine is used for VCM, and 30% for other organic compounds, as shown in Table 6.2.[2] Vinyl chloride monomer (VCM) is the largest single consumer of chlorine. Production and the price of chlorine are thus influenced greatly by the market situation of such materials containing chlorine.

Although VCM can be produced by the reaction of HCl with acetylene:

$$C_2H_2 + HCl = CH_2CHCl \tag{6.1}$$

a major part of VCM is being produced by the ethylene dichloride (EDC) route, i.e., ethylene is chlorinated to EDC, and is converted into VCM and HCl by pyrolysis:

$$C_2H_4 + Cl_2 = C_2H_4Cl_2 \tag{6.2}$$

$$C_2H_4Cl_2 = CH_2CHCl + HCl \tag{6.3}$$

$$\overline{C_2H_4 + Cl_2 = CH_2CHCl + HCl} \tag{6.4}$$

As is shown in Eq. (6.4), half of the chlorine supplied is converted into HCl, and should be recycled efficiently. The oxychlorination process and the mixed gas process $(C_2H_4 + C_2H_2)$ are examples of innovative approaches of this task.[3]

The Deacon process is very important both in a fundamental sense and industrially, i.e., HCl is oxidized by oxygen and/or air with $CuCl_2$ as the catalyst:

$$4HCl + O_2 = 2Cl_2 + 2H_2O \tag{6.5}$$

TABLE 6.1

End-use Pattern of Chlorine in Japan[a]

	1979	1980
Chlorine gas		
Liquefied chlorine	777,520 (27%)	751,500 (28%)
Hydrochloric acid	193,074 (7%)	200,942 (7%)
Hypochlorites	48,613 (2%)	49,023 (2%)
Miscellaneous	1,843,607 (64%)	1,700,655 (63%)
TOTAL	2,863,059 (100%)	2,702,415 (100%)
Liquefied chlorine		
Chemical industries		
Organic chemicals	271,114 (35%)	288,396 (39%)
Inorganic chemicals	93,259 (12%)	90,369 (12%)
Plastics	14,779 (2%)	10,801 (1%)
Miscellaneous	71,249 (9%)	42,390 (6%)
SUBTOTAL	450,401 (58%)	431,956 (58%)
Pulp and paper	228,103 (29%)	229,170 (31%)
Water treatment	75,242 (10%)	69,936 (9%)
Miscellaneous	20,012 (3%)	17,907 (2%)
TOTAL	773,758 (100%)	748,969 (100%)

[a] Unit $= t/y$.

While selection of the materials for equipment and apparatus is difficult because of the serious corrosion environment, a large plant using a modified Deacon process is being operated in the United States.[4,5]

Chlorine can be recovered by electrolysis of HCl solution, which is also an old process. A commercial cell was designed and operated in Germany

TABLE 6.2

End-use Pattern of Chlorine in the
United States in 1980

Plastics	20%
Pulp and paper	13%
Chlorinated methanes	13%
Inorganic chemicals	11%
Propylene oxide	10%
Other organic chemicals	21%
Water treatment	6%
Miscellaneous	6%

during World War II.[6] HCl electrolysis is of growing importance today because it is an economical procedure for the recovery of chlorine from byproduct HCl at the chlorination plant.[7] Byproduct HCl contains some organic impurities, which can cause serious trouble for electrolysis. Therefore, pretreatment or purification of HCl gas and/or solution should be made before sending it to the cell. Figure 6.1 shows an example of a flowsheet for HCl electrolysis.[8]

Byproduct hydrochloric acid from processes for the chlorination of organic compounds contains a trace of organic impurities. These impurities seriously affect the electrocatalytic activity of the electrodes in the HCl cell, resulting in high cell voltage. Organic impurities must be treated completely. The purified HCl gas is absorbed by weak acid which is sent back from the electrolyzer. The concentrated HCl solution is then brought to electrolysis again.

HCl electrolyzers have been operated commercially in Germany for many years. I. G. Farbenindustrie operated HCl cells in two plants located at Bitterfeld and Wolfen during 1932–1944. This technology has been continued after World War II by Farbwerke Hoechst AG and Friedrich Uhde GmbH. The results obtained at Farbwerke Hoechst AG (1200 t-Cl_2/day) and the Farbenfabriken Bayer AG Leverkusen plant (90 t-Cl_2/day) since 1964 have been published.[8]

Chlorine and hydrogen are contaminated when HCl gas or solution contains organic impurities. Those materials deposit on the electrode surface and the catalytic activity of electrode fails. Also, impurities plug the diaphragm and the *IR* drop through it increases.

There is no direct effect of methane in HCl coming from a methane

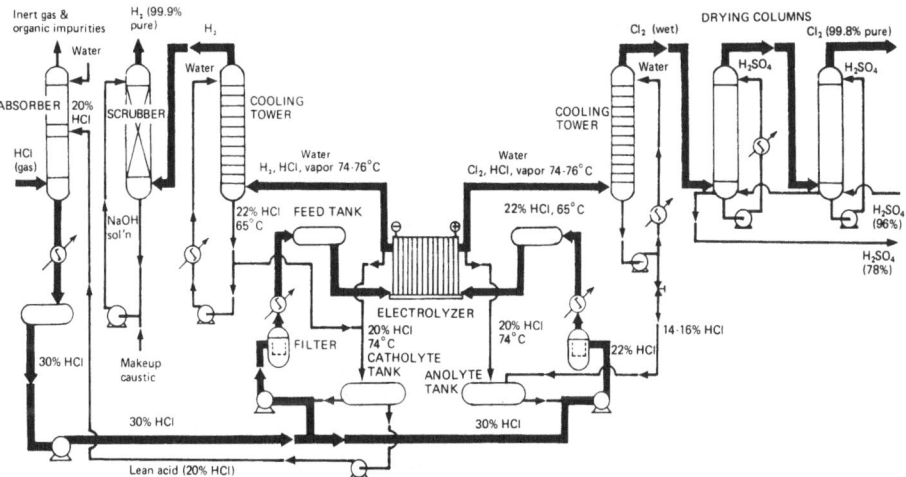

FIGURE 6.1. Flowsheet for electrolysis of hydrochloric acid (Uhde cell).

chlorination plant, but hydrogen gas is contaminated because the solubility of chlorinated methanes in HCl solution is large. On the other hand, methane is chlorinated at the anode to form carbon tetrachloride, which contaminates chlorine gas. The terminal voltage becomes high rapidly when HCl solution contains benzene, chlorobenzenes, phenol, chlorophenols, etc., due to deposition or blocking of chlorinated compounds on the anode and the diaphragm. The chlorinated substances consist of hexa- and penta-chlorobenzenes, tetra- and pentachlorophenols and hexachlorocyclohexane. These compounds intercalate into the graphite pores and finally destroy the graphite electrode.

The organic impurities can be removed by adiabatic absorption followed by adsorption with activated carbon, whereas the purification procedure could be changed by the composition of HCl to be treated.

Hydrochloric acid obtained from the fluorination plant contains hydrogen fluoride and chlorofluorohydrocarbons, and is treated prior to electrolysis.

On the other hand, inorganic impurities, which can form multivalent ions, decreases the current efficiency of electrolysis because the oxidation-reduction processes of those ions may occur at the anode and the cathode together with the main reactions on the respective electrodes. Graphite anodes degraded seriously when HCl solution contains SO_4^{2-}. Generally, the effect of organic impurities on the performance of the HCl cell is greater than that with inorganic impurities. For example, 40–50 g/liter of SO_4^{2-} is tolerated in practice, compared to only 5 ppm of orthodichlorobenzene.[8]

6.1. HCl ELECTROLYZERS

Figure 6.2 shows the DeNora HCl electrolyzer, Model 260D-50C. The capacity is 8 tons of chlorine per day at 5 kA.

The electrolytic cells can be classified into two types: monopolar and bipolar. The monopolar cell consists of at least one anode and cathode pair in a cell, and each pair of electrodes is connected with the outside line of the dc power source directly, i.e., the elements of a cell are connected in parallel. Of course, series connection is also possible and common with higher string voltage. The Hooker diaphragm-type chlorine cell is an example of the monopolar cell. On the other hand, several elements are fitted in series in a bipolar cell, and the dc power is supplied at both ends of the cell (see Fig. 6.3). Of course, both cell types are useful depending on the application, and the cell configuration should be well discussed before design of the system. Since the bipolar cell is simple, it is normally used for HCl cells as well as water electrolysis cells. The cell element consists of the frame made of insulating material such as rubber-lined steel, synthetic resin moldings,

FIGURE 6.2. The DeNora HCl cell. Model 260D-50C. Capacity: 8 t-Cl$_2$/day at 5000 A. (Courtesy of Oronzio DeNora Impianti Elettrochimici.)

and fiberglas-reinforced plastics (FRP). The anode and the cathode are positioned in a frame to make a cell element. The diaphragm is positioned between the two electrodes. Several elements are fitted together by means such as tie rods. Since the configuration of the bipolar cell resembles a filter press, the cell is often called the "filter-press type cell." The Uhde HCl cell and the DeNora HCl cell are both of the filter-press type, and are being operated in Europe and the United States.

The reversible potentials for chlorine and hydrogen electrodes, and the decomposition voltage for HCl electrolysis have been discussed previously in detail (see Sec. 2.2.1), where calculation has been made for 25°C because of a lack of thermodynamic data. Since the equilibrium constant K is a function

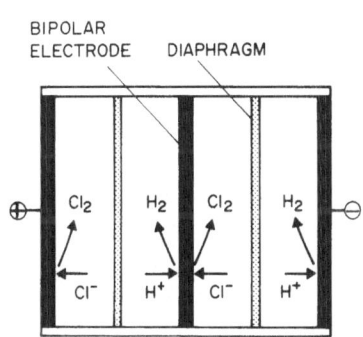

FIGURE 6.3. Bipolar-type HCl electrolysis cell.

of temperature, the reversible potential and the decomposition voltage will change with the operating temperature of the cell. According to Gallone and Messner, the reversible potentials for the chlorine electrode and hydrogen electrode in 20% HCl at 80°C are +1.16 V vs. SHE and +0.07 V, respectively.[9] The overvoltages of the hydrogen electrode and chlorine electrode are shown by Eq. (6.6) and (6.7).

$$\eta_A = 0.3 + 0.07 \log i \tag{6.6}$$

$$-\eta_C = 0.23 + 0.06 \log |-i| \tag{6.7}$$

Thus, we can obtain the voltage balance of HCl electrolysis:

$$V_T = E_d + \eta_A + (-\eta_C) + IR \tag{6.8}$$

$$E_d = E_A - E_C \tag{6.9}$$

Voltages of the hydrogen electrode and chlorine electrode are shown by Eqs. (2.33) and (2.44), respectively. With some assumptions, the voltage balance is calculated as follows:

- Basis: at 20 A/dm² in 20% HCl at 80°C
- Reversible potential: Chlorine electrode, E_A +1.16 V
                          Hydrogen electrode, E_C +0.07 V

- Decomposition voltage: $E_d = E_A - E_C$ 1.09 V
- Overvoltage: Chlorine, $\eta_A = 0.3 + 0.07 \log i_A$ 0.35 V
                Hydrogen, $-\eta_C = 0.23 + 0.06 \log |-i_C|$ 0.27 V
- IR-drops: $\sum IR$ 0.74 V
- Terminal voltage: V_T 2.45 V

The terminal voltage is represented by the sum of the decomposition voltage, the overvoltages on the anode and the cathode, and the ohmic voltage drops. In order to evaluate the sum of ohmic drop, $\sum IR$, assume that the conductivity of 20% HCl at 80°C is 1.4 mho/cm (see Fig. 4.6) and that the anode-to-cathode gap is 1 cm. The solution IR drop is then calculated to be $0.2(A/cm^2) \times 1(cm)/1.4(mho/cm) \simeq 0.14$ V, and is considerably smaller than $\sum IR$ shown above. One of the major reasons is the increase in the resistivity of electrolytic solution due to dispersion of hydrogen and/or chlorine bubbles generated from the respective electrodes. These phenomena are called "bubble effects" and are important for electrochemical engineering (see Sec. 4.1.4).

The IR drop is considerable through the diaphragm, which is made of corrosion-resistant material such as PVC fabric.

Graphite is resistant to HCl and dissolved chlorine, and is used for the chlorine electrode. Generally, the graphite blade is 2–3 cm thick to obtain mechanical strength. The electrical resistivity is about $800\,\mu\Omega$ cm, which is high compared to $1.7\,\mu\Omega$ cm for copper and $10\,\mu\Omega$ cm for iron.[10] Therefore, there are some *IR* drops in the graphite anode and cathode.

The contact resistance of metallic parts is also a factor. Now we tabulate those factors as follows:

- Solution *IR*, including bubble effects: 0.27 V
- Diaphragm: 0.30 V
- Graphite anode and cathode: 0.12 V
- Others, including contact resistance: 0.05 V
- Sum of *IR* drops, $\sum IR$:   0.74 V

The decomposition voltage is, of course, constant under given conditions of cell operation, but other factors of the terminal voltage depend greatly on the current density. A simple equation [also Eq. (4.36)] is useful in practice:

$$V_T = A + Bi \qquad (6.10)$$

where A is the apparent decomposition voltage and B is the slope of the

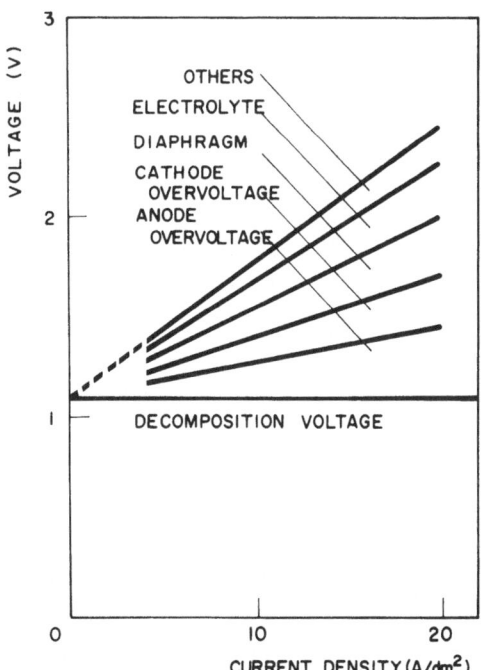

FIGURE 6.4. Breakdown of the terminal voltage of a HCl electrolyzer operated with 20% HCl at 80°C.

voltage vs. current density curve. These factors are important for cell design and operation of practical cells. Figure 6.4 shows an example of the voltage breakdown of a HCl cell containing 20%HCl at 80°C.

6.2. RECOVERY OF CHLORINE FROM HYDROCHLORIC ACID

The demand for recovery of chlorine from byproduct HCl is increasing with the increase of the chlorination of hydrocarbons. There are several processes. Production of chlorine from hydrochloric acid is not new. Scheele first discovered oxidation of HCl with MnO_2 about 200 years ago.[11] Deacon modified Scheele's experiment and developed the oxidation process of HCl using oxygen or air, although the start-up of commercial plants using this process is fairly recent (see Table 6.3).

Electrolysis of HCl solution is a relatively simple process. It has been commercialized with the development of chlorination of hydrocarbons in Europe, although capacity is not large even at present.

There are two ways for recovery of chlorine from HCl: electrolysis and oxidation. Table 6.3 describes several processes published or now operated and includes the oxychlorination process of hydrocarbons for reference.

The electrochemical processes are classified into two methods: direct and indirect electrolysis. Direct electrolysis has been described in the preceding section. The DeNora cell and the Uhde cell are examples in operation. The flowsheet for this process is simple (see Fig. 6.1), but the energy consumption is somewhat large, the largest factor being the decomposition voltage of about 1.3 V. Alternative cathode reactions whose reversible potential is more noble than that of the hydrogen electrode process have been investigated to replace hydrogen evolution and thus reduce the cell voltage. Use of the oxygen-depolarized cathode is an example of such indirect electrochemical processes.

Figure 6.5 is the flowsheet of the Westvaco process for the electrolysis of the mixed solution of HCl and $CuCl_2$, which has been extensively studied. The mixed solution of electrolyte is sent from its head tank to the cell, and chlorine gas generated at the anode is recovered. Cupric chloro-complex ions are reduced at the cathode. The catholyte effluent containing HCl, CuCl, and $CuCl_2$ is pumped to the multistage column where cuprous ions in the solution are oxidized by air or oxygen. The electrolytic solution is then brought to the cell for electrolysis. The terminal voltage of the Westvaco cell is advantageously low as shown in Table 6.4, which summarizes operating conditions and performance.[12,13]

Schroeder investigated electrolysis of mixed solution consisting of HCl and $NiCl_2$ with graphite electrodes.[14] He obtained chlorine at the anode and a nickel electrodeposit at the cathode. Electrolysis is interrupted when

TABLE 6.3

Recovery of Chlorine from Byproduct Hydrochloric Acid

Electrolytic processes
  Direct electrolysis
    Uhde[7,8], DeNora[9] $2HCl \rightarrow H_2 + Cl_2$
  Indirect electrolysis

    Westvaco process[12,13] $CuCl_2 \xrightarrow{\text{electrolysis}} CuCl + \frac{1}{2}Cl_2$

                                 $CuCl + HCl + \frac{1}{4}O_2 \xrightarrow{\text{oxidation}} CuCl_2 + \frac{1}{2}H_2O$

    Schroeder process[14] $NiCl_2 \xrightarrow{\text{electrolysis}} Ni + Cl_2$

                                 $Ni + 2HCl \xrightarrow{\text{oxidation}} NiCl_2 + H_2$

    Kyoto process[15,16] $2HCl + \frac{1}{2}O_2 \xrightarrow{\text{electrolysis}} H_2O + Cl_2$

Oxidation processes
  Deacon Process, Airco,
    fluidized bed[17] $4HCl + O_2 \xrightarrow[\text{CuCl}_2]{450-650°C} 2Cl_2 + 2H_2O$

  Modified Deacon process,
    Kel-Chlor process[4,5,18] $HCl + NOHSO_4 = NOCl + H_2SO_4$
                                 $2NOCl = 2NO + Cl_2$
                                 $O_2 + 2NO = 2NO_2$
                                 $NO_2 + 2HCl = NO + Cl_2 + H_2O$
                                 $NO + NO_2 + 2H_2SO_4 = 2NOHSO_4 + H_2O$
                                 $NOCl + H_2SO_4 = NOHSO_4 + HCl$

  Chloride melts process,
    I. G. Farbenindustrie[17] $FeCl_3 - KCl(\text{melts}) + \frac{3}{4}O_2 \rightarrow \frac{1}{2}Fe_2O_3 + \frac{3}{2}Cl_2$
                                 $Fe_2O_3 - KCl + 6HCl \rightarrow 2FeCl_3 + 3H_2O$

                                 $S_2Cl_2 + \text{dilute } Cl_2 \xrightarrow{\text{absorption}} 2SCl_2$

                                 $2SCl_2 \xrightarrow{\text{desorption}} Cl_2 + S_2Cl_2$

  Metal chloride decomposition
    process
    Grosvenor–Miller process[17] $Fe_2O_3 + 6HCl \xrightarrow{250-300°C} 2FeCl_3 + 3H_2O$

                                 $2FeCl_3 + \frac{3}{2}O_2 \xrightarrow{475-500°C} Fe_2O_3 + 3Cl_2$

    Chloro-sulfonic acid process[17] $HCl(g) + SO_3(g) \xrightarrow{70-100°C} HSO_3Cl(l)$

                                 $2HSO_3Cl \xrightarrow[\text{HgCl}_2]{160°C} SO_2Cl_2(g) + H_2SO_4$

                                 $SO_2Cl_2(l) \rightarrow SO_2 + Cl_2$

                                 $SO_3 + 2HCl \rightarrow H_2SO_4 + Cl_2 + SO_2$

Oxychlorination processes[17]

                                 $CH_2CH_2 + HCl + \frac{1}{2}O_2 \rightarrow CHClCH_2 + H_2O$
                                 $CH_2CH_2 + 2HCl + O_2 \rightarrow CHClCHCl + 2H_2O$
                                 $CH_2CH_2 + 3HCl + \frac{3}{2}O_2 \rightarrow CHClCCl_2 + 3H_2O$
                                 $CH_2CH_2 + 4HCl + 2O_2 \rightarrow CCl_2CCl_2 + 4H_2O$

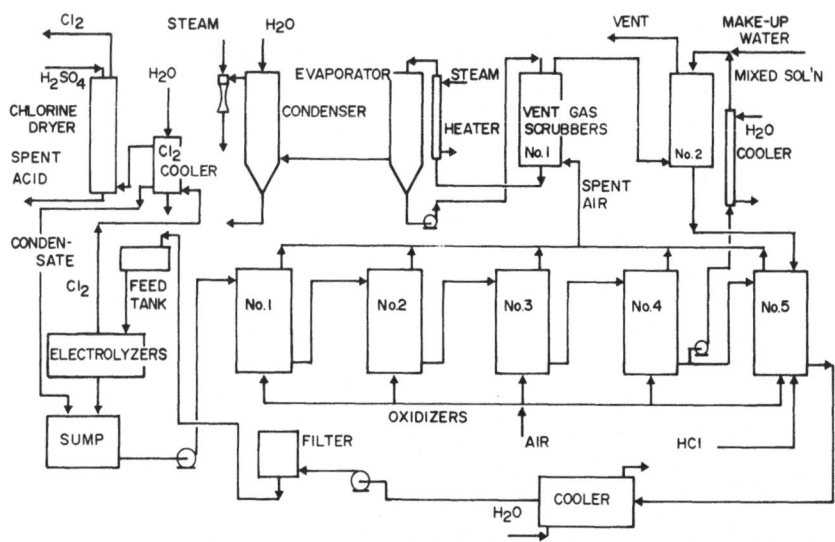

FIGURE 6.5. Flowsheet of the Westvaco process.

the nickel deposit becomes thick and deposited nickel is dissolved with oxygenated hydrochloric acid solution. Practical application is considered to be hindered by the intermittent nature of the process.

Hine and coworkers at Kyoto University investigated electrolysis of the mixed solution of $CuCl_2$ and HCl with an oxygen depolarized cathode, where the expected cathode reaction is

$$4H^+ + O_2 + 4e = 2H_2O \qquad (6.11)$$

because of its noble potential.[15,16] Reaction 6.11 is the same as for the cathode process of the acid-fuel cell. Since the performance of the fuel cell oxygen cathode was insufficient, cupric chloride was added to the solution and oxygen or air was brought to the cathode compartment. Therefore, the cathode reaction is the same as that of the Westvaco process, but cuprous ions formed at the cathode are oxidized immediately in the bulk of catholyte of the compartment (see Sec. 4.3.3).

All the indirect electrochemical methods consist of a combination of electrolysis and oxidation of metals or metal chlorides of lower valencies, whereas the overall reaction is similar to that for the Deacon process. That is, processes can be candidates for energy saving or recovery of electricity by application of fuel cell concepts for conversion of the free energy for water formation from H_2 and O_2.

Nonelectrochemical processes, including oxidation of HCl and oxychlorination of hydrocarbons, are also applications or modifications of

TABLE 6.4

Performance of the Westvaco 4kA Cell

Electrolytic solution		
Flow velocity	gal/hr	70
Flow rate	gal/hr ft^2	7.0
Composition: $CuCl_2$	%	15
HCl	%	20
Temperature of feed	°C	64
Operating temperature	°C	80
Amperage	kA	4
Cathode current density	A/ft^2	400
Current efficiency	%	82
Terminal voltage	V	1.80
dc power consumption	kWh/lb Cl_2	0.80
Production of chlorine	lb/hr	9.6
Composition of catholyte effluent		
CuCl	%	3.7
$CuCl_2$	%	10.2
HCl	%	20.2

the Deacon process. For example, HCl is oxidized by NO_2 and/or SO_3, and the NO or SO_2 formed are oxidized by air or oxygen and these gases are recycled in the oxidation process. On the other hand, oxygen is brought to the chlorination reactor of the oxychlorination process so as to oxidize by-product HCl simultaneously. Details of the nonelectrochemical methods can be found in the original papers and/or reference books.[18,19]

REFERENCES

1. *Soda to Enso* (Soda and Chlorine) **31**, 209 (1980); and **32**, 273 (1981).
2. E. J. Rudd and W. E. Darlington, *Report of the Electrolytic Industries in 1980*, presented at Electrochemical Society Minneapolis, Minnesota, May 1981.
3. *Kagaku Purosesu Shusei* (Collection of Chemical Processes), p. 496, Tokyo Kagaku Dojin, Tokyo (1970).
4. C. P. van Dijk and W. C. Schreiner, *Chem. Eng. Prog.* **69**(4), 57 (1973).
5. L. E. Bostwick, *Chem. Eng.*, p. 86 (October 11, 1976).
6. W. C. Gardiner, *Chem. Eng.*, p. 100 (January 1947).
7. F. B. Grosselfinger, *Chem. Eng.*, p. 172 (September 14, 1964).
8. S. Payer and W. Strewe, *Chlorine Bicentennial Symposium*, p. 257, Electrochemical Society (1974).
9. P. Gallone and G. Messner, presented at Electrochemical Society meeting, San Fransisco, CA, May 1965.
10. *Handbook of Chemistry and Physics*, 44th ed., p. 2667, Chemical Rubber, Cleveland (1963).

11. I. Eiden and L. Lundevall, *Chlorine Bicentennial Symposium*, p. 20, Electrochemical Society (1974).
12. C. P. Roberts, *Chem. Eng. Prog.* **46**(9), 456 (1950).
13. U.S. Patents 2,468,766 and 2,470,073.
14. D. W. Schroeder, *I & EC, Process Des. and Dev.* 1(2), 141 (1962).
15. F. Hine, S. Yoshizawa, K. Yamakawa, and Y. Nakane, *Electrochem. Technol.* **4**, 555 (1966).
16. F. Hine and K. Yamakawa, Abstract No. 258, Electrochemical Society meeting, Los Angeles, CA, May 1970.
17. A. Redniss, HCl oxidation processes, in *Chlorine*, edited by Sconce, ACS Monograph 154, Reinhold, New York (1962).
18. A. G. Oblad, *Ind. Eng. Chem.* **61**(7), 23 (1969).
19. *Harogen-ka*, *Datsu-harogen*, *Kuroru-mechiru-ka* (Halogenation, Dehalogenation, and Chloromethylation), Kagaku-Kogyo Sha, Tokyo (1964).

AMALGAM-TYPE CHLOR-ALKALI INDUSTRY

Chlorine and caustic soda, which are produced simultaneously by electrolysis of sodium chloride solution, together with sulfuric acid and ammonia, are the most important industrial inorganic chemicals. Table 7.1 shows the world production of caustic soda.[1] Figure 7.1 represents the growth of production of chlorine and caustic soda in the United States and Japan. For the United States, the production is shown in short tons of chlorine and by metric tons of caustic soda in Japan. Fortunately, those units are roughly equivalent to each other: $(Cl_2/NaOH) \times$ (metric ton/short ton) $= 0.888/0.907 = 0.98$. The production of chlorine and caustic soda increased at a healthy rate until the year 1970. The growth rate has, however, decreased with the decline of the world economy due to the oil crisis. The decline is projected to continue for several years, with a gradual recovery.[2-4]

Table 7.2 represents the end-use pattern of chlorine and caustic.[4,5] Since a large amount of chlorine is consumed by chlorination of hydrocarbons, the chlor-alkali industry is seriously affected by the market for these chlorine compounds. The rayon industry was once the largest consumer of caustic soda in Japan. Since it has a variety of users today, the effects of market fluctuations seem to be minimized.

Because the United States and Canada have abundant crude salt, fuel, and chrysotile asbestos, diaphragm cell plants have a large share of production, as shown in Table 7.3, whereas some amalgam cell plants are being operated to produce pure caustic.[5,7] Almost all the plants in Europe are operating amalgam cells.[8]

The Japanese chlor-alkali industry has experienced serious problems with the so-called "third" and "fourth" Minamata diseases in 1973. Several plants using amalgam cells were switched off for a week or so because sludge containing a lot of mercury was found near the plant site. It should be noted that mechanisms for the conversion of metallic mercury and inorganic compounds to harmful organics have still not been established.

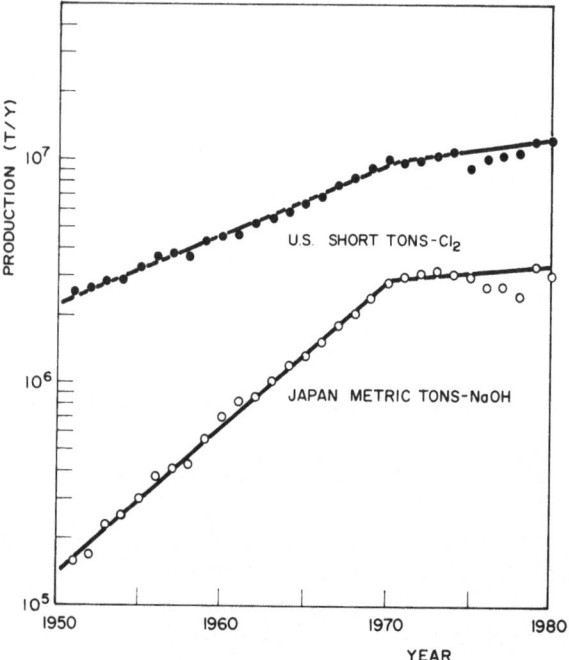

FIGURE 7.1. Annual production of chlorine and caustic soda.

The Japanese EPA ordered the conversion of all amalgam-type chlor-alkali plants to the diaphragm process by 1978, while the conversion program was delayed by the slump of economy, as shown in Table 7.3 and Fig. 7.1. The production capacity in Japan was the second largest after the United States, and was estimated to be about 3.9 million metric tons of caustic in the year 1974, 100% NaOH basis, of which 95% was made by

TABLE 7.1

Production of Caustic Soda in Eight Major Countries[a]

	1970	1975	1979
U.S.A.	9.2	8.4	11.1
West Germany	1.7	2.5	3.4
U.S.S.R.	1.8	2.4	3.0
Japan	2.6	2.9	2.9
France	1.1	1.1	1.4
United Kingdom		1.0	1.2
Canada	0.9	0.8	1.1
Italy	1.0	1.0	1.1

[a] 100% NaOH basis. Unit = million metric tons per year.

TABLE 7.2
End-use Pattern of Chlorine and Caustic Soda

	U.S.A. (1974)	Japan (1980)
Chlorine		
Vinyl chloride	20%	29%[a]
Other organic compounds	30%	26%[b]
Pulp and paper	15%	8%
Inorganic chemicals	11%	9%
Others	17%	28%
Caustic soda		
Inorganic chemicals		17%
Pulp and paper		11%
Chemical fibers		7%
Alumina		7%
Other domestic use		52%
Export		6%

[a] Including vinylidene chloride.

[b] Including propylene oxide.

amalgam cells, but the share in the world production declined in 1979, as shown in Table 7.1. It should be pointed out that one-third of the amalgam-type chlorine cell plants are very new and are being operated with fairly good performance and extremely low consumption of mercury, say 2 g/t-NaOH or less, which would be the best level in the world. In order to comply with the EPA directive, chlorine plants must spend more than 2.5 billion dollars for conversion of mercury cells.

TABLE 7.3
Production and Capacity of Chlorine and Caustic

	Chlorine produced in 1974		Capacity in Japan		
	U.S.A.	Canada	1974	1980	1984[a]
Diaphragm process	69.9%	64.5%	4.6%	65.8%[b]	100%[b]
Amalgam process	24.8%	35.5%	95.4%	34.2%	0%
Others	5.3%				
TOTAL	100.0%	100.0%	100.0%	100.0%	100.0%
Capacity (10^6 t/y)	12.1	1.2	3.8	4.0[c]	

[a] Projected.

[b] Including ion exchange membrane cell process.

[c] Caustic soda, 100% NaOH basis.

Operation of amalgam-type chlor-alkali plants rather than diaphragm cell plants is preferred in Japan because a lot of high-grade caustic is required for use in the rayon, pulp and paper, and miscellaneous chemical industries. The Japan Soda Industry Association distributed samples of diaphragm cell caustic to users to check its applicability in November 1974. Results of the effort showed that rayon, cellophane, and flavor enhancer would be seriously affected by the NaCl in diaphragm caustic. Some inorganic chemicals such as sodium silicate, sodium phosphate, sodium sulfite and hyposulfite, sodium cyanide, and sodium bichromate could not pass Japan Industrial Standards (JIS), if produced with regular-grade diaphragm caustic. Petrochemical industries, including petroleum refiners, are worried by corrosion problems caused by the chloride in diaphragm caustic, which may give rise to serious accidents.

Thus, the Japanese market required high-grade caustic of 700 thousand tons per year, which is about 23% of the total consumption of alkali. Recently the ion-exchange membrane cell technology has been developed, and several plants using the membrane cells have already been started up to supply pure caustic to the market. It is estimated that the membrane cells will be used for the process conversion of remaining amalgam cell plants in the near future.

7.1. FLOWSHEET

The amalgam-type chlor-alkali plant consists of a brine treatment, electrolysis, amalgam decomposition, and products polishing. Those processes are, of course, closely related and connected to each other via flows of energy and materials, as shown in Figs. 7.2 and 7.3.

In the brine treatment yard, crude salt is dissolved with depleted brine returned from the amalgam cell. Impurities, mostly magnesium and calcium, are precipitated by addition of a slight excess of sodium hydroxide and soda ash and are separated by a thickener. The effluent is further polished by the brine filter and the brine is slightly acidified (pH = 2–3) prior to electrolysis. This procedure is important to keep the performance of the cell high. The depleted brine (10–15% conversion) is brought to a reservoir, where chlorine gas is separated. The brine is further dechlorinated if necessary, then is sent to the treatment yard. On the other hand, sludge or underflow of the precipitator or thickener is dehydrated with a filter press to recover salt, and washed out with fresh water prior to being purged.

The cell gas or wet chlorine from the cell is cooled to ambient temperature by a plate-type titanium heat exchanger to demist and dehydrate. The gas is dried by a multistage acid tower containing concentrated sulfuric acid, and is brought to liquefaction. Caustic soda solution,

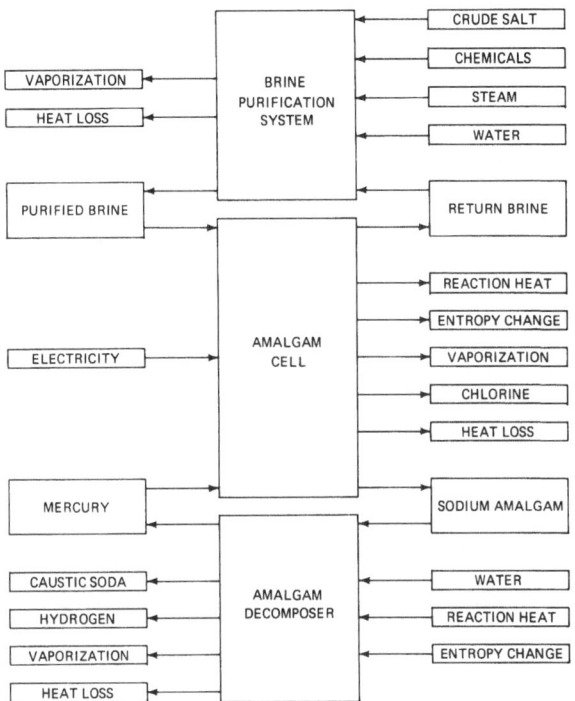

FIGURE 7.2. Flow chart of materials and enegy in the amalgam cell plant.

mostly 50% by weight in concentration, produced by the amalgam decomposer is cooled to 50–60°C, and is polished by a leaf-type filter using activated carbon to aid removal of traces of mercury and carbon before storage or shipping.

It is well known that impurities in the brine affect the cell performance in electrochemical processes in general. The effect of contamination on the performance of the amalgam cell is most serious because this process is based on the concept that the hydrogen overvoltage on fresh mercury or sodium amalgam cathode is very high, and hence no formation of hydrogen occurs at all. This is valid only for electrolysis of a highly pure solution of sodium chloride; thus, brine treatment is essential for the amalgam process plant.

The return brine from the cell may contain a trace of mercury in either metallic or ionic form, and mercury is a concern if discharged with the sludge of the precipitator to the environment. Caustic soda and hydrogen produced in the amalgam decomposer also contains mercury. Mercury was

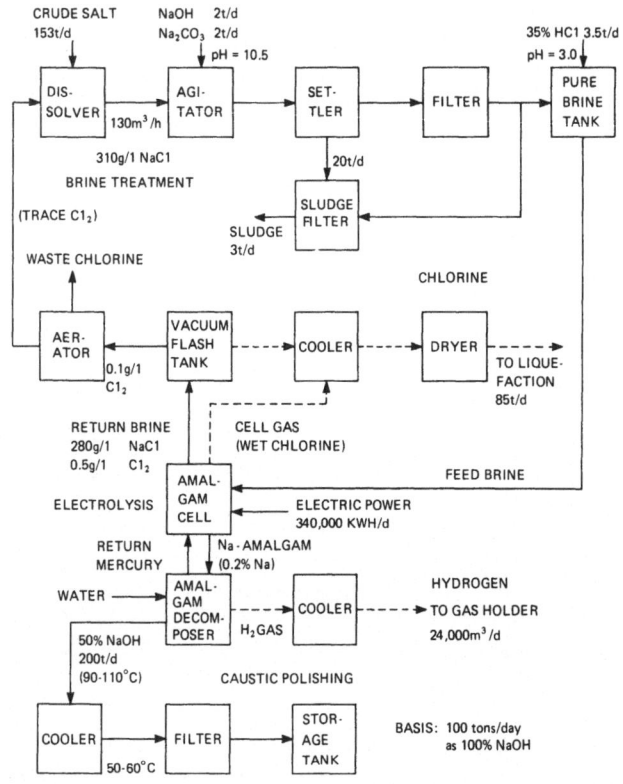

FIGURE 7.3. Flowsheet and material balance in the amalgam cell plant.

formerly detected in the atmosphere in the cell room. Abatement of discharge and leakage of mercury from the amalgam-type chlorine plant has been investigated by a joint group of Japanese chlor-alkali industries. (See Sec. 15.2.)

7.2. AMALGAM CELL

Figure 7.4 illustrates an example of the amalgam cell and the decomposer. Saturated NaCl solution and mercury are brought to the cell from the top end (left) and flow along to the outlet (right). Depleted brine and chlorine are separated, and the brine is sent back to the treatment yard. Sodium amalgam is decomposed in a column-type amalgam decomposer, and mercury is recycled to the cell top.

Conversion of the brine is normally 10–15%, and the sodium amalgam concentration is about 0.2% by weight at the outlet end box.

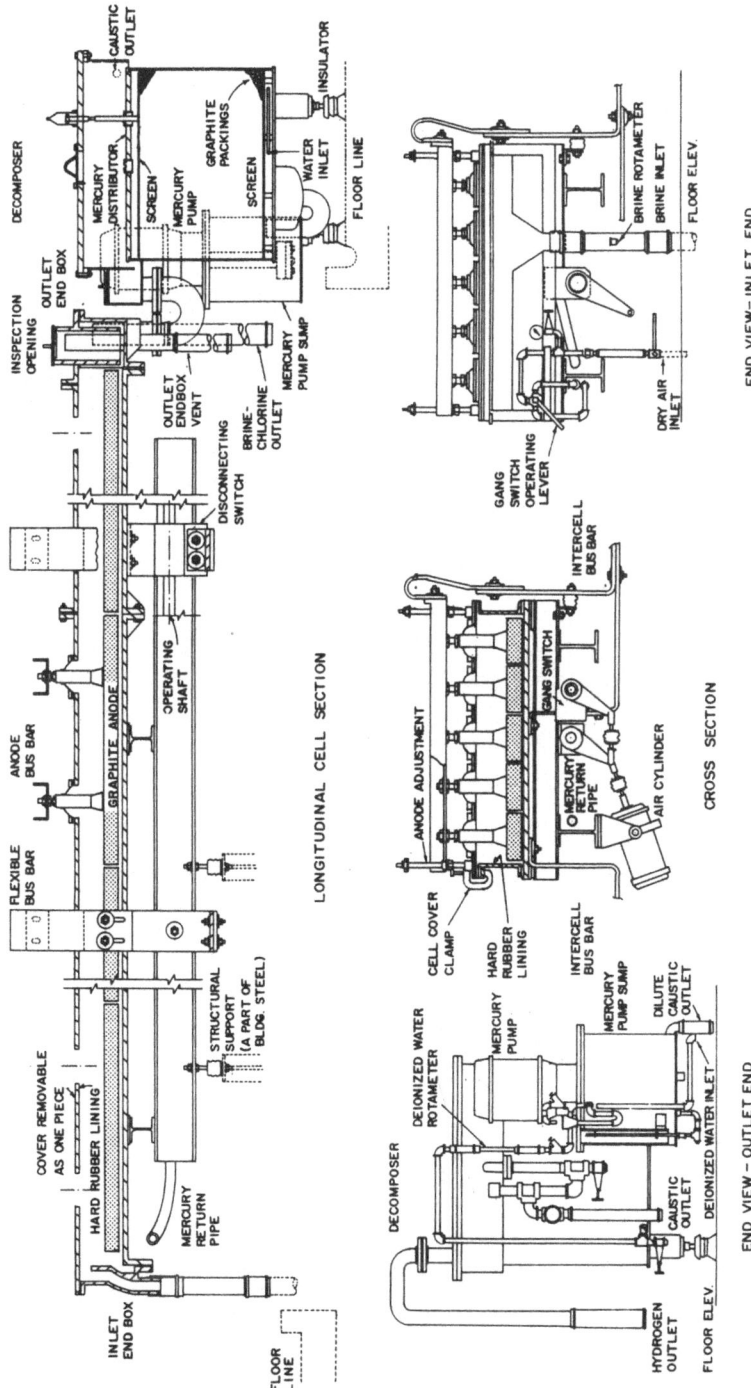

FIGURE 7.4. Olin-type E-11 amalgam cell. (By permission of Olin Corporation.)

The electrochemical processes in the amalgam cell are as follows:

At the anode: $Cl^- = \frac{1}{2}Cl_2 + e$ (7.1)

At the cathode: $Na(Hg) = Na^+ + (Hg) + e$ (7.2)

Overall: $Na^+ + Cl^- + (Hg) = Na(Hg) + \frac{1}{2}Cl_2$ (7.3)

where Na(Hg) denotes the sodium amalgam.

The reversible potentials for the chlorine electrode and the sodium amalgam electrode have been discussed in Secs. 2.2.3 and 2.2.4, respectively.

Now we may discuss the voltage balance of the amalgam cell with those data. Figure 7.5 shows an example obtained in a cell using graphite anodes.[9] At 100 A/dm^2

Reversible potential: Anode:	1.34 V vs. SHE	
Cathode:	−1.76 V	
Decomposition voltage:	3.10 V	60.2%
Anode overvoltage:	1.50 V	29.1%
Solution *IR* drop including bubble effects:	0.44 V	8.5%
Cathode overvoltage:	0.06 V	1.2%
IR drops in hardware:	0.05 V	1.0%
(Total) Cell voltage:	5.15 V	100.0%

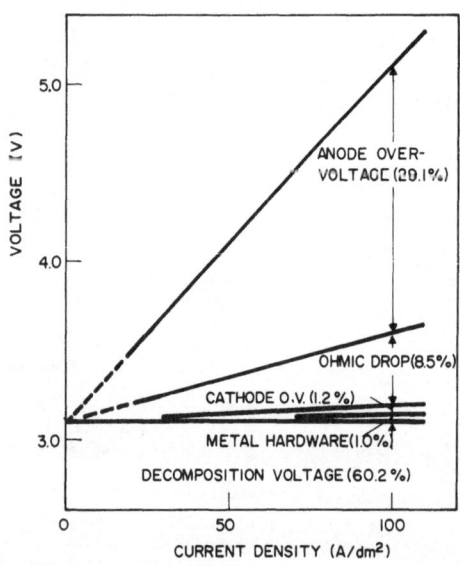

FIGURE 7.5. An example of the voltage balance of an amalgam cell.

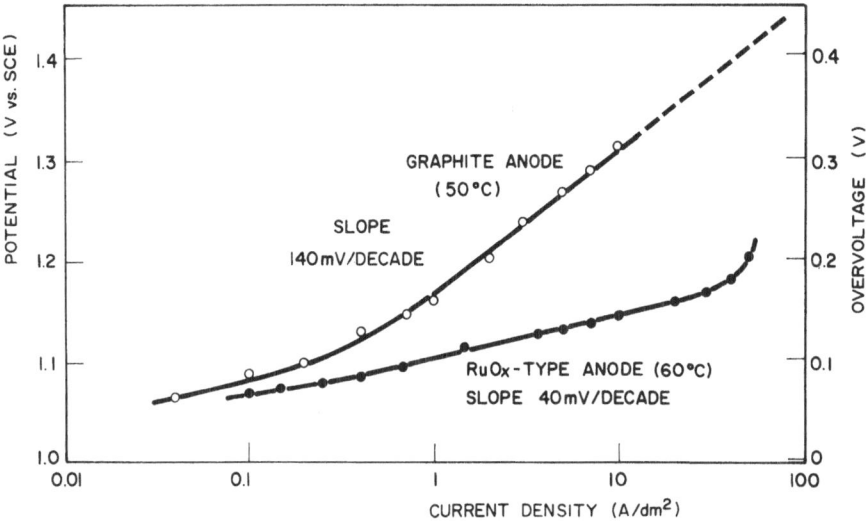

FIGURE 7.6. Chlorine overvoltage on graphite anode and RuO_x type metal anode in 5 M NaCl (pH = 1).

Since the anode overvoltage is a large factor of the cell voltage, it should be minimized to reduce the electric power consumption.

By extrapolation of the polarization curve of a graphite anode in 5 M NaCl at 50°C shown in Fig. 7.6,[10] the chlorine overvoltage at 100 A/dm² should be some 0.45 V, which disagrees with the table shown above. A major reason is the existence of "bubble effects." Chlorine bubbles

GRAPHITE ANODE

BUBBLES

BRINE FLOW

AMALGAM CATHODE

(A) FLAT PLATE ANODE

GAS HOLE

ANODE

BUBBLES

BRINE FLOW

AMALGAM CATHODE

(B) PERFORATED ANODE

FIGURE 7.7 Blocking of anode surface by gas bubbles.

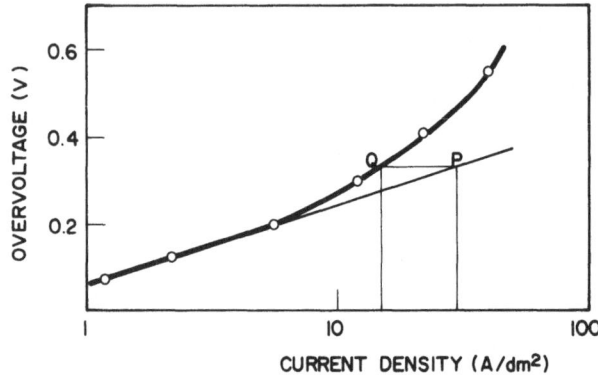

FIGURE 7.8. An example of chlorine overvoltage on a graphite anode.

generated by electrolysis cover the working surface of the anode, as shown in Fig. 7.7(A). Also, the *IR* drop between two electrodes increases significantly at high current densities because gas bubbles disperse in the brine near the anode. As a result, the polarization curve does not follow a straight line (Tafel line) on the η vs. $\log i$ diagram, as shown in Fig. 7.8. It is estimated that the current density at the working surface (Q) becomes high compared to the average or superficial current density (P).[11]

For reducing the bubble effects, the graphite anodes have been machined as illustrated in Fig. 7.7(B). Chlorine bubbles escape from the anode-to-cathode gap through the grooves and the holes.[12]

The chlorine overvoltage on the graphite anode is still a large contribution to the terminal voltage even if the bubble effects are eliminated. About fifteen years ago, Beer developed an excellent anode material consisting of the mixed oxides of Ru and Ti thermally deposited on a Ti substrate which is corrosion resistant and shows a very low chlorine overvoltage under operating conditions of a chlorine cell,[13] as shown in Fig. 7.6.

With a metal substrate, the area of gas holes can be optimized for reducing the voltage drop. Although expensive compared to graphite, Beer's anodes are being used by chlorine industries all over the world due to a significant saving of electricity and hence manufacturing cost.

Figure 7.9 shows the reversible potentials of the chlorine electrode and of the amalgam electrode in the chlor-alkali cell as functions of position along the cell. Since the brine concentration decreases gradually by only about 10–15% with flow-down from the inlet to the outlet, the reversible potential of the anode reaction is kept almost constant. At the cathode, on the other hand, sodium deposits on and dissolves in mercury, and the sodium concentration in the amalgam cathode increases from 0.002% at the top end to, say, 0.2% at the bottom and of the cell. The reversible potential

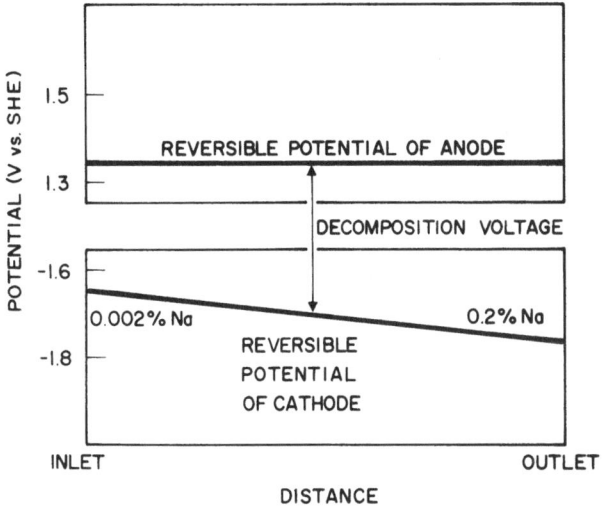

FIGURE 7.9. Potential distribution in the amalgam cell.

of amalgam electrode thus changes significantly with the cell length. Since the terminal voltage

$$V_T = E_A - E_C + \eta_A + (-\eta_C) + \sum IR \qquad (7.4)$$

is assumed to be almost independent of the cell length, the sum of η_A, η_C, and $\sum IR$ will change along the flow direction of the mercury cell:

$$\eta_A + (-\eta_C) + \sum IR = (V_T - E_A) + E_C(x) = f(x, i) \qquad (7.5)$$

Therefore, the current density decreases gradually in the direction of mercury flow from the top to the end of cell. There is a correspondingly different consumption of graphite anodes located at different positions in the amalgam cell which is a function of ampere-hours or current density.

Degradation of the graphite anodes during electrolysis is a problem, and the cell voltage increases gradually. Figure 7.10 shows the experimental results on the changes of the thickness of graphite anode plate and of the terminal voltage of a commercial cell over a 15-month period.[12] The cell voltage zig-zagged since the anode-to-cathode gap was adjusted every day, but the average value rose with time. Many devices for anode adjustment have been investigated and some patents have also been issued. Recently, computerized control systems for adjusting the anode-to-cathode space have been developed and employed universally. This technique is based on data analysis of the current load, the cell voltage, and other factors, and on the feedback of these data to the computer system so as to minimize the cell voltage.

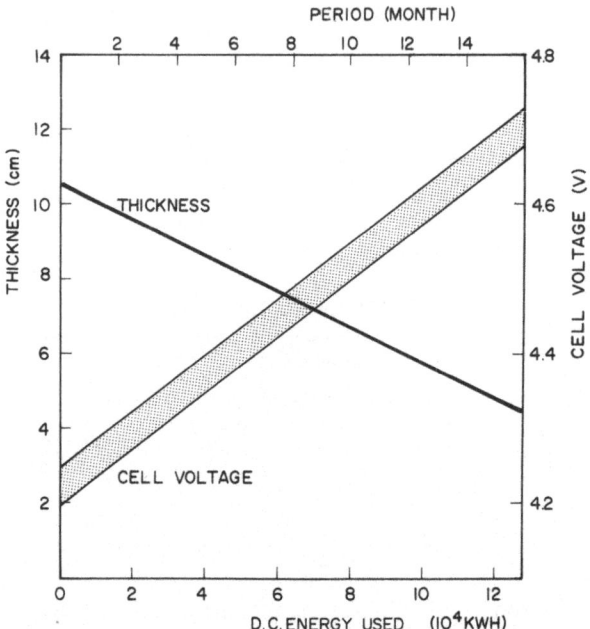

FIGURE 7.10. Thickness of graphite anode and the cell voltage.

Problems of gradual change of the electrode gap have been solved completely by Beer by the invention of very stable anode materials consisting of the mixed oxides of RuO_2 and TiO_2 on a Ti substrate. This oxide-coated Ti electrode has been further investigated and commercialized by the Italian engineering company DeNora, and is named the Dimensionally Stable Anodes (DSA). The DSA is now widely used in amalgam-type chlorine plants (see Fig. 7.11). Those anodes also require a device for anode adjustment in order to prevent short-circuiting.

Figure 7.12 illustrates the cell room in a modern amalgam cell plant. The control room can be seen to the rear. A large movable overhead crane is used for installation of cells and maintenance.

Table 7.4 shows the dimensions of typical amalgam cells. The capacity of these cells is typically 150 kA or greater. Cells rated at 450–500 kA are also being operated. The current density is also large: 70–130 A/dm². The current load and the current density are, respectively, 10 to 30 times and 3 to 7 times those of cells operated thirty years ago, but the cell voltage has actually been reduced. Hence, the consumption of electric energy per unit ton of chlorine has been almost unchanged.

There are two methods for scaling-up the capacity of electrolyzers: increase in geometry and increase in current density. Table 7.5 shows the

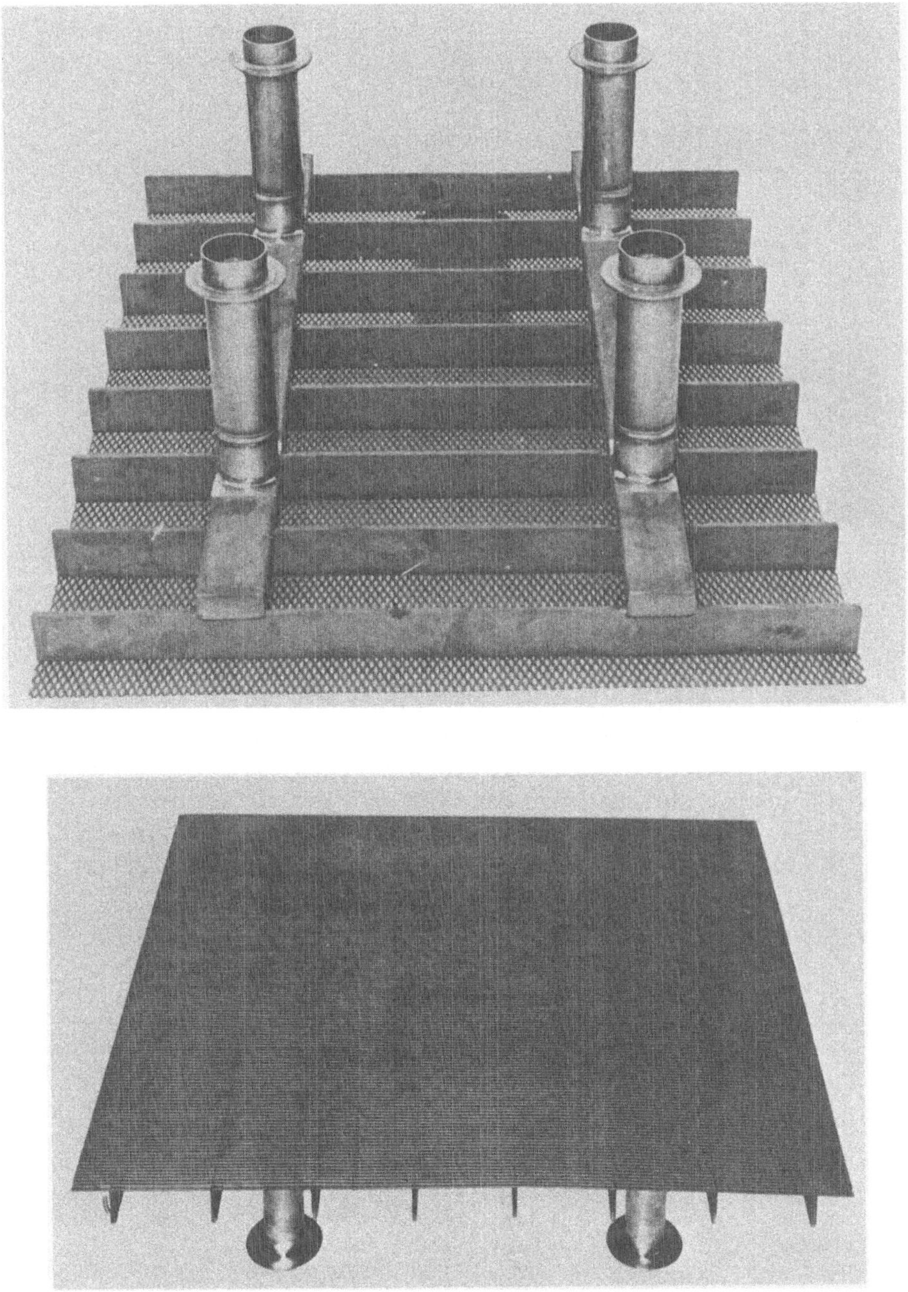

FIGURE 7.11. RuO_2-TiO_2-type oxide-coated titanium anodes for the amalgam-type chlorine cells. (a) Expanded metal-type DSA; (b) spaghetti-type DSA. (From Oronzio DeNora Impianti Elettrochimici.)

FIGURE 7.12. View of an amalgam cell plant. (Courtesy of Dynamit Nobel AG.)

dimensions of various models of Olin mercury cells where, for example, Models E-X1, EX-2, and EX-3 are test cells.[14] The cathode area of the largest cell, E-1014, is about 6 times that of E-X1, and the current density is about 5 times, so that the production capacity is as large as 30 times. The last line in this table shows the ratio of the cathode length to the width, which decreases gradually with increase of the geometry because of difficulty

TABLE 7.4

Dimensions of Various Amalgam-type Chlorine Cells

		Olin E-812	DeNora 24H5	Hoechst-Uhde 30M^2	Krebs-BASF 2T200-8	Solvay V200-F	Kureha HD-4
Amperage	kA	288	250	240	200	160	150
Cathode area	m^2	28.8	34.2	31.5	25.0	30.0	19.8
Width	m	1.94	1.77	2.25		1.29	1.20
Length	m	14.8	19.3	14.0		23.26	16.5
Current density	A/dm^2	100.0	73.1	76.2	80.0	53.4	75.7
Power consumption	dc kWh/t-Cl$_2$	3,679	3,380	3,708	3,155	3,053	3,096
Slope of cathode	mm/m	6	7–8	5–6		6–7	18.8
Number of anodes		96	120	192	144	180	64
Mercury inventory	t/cell	3.8	3.6	3.6	2.8	3.7	1.7

TABLE 7.5
Changes of Olin Mercury Cells

	E-X1	E-X2	E-X3	E-4	E-8	E-11 E-510	E-812	E-1014
Cathode area (m^2)	0.66	3.08	3.23	4.03	6.3	15.0	28.8	42.0
Amperage (kA)	1.4	10	12	15	27	76	240	420
Current density (A/dm^2)	21	32.5	37.2	37.2	43	50.6	83.3	100
Cathode width (m)	0.27	0.457	0.457	0.457	0.615	1.22	1.94	2.42
Cathode length (m)	2.45	6.74	7.03	8.8	10.2	12.3	14.8	17.3
Ratio of length to width	9.1	14.8	15.4	18.5	16.6	10.1	7.6	7.15

in maintaining uniform flow of the mercury layer on a wide cathode plate. Consequently, the ratio of length to width is an important design factor, especially for large-scale cells.

The advantage of scale-up has been sought by all process industries for the last thirty years from the economic point of view, and the chlor-alkali industry has followed the worldwide tendency. However, we need to reconsider those concepts for chlor-alkali cells for several reasons. Figure 7.13 illustrates the change of the floor space with increase of the cathode area.[15] It is clear that reduction of the ratio of floor space to cathode area is almost

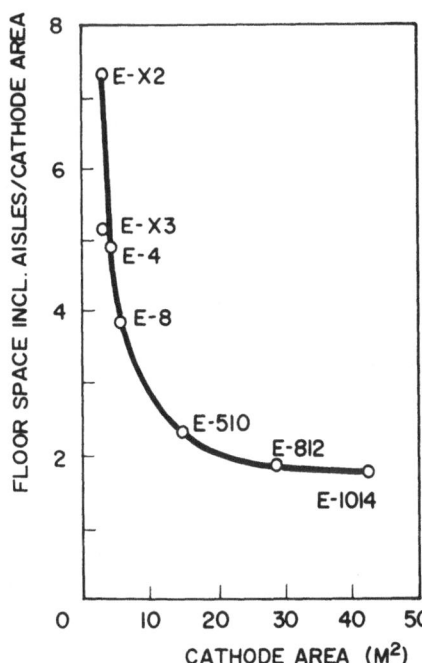

FIGURE 7.13. Relationship between floor space and cathode area of the Olin amalgam cells.

limited at present, so that no saving on floor space could be achieved unless the design concept is changed extensively. A large coefficient of the Williams equation showing the relationship between the investment and the plant size is also a problem, as described in Chap. 14, for the amalgam cell plant and almost all electrochemical process plants as well.

The production capacity of electrolyzers could be increased by an increase of the current density. Such an approach should be considered carefully in relation to the sky-rocketing rise of the cost of electric power. The optimum current density may become rather low in the future.

For the optimum size of electrolyzer to be installed in a plant of a given capacity of production, Burt and Clarke of the Olin Corporation have suggested[15]:

> The industry trend has been to steadily increase cell sizes. Indeed, there is a tendency to believe that if a big cell is good, a bigger one must be better. These studies fail to confirm such a viewpoint unless a rather unlikely combination of economic factors are present. None of the situations studied thus far have required a cell as large as the E-1014. Of the other two sizes studied in detail, the E-510 is preferred below capacities of 300 t/d, the E-812 is preferred between 300 and 2000 t/d. This surprising, to us, at least, inability of large cells to be competitive with smaller cells can be traced to several causes, including:
>
> - Derating of a large cross-section bus to avoid excessive internal temperatures.
> - Shorting switch requirements apparently at the limit of the state of the art.
> - Limited number of shops capable of fabricating cell components, limiting or precluding competitive bidding.
>
> Another factor, not included in the economic studies, is the risk of using a newer, less proven design. While future developments or special situations may affect the relative economics, we concluded that 25–30 m^2 cells are generally the best choice for economically optimum chlorine plants with capacities above 300 t/d.

7.3. AMALGAM DECOMPOSITION

Sodium amalgam produced in the amalgam cell by electrolysis is immediately brought to the amalgam decomposer, where it reacts with water to form sodium hydroxide solution and hydrogen gas. Mercury after decomposition is recycled to the cell top (see Fig. 7.4).

The overall process of amalgam decomposition is

$$Na(Hg) + H_2O = NaOH + \tfrac{1}{2}H_2 + (Hg) \tag{7.6}$$

This reaction is stimulated by contact of the sodium amalgam with a graphite piece, at which hydrogen evolution takes place. The amalgam decomposition may be considered as an electrochemical process consisting

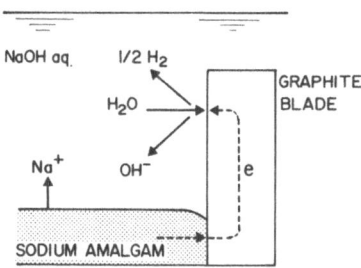

FIGURE 7.14. Schematic profile of amalgam decomposition.

of an amalgam anode and a hydrogen cathode on the graphite lumps under short-circuiting conditions in caustic soda solution (see Fig. 7.14), that is,[16]

At the anode: $Na(Hg) \rightarrow Na^+ + (Hg) + e$ (7.7)

At the cathode: $\frac{1}{2}H_2 + OH^- \leftarrow H_2O + e$ (7.8)

The EMF for reaction (7.6) is

$$E_f = E_f^0 + \frac{RT}{F} \ln \frac{a_{Na(Hg)} \ a_{H_2O}}{a_{NaOH} \sqrt{p_{H_2}}} \qquad (7.9)$$

where E_f^0 is the standard EMF and is a function of temperature, $a_{Na(Hg)}$ is the activity of sodium amalgam which depends on the amalgam concentration and temperature, a_{H_2O} and a_{NaOH} are the activities of H_2O and of NaOH, respectively, (also functions of the NaOH concentration and temperature), and p_{H_2} is the partial pressure of hydrogen, which is assumed to be unchanged for simplicity. Therefore, E_f is a function of the amalgam concentration and the NaOH concentration at given temperatures. E_f is compensated by the sum of the overvoltages of reactions (7.7) and (7.8) and the solution IR, in which the overvoltage of the amalgam anode is very small. The hydrogen overvoltage on the graphite cathode is almost independent of the caustic concentration, and is a linear function of current density up to more than 5 A/dm². Therefore, the voltage balance in the amalgam decomposition tower shown in Fig. 7.15 is represented by

$$E_h = E_f - \Delta E_h = -\eta_C + IR(\text{solution}) \qquad (7.10)$$

where E_h is the effective EMF, ΔE_h is the intersection of the E vs. i plot $(i = 0)$, η_C is the hydrogen overvoltage, and IR (solution) is the solution IR drop. The current per unit height of tower is

$$I = K_H FS(-\eta_C) \qquad \text{(cathode process)}$$

$$= K_F FS\kappa[E_h - (-\eta_C)] \qquad \text{(solution } IR)$$

$$= KFSE_h \qquad \text{(overall)} \qquad (7.11)$$

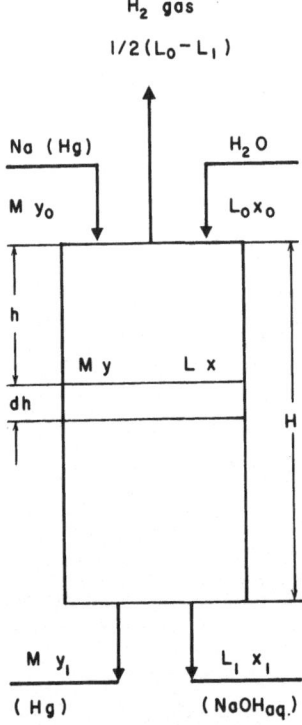

H$_2$ gas

1/2 (L$_0$ - L$_1$)

Na (Hg)   H$_2$O

M y$_0$ L$_0$x$_0$

h

M y   L x

dh   H

M y$_1$ L$_1$ x$_1$

(Hg)   (NaOH$_{aq}$.)

FIGURE 7.15. Material balance in the amalgam decomposer.

where S is the sectional area of the tower, F is Faraday's constant, K_H, K_F, and K are the coefficients, and κ is the conductivity of the caustic soda solution. Thus we obtain

$$\frac{1}{K} = \frac{1}{K_H} + \frac{1}{\kappa K_F}$$

(7.12)

The material balance in a small-section dh is

$$dw = -M\, dy = d(Lx) = -dL$$

(7.13)

By Faraday's law, we have

$$w = \frac{I}{F}\, dh$$

(7.14)

From Eqs. (7.13) and (7.14)

$$d(Lx) = KE_h S\, dh$$

(7.15)

$$-M\, dy = KE_h S\, dh$$

(7.16)

where M is the mercury flow (kg mol/hr), L is the water flow (kg mol/hr), y is the Na concentration (kg mol Na/kg mol Hg), and x is the NaOH concentration (kg mol NaOH/kg mol H_2O). By integration

$$SH = \int_{L_0 x_0}^{L_1 x_1} \frac{d(Lx)}{KE_h} \tag{7.17}$$

$$\frac{SH}{M} = -\int_{y_0}^{y_1} \frac{dy}{KE_h} = -\frac{1}{K} \int_{y_0}^{y_1} \frac{dy}{E_h} \tag{7.18}$$

where 0 and 1 denote the top and the bottom of tower, respectively, and H is the height of tower.

The required volume of the amalgam decomposer may be calculated by Eq. (7.18). However, the more important problem in the design of a decomposer is to determine the required height of the packing and the diameter of the column instead of its volume. As a general approach to the study of packed columns, Chilton and Colburn have applied the idea of the number of transfer units (NTU) and the height of transfer unit (HTU) in a common packed column.[17] The height H is represented by the product of both values:

$$H = \text{HTU} \cdot \text{NTU} \tag{7.19}$$

HTU has the dimension of length and NTU is usually dimensionless. However, if the NTU and the HTU of the amalgam decomposer are considered, the values have somewhat different dimensions because of the different mechanism of the mass transfer in electrochemical processes.

Because K is independent of y owing to negligibly small polarization of the anode, Eq. (7.18) may be transformed into the following equations:

$$\text{NTU} = -\int_{y_0}^{y_1} \frac{dy}{E_h} \tag{7.20}$$

$$\text{HTU} = M/KS \tag{7.21}$$

The value of NTU may be evaluated by Eq. (7.20) if the concentration of caustic soda solution and the operating temperature are kept constant.

An empirical relation has been obtained:

$$E_h = -0.910x + 11.0y - 270.0y^2 + 0.9060$$
$$+ 0.0005(70 - t) - 15.977t^{-0.928} \tag{7.22}$$

at $t°C$, under the conditions of

$$\text{NaOH concentration} = 10\text{--}50\,\%(\text{wt})$$

$$\text{Amalgam concentration} = 0\text{--}0.25\,\%(\text{wt})\,\text{Na}$$

$$\text{Temperature} = 50\text{--}90°C$$

These data were obtained in the laboratory and then confirmed on two experimental decomposers at a plant. The laboratory one was 75 mm in diameter and 80 mm high and the other was 100 mm in diameter with 100 mm height of packing. It was connected to an electrolyzer operating at 60,000 A.

Substituting Eq. (7.22) into Eq. (7.20)

$$\text{NTU} = \frac{2}{F(x,\,t)} \left[\tanh^{-1} \frac{11 - 540y}{F(x,\,t)} \right]_{y_0}^{y_1} \tag{7.23}$$

$$[F(x,\,t)]^2 = 1099.48 + 0.54(70 - t) - 17280\,t^{-0.928} - 982.8x \tag{7.24}$$

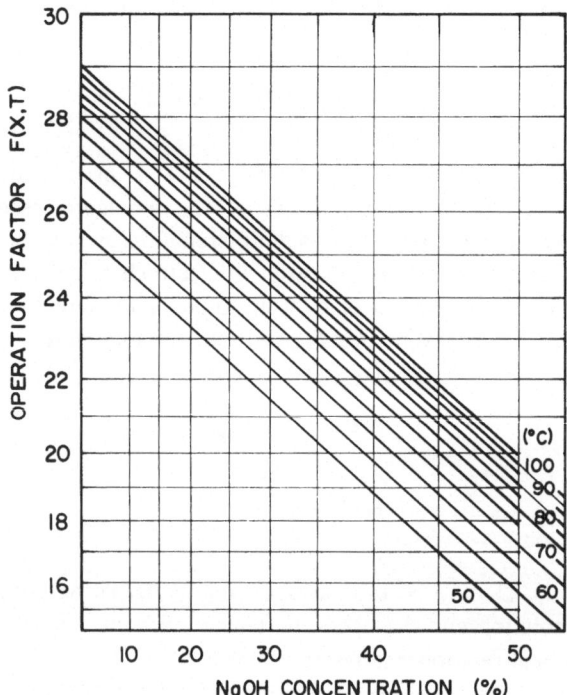

FIGURE 7.16. The operation factor diagram as a function of the concentration of the caustic soda solution at various temperatures.

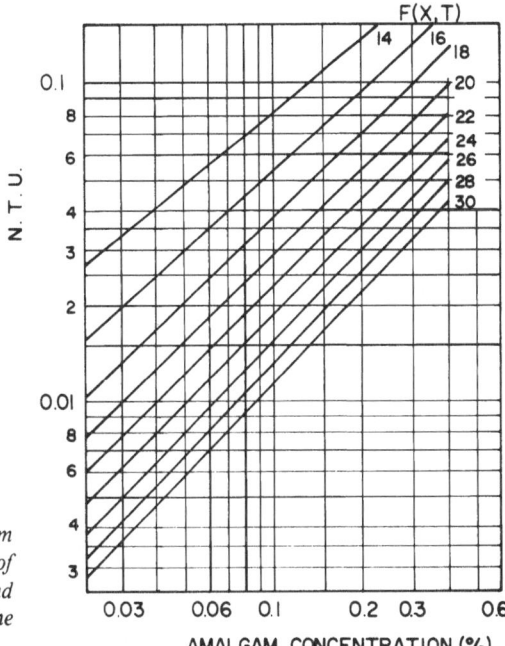

FIGURE 7.17. NTU of the amalgam decomposition tower as the function of the concentration of the amalgam and of the operation factor under the conditions of $y_1 = 0$.

where $F(x, t)$ is a function of x and t. The value of $F(x, t)$ is defined by operating conditions such as the concentration of caustic soda and temperature, so that it may be called the "operation factor." The value of $F(x, t)$ calculated by Eq. (7.24) is shown in Fig. 7.16 as a function of the concentration of caustic and operating temperature. The value of this function $F(x, t)$ can be substituted in Eq. (7.23) to calculate the value of NTU, or Fig. 7.17 may be used. Figure 7.17 shows the value of NTU as a function of the concentration of the amalgam and of the operation factor for the condition of $y_1 = 0$, which means that the concentration of the amalgam at the bottom of the column is quite low.

Next, the value of HTU must be determined. The value of HTU may be represented as a function of the flow of mercury and of the specific conductivity of caustic in the tower, because both K_H and K_F are functions of the flow of mercury for the unit sectional area of the tower, M/S.

It is desirable to operate the decomposer under such conditions that the overall coefficient for mass transfer K is a maximum and that this depends on the rate of flow of mercury. This relation was measured experimentally, as shown in Fig. 7.18, with 25% caustic soda solution at 50°C, using spheres 10 mm in diameter as packing. According to this result, a suitable flow may be in the range between 300 and 450 kg mol/hr m², or 80–110 liters/min m².

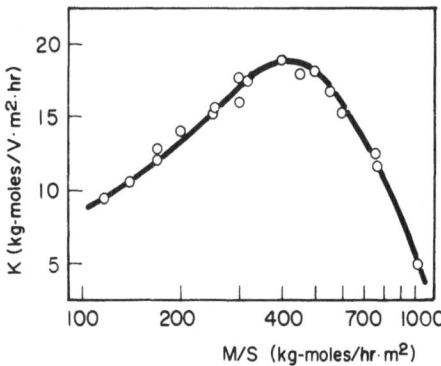

FIGURE 7.18. A relation between K and M/S in 25% caustic soda solution at 50°C. Spheres 10mm in diameter used as packing.

Values of HTU as a function of the flow velocity of mercury, M/S, and the specific conductivity of caustic soda κ are shown in Fig. 7.19. Although it was figured for wide ranges of M/S, in practice, the value of HTU is available for only 150–400 kg mol/hr m^2 of M/S, because poor distribution occurs at low ranges of M/S and it may flood at large mercury flow rate. An example of the flooding phenomena of the mercury is shown by the dotted line labeled $\kappa = 62$ in Fig. 7.19, which occurred in the operating tower. There was reasonable agreement with the solid line, $\kappa = 60$, up to $M/S = 300$, above which the tower is stated to have flooded. Now it is believed the tower may be operated safely up to 400 kg mol/hr m^2 of mercury flow.

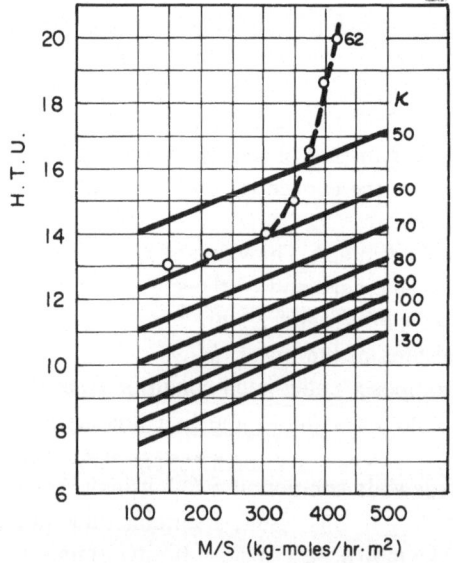

FIGURE 7.19. HTU of the amalgam decomposer as a function of the flow velocity of the mercury and the specific conductivity of the caustic soda solution. (κ = conductivity of NaOH solution in mho/m.)

Here we may have a sample calculation under the conditions of:

Total current of the electrolyzer	100 kA
Concentration of the amalgam	
At the top	0.21%
At the bottom	0.01%
Concentration of the caustic	50%
Operating temperature	90°C

7.3.1. Minimum Sectional Area

The flow of mercury is 52.7 liters/min or 213 kg mol/hr. A value of 99 liters/min m^2 (400 kg mol/hr m^2) is assumed for M/S. Accordingly, the minimum area should be

$$S = \frac{52.7}{99} = 53.3 \times 10^{-2} \ (\text{m}^2)$$

Therefore, the diameter of the column is about 0.83 m.

7.3.2. Minimum Height

The specific conductivity of 50% caustic soda solution at 90°C is 100.2 mho/m, and hence, from Fig. 7.19, the value of HTU at $M/S = 400$ is about 11.10. From Fig. 7.16, the operation factor is obtained: $F(x, t) = 19.5$. Therefore, the NTU is 5.8×10^{-2}, which is obtained from Fig. 7.17. Thus, the minimum height is as follows:

$$H = \text{HTU} \cdot \text{NTU} = 11.10 \times 5.8 \times 10^{-2} = 0.644 \ (\text{m})$$

Amalgam decomposition is an exothermic reaction, and the reaction heat is counted to be about 20% of total heat. The largest source is the sensible heat of mercury as shown in Table 7.6.[18] Therefore, the operating temperature ($t°C$) of the amalgam decomposition tower depends on the flow rate and the temperature of mercury or amalgam, $t_{\text{Na-Hg}}°C$. An empirical equation obtained by Hine is as follows:[19]

$$t = [t_{\text{Na-Hg}} + (377 - 2.48 \, p) \, \Delta q] \left[1 - \frac{55.4}{p} \Delta q \right] \tag{7.25}$$

$$\Delta q = q_0 - q_1$$

where p is the NaOH concentration in percent by weight, and q is the Na concentration in the amalgam in percent by weight. If $t_{\text{Na-Hg}} = 80°C$, $p = 50\%$, and $\Delta q = 0.2\%$, the operating temperature is estimated to be about 102°C.

TABLE 7.6

An Example of Heat Balance in the Amalgam Decomposer

	kcal/hr	%
Input		
Sensible heat of mercury flow	28,118	78.2
Sensible heat of water	702	2.0
Heat of reaction	7,171	19.8
TOTAL	35,991	100.0
Output		
Sensible heat of mercury flow	33,364	92.7
Sensible heat of caustic soda	2,363	6.6
Sensible heat of hydrogen gas	95	0.2
Latent heat of water vapor	129	0.4
Heat loss	40	0.1
TOTAL	35,991	100.0

7.3.3. Notes for Convenience

$$M = 4.04Q \qquad M = \text{Hg, kg mol/hr} \tag{7.26}$$
$$Q = \text{Hg, liters/min}$$

$$x = \frac{0.45p}{100 - p} \qquad x = \text{NaOH, kg mol/kg mol } H_2O \tag{7.27}$$
$$p = \text{NaOH in percent by weight}$$

$$y = 8.75 \times 10^{-2}q \qquad y = \text{Na, kg mol/kg mol Hg} \tag{7.28}$$
$$q = \text{Na in percent by weight in the amalgam}$$

$$Q = \frac{0.1056J}{q_0 - q_1} \qquad J = \text{total current of the attached electrolyzer in kA} \tag{7.29}$$

7.4. EFFECTS OF IMPURITIES AND NECESSITY OF BRINE PURIFICATION

The basic concept for the amalgam process is clearly related to the characteristics of the mercury and/or sodium amalgam cathode, where the hydrogen overvoltage is extremely high, and hence hydrogen evolution is retarded even at less noble potentials where sodium ions may be reduced cathodically:

$$Na^+ + (Hg) + e = Na(Hg) \qquad (7.30)$$

The rate of this process is fast, but is mainly controlled by diffusion of sodium in mercury. Operation of the amalgam cell becomes impossible if for any reason the cathode process is affected.

Impurities such as Mg and Ca and other metal ions in the brine are reduced at the cathode. Some of them may deposit as a colloidal film, thereby disturbing the cathodic deposition of Na^+ and diffusion of Na into the bulk of mercury.

Generally, crude salt contains some Mg and Ca as chlorides and sulfates. Those impurities can be precipitated and removed by addition of caustic soda and soda ash into the saturated brine, followed by precipitation. However, other contamination may come from various routes to disturb the cathodic reduction of Na^+, although a part of those impurities might be removed with precipitates of Mg and Ca in the brine treatment stage. Vanadium, probably from graphite anodes, is one of the most harmful impurities, as shown in Table 7.7. Raw materials used in graphite manufacture such as petroleum coke contain a lot of V, and some V remains in the product even after heating at very high temperatures, such as at 2000°C. Consequently, the V content in the graphite anodes for the amalgam cell is limited to a very low level.

TABLE 7.7

Effect of Metallic Impurities on Hydrogen Evolution at the Cathode of a Mercury Cell Operated at $40A/dm^2$ and at 80°C [a]

Metal	mg/liter
Ca	100.0
Mg	1.0
Fe	0.3
Ti	0.1
Mo	0.001–0.010
Cr	0.001–0.010
V	0.001–0.010
Very harmful	V, Mo, Cr, Ti, Ta, (Mg + Fe)
Moderately harmful	Ni, Co, Fe, W
Slightly harmful	Ca, Ba, Cu, Al, Mg, graphite
No effect	Ag, Pb, Zn, Mn

[a] The maximum allowable concentration of impurities in the brine, for 0.3% H_2 in Cl_2 gas, is shown.

There has been an experience of serious troubles with Al, which comes from diatomaceous earth filter aid. The hydrogen content in the cell gas rose to the harmful explosive limit as the result of aluminum in the brine polishing filter material being extracted by alkaline brine.[20] There are many experiences of troubles with impurities, and many publications in this field can be found.[21-25] Table 7.7 is a summary of effects of impurities on the amalgam cell by MacMullin.[26] The data in this table are, however, related to cell operation at relatively low current densities: 40 A/dm^2 compared to 100 A/dm^2 of modern amalgam cells. Influences of brine impurities at high-current densities are more serious, and hence the limit of contamination should be lowered.

Effects of impurities in the brine on the amalgam cathode can be classified in two types: one, the direct discharge of hydrogen ions or the decrease of hydrogen overvoltage at the amalgam cathode and the other, the decomposition of sodium amalgam at the cathode surface by local cell action. Magnesium ions may form a colloidal precipitate at the surface of the amalgam cathode which may stimulate discharge of hydrogen ions at relatively high-current densities. On the other hand, some ferric and calcium ions may deposit in the amalgam and cause decomposition of amalgam or hydrogen evolution. Few cases, V^{5+} and Mg^{2+} for example, are of the first type, and some impurities are of the second type. Impurities such as Ag, Pb, and Zn may form a relatively uniform amalgam, and may not affect the hydrogen formation reaction on the amalgam cathode. Others such as Fe, Ni, and C do not dissolve in mercury, but form the amalgam butter on which hydrogen discharge is stimulated.

REFERENCES

1. *Soda News*, Japan Soda Industry Association, No. 4 (1981).
2. K. Hass, *Chem. Ing. Techn.* **47**, 121 (1975).
3. *Annual Report of the Electrolytic Industries*, Electrochemical Society, 1970–1979.
4. Data published by Japan Soda Industry Association.
5. R. S. Karpiuk and S. D. Argade, *J. Electrochem. Soc.* **122**, 310C (1975).
6. Reported at Annual Meeting, Japan Soda Industry Association, June 1981.
7. F. Hine, *Kagaku* (Chemistry) **29**, 427 (1974).
8. K. Hass, *Ullmanns Encyklopädie der technischen Chemie*, Band 9, p. 319 (1975).
9. F. Hine, *Electrochem. Technol.* **8**, 69 (1968).
10. F. Hine and M. Yasuda, *J. Electrochem. Soc.* **121**, 1289 (1974).
11. F. Hine, *Denki Kagaku* (J. Electrochem. Soc. Jpn.) **39**, 60 (1971).
12. F. Hine, S. Yoshizawa, S. Okada, and T. Uesugi, *Kogyo Kagaku Zasshi* (J. Chem. Soc. Jpn, Industrial Chem. Section) **58**, 554 (1955).
13. H. B. Beer, British Patent 6490/67 (1967).
14. W. C. Gardiner and R. D. Burt, Abstr. No. 261, Electrochemical Society meeting, Los Angeles, CA, May 1970.
15. R. D. Burt and W. McM. Clarke, AIChE meeting, Puerto Rico, May 1970.

16. F. Hine, *Electrochem. Technol.* **2**, 79 (1964).
17. T. H. Chilton and A. P. Colburn, *Ind. Eng. Chem.* **27**, 255 (1935).
18. T. Sugino, Ph.D. thesis, Kyoto University, p. 226 (1958).
19. F. Hine, unpublished data.
20. F. Hine and T. Nozu, *Soda to Enso* (Soda and Chlorine) **26**, 101 (1975).
21. G. Angel and T. Lunden, *J. Electrochem. Soc.* **99**, 435, 442 (1952); **100**, 39 (1953); **102**, 124, 243 (1955); and **104**, 167 (1957).
22. W. E. Cowley, B. Lott, and J. H. Enteisle, *Trans. Inst. Chem. Eng.* **41**, 372 (1963).
23. K. Hass, *Electrochem. Technol.* **5**, 246 (1967).
24. F. Hine, S. Matsuura, and S. Yoshizawa, *Electrochem. Technol.* **5**, 251 (1967).
25. F. Hine, M. Yasuda, F. Wang, and K. Yamakawa, *Electrochim. Acta* **16**, 1519 (1971).
26. R. B. MacMullin, *Chlorine*. ACS Monograph No. 154, p. 151 (1962).

CHLOR-ALKALI INDUSTRY USING DIAPHRAGM CELLS

The diaphragm-type chlor-alkali plant consists of the brine treatment yard, the electrolytic cell room, the caustic evaporation system, and the gas processing system, both chlorine and hydrogen, as illustrated in Fig. 8.1., where the material balance is also shown.

Crude salt is dissolved with water, and the concentrated NaCl solution is sent to the electrolyzer after purification. The procedure for brine purification is almost the same as for the amalgam-type chlor-alkali plant, but the brine is not returned to the brine treatment yard after electrolysis as in the amalgam cell plant. Instead, salt recovered from the caustic evaporator is recycled to the brine treatment yard as slurry. In the amalgam cell plant, the feed brine is acidified with HCl to pH = 3–4 prior to electrolysis. In contrast, neutralized or alkaline solution is fed to the diaphragm cell in most plants to protect the asbestos mat from chemical attack.

The catholyte effluent or the cell liquor, consisting of about 12% NaOH and 15% NaCl, is treated by the evaporator to separate crystallized salt from caustic soda. A caustic evaporator of double or triple effects is generally used, but quadruple effect evaporators are being used in very large plants to minimize steam consumption.[1] Table 8.1 shows an example of the operating conditions of a triple-effect evaporator.[2] Figure 8.2 illustrates an example of a triple-effect evaporator in a chlor-alkali plant. Diaphragm-cell caustic contains NaCl of about 1% by weight, which is about equal to the solubility of NaCl in 50% caustic soda solution at room temperature. This contamination causes certain problems such as insufficient quality for further processing, local corrosion of equipment, etc. Although there are several processes for purification of diaphragm-cell caustic,[3,4] they add an extra cost of ten dollars or more per ton of caustic.

Although the graphite anode resists attack by the chlorine electrode process, it is gradually corroded by the oxygen electrode reaction, and is converted into CO and CO_2.[5,6] Consumption of the graphite anode is also troublesome in the amalgam cell. Adjustment of the anode-to-cathode gap, or

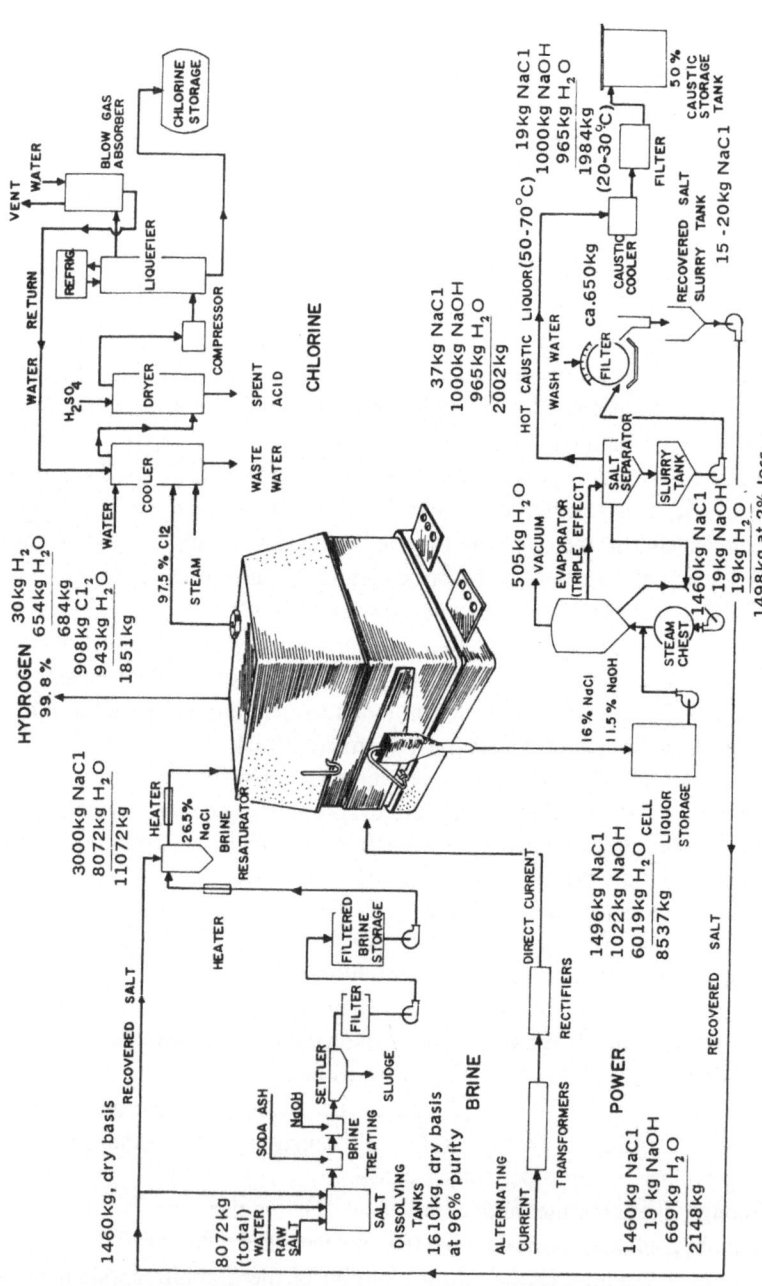

FIGURE 8.1. Flowsheet of a typical diaphragm-type chlor-alkali plant.

TABLE 8.1

Typical Operating Conditions of Triple-effect Caustic Evaporation System

	Third effect	Second effect	First effect
Solution composition (%)			
NaOH	16	26	45
NaCl	14	9	5
Temperature (°C)			
Vapor phase	46	71	110
Liquid phase	57	88	150
Pressure, psia	1.45	4.5	20.7
kg/cm^2	0.102	0.316	1.46

the anode adjustment, must be made frequently to minimize the cell voltage or the power consumption per ton of caustic, and is a troublesome feature for operation of the amalgam cell using graphite anodes. The average consumption of graphite anodes in the amalgam cell is relatively small, say 2 kg/ton NaOH, depending on the brine concentration and the pH.

In the anode compartment of the diaphragm cell, the brine concentration is weak compared to the amalgam cell, i.e., 50% conversion in the

FIGURE 8.2. Evaporation system for diaphragm-cell caustic soda. (Courtesy of Central Chemical Co., Japan, and Hooker Chemicals and Plastics Corporation.)

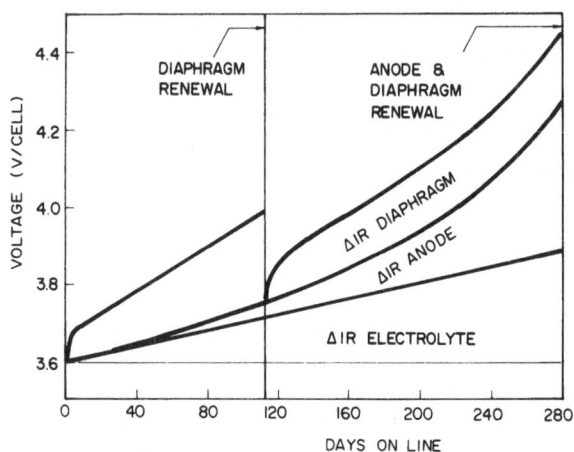

FIGURE 8.3. Variation of voltage of a diaphragm cell.

diaphragm cell vs. 10% in the amalgam cell, the solution pH is high due to back migration of hydroxyl ions through the diaphragm, and the concentrations of hypochlorite and chlorate ions are high. All of these factors are detrimental in terms of contributing to a graphite consumption of about 5–6 kg/ton NaOH. Also, the interelectrode gap cannot be adjusted during operation because of the cell configuration. As a result, the *IR* drop between the anodes and cathodes increases with time after start-up, as shown in Fig. 8.3.[7] Clogging of the asbestos diaphragm with graphite particles and the formation of precipitates of Mg and Ca compounds in the brine are also factors which increase the *IR* drop through the diaphragms.[8] The interelectrode gap is, of course, kept at a suitable dimension if the DSA is used, and no adjustment is required. Use of high-purity brine minimizes gradual change of the diaphragm *IR* drop.

8.1. DIAPHRAGM-TYPE CHLORINE CELLS

Diaphragm cells have been modernized extensively in the last few years since the DSA has been introduced in the market. An example of modern cells is shown in Fig. 8.4. The DSAs are mounted on the rubber-lined copper plate bed, and the cell cover is made of the fiberglas-reinforced plactic (FRP), bisphenolic resin, instead of the heavy concrete cover of older diaphragm cells of the type shown in Fig. 8.5.

The cathode has also been modernized. Figure 8.6 shows two types of cathode for Hooker cells. Type S-4 is used with graphite anodes in older cells, and Type H-60 with the DSA in new cells. The type H-60 cathode is

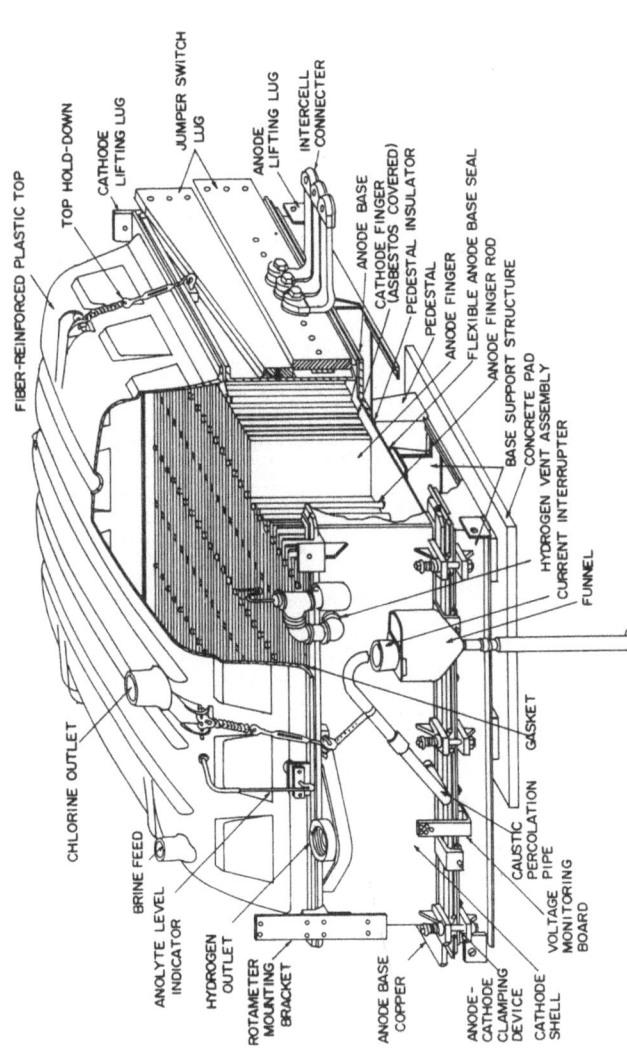

FIGURE 8.4. Hooker-type H-4 diaphragm cell. Nominal current rating: 150 kA. (By permission of Hooker Chemical and Plastics Corporation.)

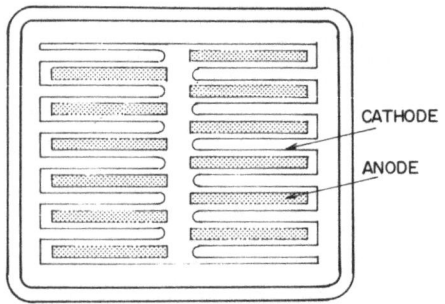

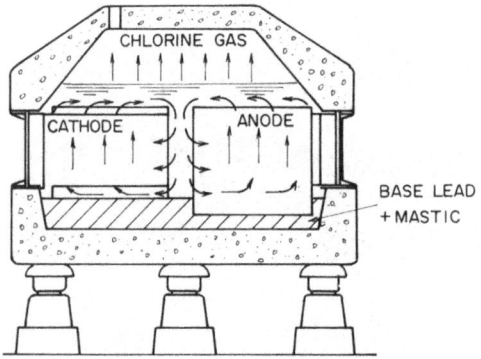

FIGURE 8.5. Hooker S-type diaphragm cell.

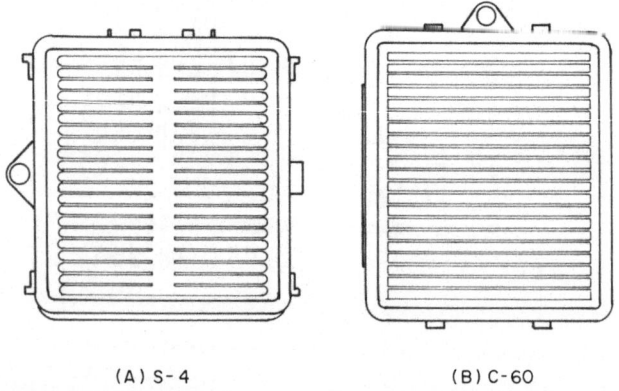

(A) S-4   (B) C-60

FIGURE 8.6. Two types of cathode for the diaphragm-type chlorine cells.

not entirely new since a similar cathode design was first used many years ago.[9] It was soon converted to the S-4 in order to provide the brine downcomer between the cathode fingers so as to eliminate the bubble effects of chlorine generated at the graphite anodes. The two-phase flow consisting of the brine and gas goes up in the gap between the anode blades and the diaphragms by a gas lift action, and chlorine bubbles are separated from the brine at the top. The dense brine goes down through the downcomer, and circulation of electrolyte in the cell is stimulated (see Fig. 8.5). It is an effective measure for reducing anode overvoltage and solution IR drop.

Since the DSAs for the diaphragm cell are in the shape of a rectangular box as shown in Fig. 8.7, the inside space of the anode box is provided with the downcomer for brine. Consequently, the cathode tubes are simply fabricated from side to side all the way across in the type H-60 design.

The cathode finger in old-style diaphragm cells is made of specially woven steel mesh, shown in Fig. 8.8A, whereas the cathode tube in modern cells is fabricated with perforated steel plate, shown in Fig. 8.8B. The ohmic voltage drop of the perforated-plate cathode is slightly less compared to that of the older-type cathode.

Chrysotile asbestos slurry prepared with the catholyte liquor is deposited on the cathode plate under vacuum to provide the diaphragm. The

FIGURE 8.7. *The DSA assembly for diaphragm cells. (By permission of Electrode Corporation.)*

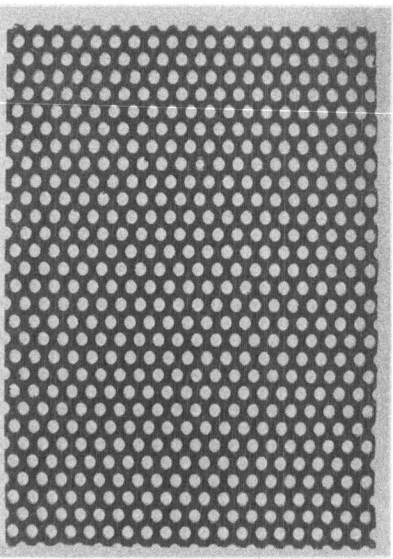

FIGURE 8.8. Cathode materials for the diaphragm-type chlor-alkali cells. (a) Steel mesh; (b) perforated plate.

asbestos diaphragm swells outward due to penetration of hydrogen evolved at the cathode and decreases the brine gap in front of the anode plate, so that the gas void fraction increases.[10] In the modern cells, a modified asbestos diaphragm is used to eliminate this problem. The HAPP of Hooker and the TAB of Diamond Shamrock are examples. Asbestos slurry containing a small amount of fluorinated polymer such as PTFE powder is vacuum-deposited on the cathode plate, dried, then heated to about 250–300°C, at which the polymer powder melts and binds the asbestos fibers. There is no problem with swelling, and cell performance is improved significantly.[11]

Substitutes for asbestos for use as diaphragms for the chlorine cell have been investigated extensively for many years. A very thin microporous PTFE sheet is a candidate, and various fibers and fabrics have also been examined.[12] However, the vacuum-deposited asbestos layer seems to be the best diaphragm material at present. The mechanism and the behavior of the asbestos diaphragm are still unclear, but the surface of asbestos fibers may be dissolved slightly in caustic alkali solution with the formation of a gelatinous layer which restricts unwanted permeation of ionic and molecular species.

Table 8.2 shows the operating conditions and the voltage balance of the Hooker H-4 cell.[9] The anode overvoltage has been reduced to an adequate level by use of the DSA. Consequently, the cathode overvoltage and the solution IR drop including the bubble effects should be minimized for further

TABLE 8.2
(A) Operating Conditions of Hooker H-4 Diaphragm Cell;
(B) Voltage Balance of Hooker H-4 Diaphragm Cell

(A)

Total current	150 kA
Current density	ca. 27 A/dm^2
Terminal voltage measured	3.788 V
Current efficiency	96.53 %
Feed brine, @ 65°C	315 g/liter
NaOH concentration in cell liquor	140 g/liter
NaCl/NaOH ratio	1.3405
NaCl concentration in anode compartment	266.10 g/liter
Temperature	
Anolyte	94.19 °C
Catholyte	100.19 °C

(B)

Reversible potential		
Anode	E_A	1.32
Cathode	E_C	−0.90
Decomposition voltage	E_d	2.25
Anode overvoltage with DSA	η_A	0.03
Cathode overvoltage	$-\eta_C$	0.30
Solution IR drop	IR (solution)	0.35
Diaphragm IR drop	IR (diaphragm)	0.60
IR drop in hard ware	IR (metal)	0.26
Terminal voltage	V_T	3.79 V

reduction of the terminal voltage. The IR drop through the diaphragm is also a factor to be optimized.

Dimensions of several commercial cells now available are summarized in Table 8.3. The Hooker cell and the Diamond Shamrock cell are monopolar, while the Glanor cell, jointly developed by DeNora and PPG, is a bipolar type, as shown in Fig. 8.9.[13]

8.2. ION-EXCHANGE MEMBRANE CELLS

In the diaphragm-type chlorine cell, Na^+, Cl^-, and OH^- permeate through the micropores of the asbestos diaphragm. The catholyte, or the cell liquor, is concentrated by evaporation to separate NaCl from caustic soda so that product caustic contains a minimum amount of salt, say 1% by weight in 50% NaOH solution. It would be possible to avoid contamination by

TABLE 8.3
Dimensions of Various Diaphragm-type Chlor-Alkali Cells

		Diamond Shamrock			Hooker		Glanor V-1144	Nippon Soda B-40	Showa Denko DS-7
		DS-31	DS-45	DS-85	H-2A	H-4			
Current	kA	40	80	150	80	150	80	40	75
Current density	A/dm^2	19.8	27.6	27.4				11.95	17.3
Capacity									
Cl$_2$	t/d	1.22	2.45	4.59	2.45	4.59	27.0		2.281
NaOH	t/d	1.38	2.76	5.18	2.76	5.18	30.3	6.80	2.592
Current efficiency	%	96.5	96.5	96.5	96.4	96.6	97	95.0	96.0
Cell voltage	V/cell	3.64	4.17	4.15	3.79[a]	3.85[a]	3.75 × 11	3.60 × 5	3.63
DC power consumption	kWh/ton Cl$_2$	2860	3270	3250			2900	2540	2521
	kWh/ton NaOH								
Life of diaphragm	days		1–2 years		180–320			240	180–350

[a] Excluding bus bar.

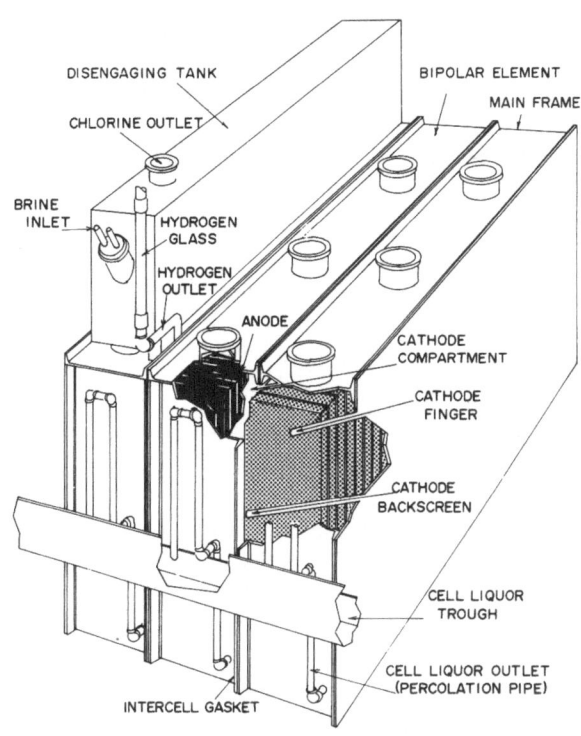

FIGURE 8.9. Type V-1144 Glanor diaphragm electrolyzer. (By permission of PPG Industries, Inc.)

NaCl in the catholyte if a suitable ion exchange membrane was used as an electrolytic separator between the anode and cathode compartments.

Organic polymer materials containing sulfonate groups and carboxyl groups have ion-exchange behavior and have found many applications.

Under the equilibrium conditions in a solution consisting of cation C and anion A, the Donnan equilibrium is established

$$[C_m] = [A_m] + [R]$$

and

$$[C_s] = [A_s]$$

where $[\ ]$ represents the concentration, R is the unfixed ion-exchangable radical, and the subscripts m and s are resin and solution, respectively. Suppose that two solutions of the $1-1$ valence salt C^+A^- of different concentrations are separated by an ion-exchange membrane, and that 1

Faraday of electricity flows through the membrane with infinitely small rate. With an assumption of sufficiently large amount of R, we have $[A_m] \ll [A_1]$ or $[A_2]$, and hence the current is almost independent of A. The subscripts 1 and 2 represent the left-hand and right-hand side of the membrane, respectively.

At the interface between solution 1 and the membrane M,

$$\Delta G_{1m} = \Delta G_1 - \Delta G_m = RT[t \ln a_{C_1} - (1 - t) \ln a_{A_1}] - \Delta G_m \qquad (8.1)$$

where t is the transference number of C. Substituting

$$a_{C_1} = a_{A_1} = a_{\pm}$$

into Eq. (8.1)

$$\Delta G_{1m} = (2t - 1) RT \ln a_{\pm} - \Delta G_m \qquad (8.2)$$

We have a similar equation at the interface between solution 2 and the membrane M. The Gibbs free energy change of the system is

$$\Delta G = \Delta G_{1m} - \Delta G_{2m} - \Delta G_L$$

where ΔG_L is the Gibbs free energy change corresponding to the liquid junction potential. Therefore, the membrane potential E_M is represented as follows:

$$E_M = -\frac{\Delta G}{F} = -(2t - 1)\frac{RT}{F} \ln a_{\pm} + (2t' - 1)\frac{RT}{F} \ln a'_{\pm}$$

$$+ \frac{\Delta G_m - \Delta G'_m - \Delta G_L}{F}$$

where the prime indicates solution 2. For simplification, we assume that

$$\Delta G_m = \Delta G'_m, \qquad \Delta G_L \ll F E_M, \qquad [A_m] \ll [A_1] \text{ and } [A_2], \qquad \text{and } t = t'$$

Thus, the above equation becomes

$$E_M = (2t - 1)\frac{RT}{F} \ln \frac{a'_{\pm}}{a_{\pm}} \qquad (8.3)$$

If the membrane has the ideal property of ion selectivity $t = 1$, Eq. (8.3) can be rewritten as follows:

$$\bar{E}_M = \frac{RT}{F} \ln \frac{a'_{\pm}}{a_{\pm}} \qquad (8.4)$$

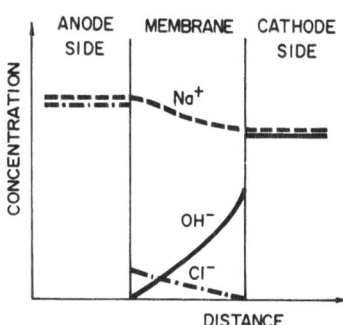

FIGURE 8.10. Estimated concentration distribution of ionic species in an ion-exchange membrane.

which is the Nernst equation. The transference number t can be estimated by measurement of membrane potential using Eq. (8.3).

The anolyte of the ion-exchange membrane cell is saturated NaCl containing dissolved chlorine, and the catholyte consists of concentrated NaOH solution. The concentration of unfixed ions in the membrane positioned between these solutions varies approximately as shown in Fig. 8.10.

A decrease in the transport number of Na^+ depends on leakage of OH^- through the membrane. The electrical field during electrolysis stimulates permeation of OH^- from the cathode compartment to the anode compartment. A decrease in the current efficiency in a practical cell, both anodic and cathodic, refers directly to a loss of OH^-, and the current efficiency decreases with an increase in the OH^- concentration in the catholyte. The concentration of NaOH to be produced is thus limited. On the other hand, the current efficiency is high when the NaOH concentration in the catholyte is low, but a lot of thermal energy for evaporation of caustic liquor is required.

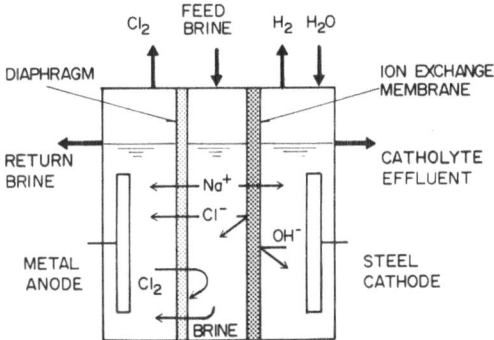

FIGURE 8.11. Three-compartment cell with a porous diaphragm and an ion-exchange membrane.

The ion-exchange membrane was first developed for the purpose of electrodialysis of brackish water rather than as a separator for electrolyzers. These membranes consist of sulfonated hydrocarbons. Several membranes are sandwiched to fabricate a stack. An anode is positioned at one end of the stack and a cathode is located at the other end. Electrochemical reactions taking place at the electrodes are not important, but the electric power supplied works as the driving force for dialysis.

In ion-exchange membrane cells, on the other hand, the main subject is electrochemical processes, and the membrane works as a separator of anolyte and catholyte. Therefore, the membrane should not be attacked by electrolytic solution, which is usually an aggressive environment. A triple-compartment cell as shown in Fig. 8.11 may be used to prevent failure of the membrane due to heavy oxidation by chlorine. A porous diaphragm or membrane prevents permeation of the anolyte containing chlorine to the center compartment. The flow of concentrated feed brine NaCl solution through the separator from the center compartment to the anode direction retards back migration of anolyte. A cation exchange membrane is arranged between the center and cathode compartments, and retards permeation of OH^- and Cl^-. This cell configuration is more complicated than the two-compartment cell shown in Fig. 8.12, and the anode-to-cathode gap is also larger. Thus, we need a membrane material that will resist attack by dissolved chlorine and concentrated caustic soda at elevated temperatures such as 100°C.

DuPont first developed chemically stable ion-exchange membranes, called Nafion, and applied them to the chlor-alkali cell. The membrane consists of sulfonated perfluorocarbons as shown in Fig. 8.13.[14] To improve the mechanical strength under working conditions, the resin film is laminated with Teflon cloth or mesh. The Nafion membranes are water permeable, which is an important property for use as an electrolytic membrane.[15] The wettability or the water content depends on the equivalent weight (EW),

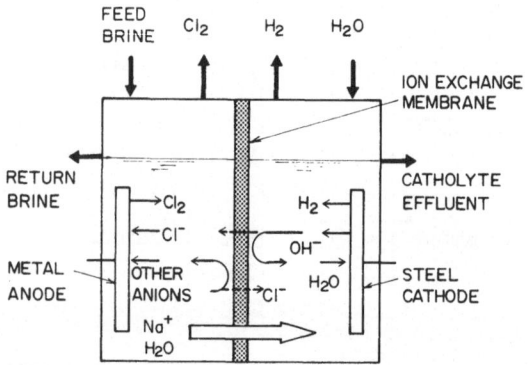

FIGURE 8.12.   *Two-compartment cell with an ion-exchange membrane.*

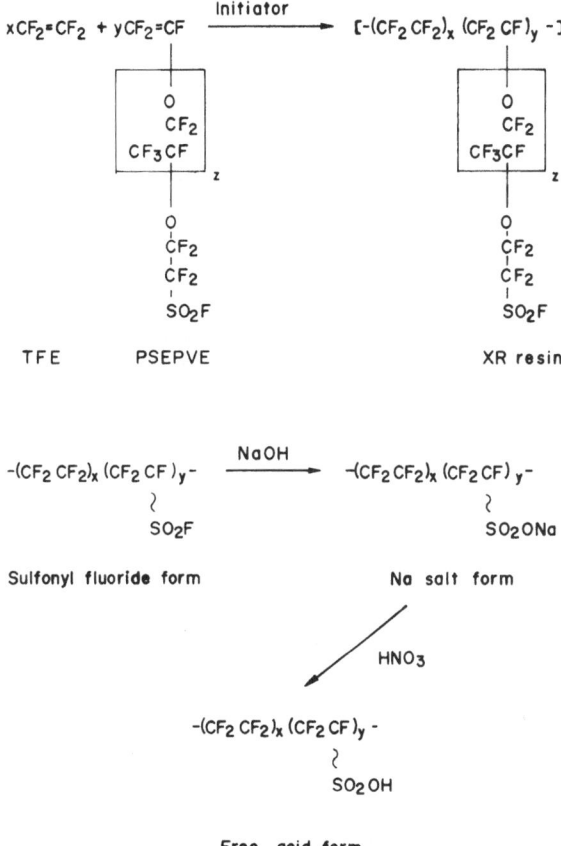

FIGURE 8.13. DuPont's NAFION perfluorosulfonic acid membrane. PSEPVE = perfluoro
[2-(2-fluorosulfonylethoxy)-propylvinyl ether].

pretreatment, solution composition, operating temperature, and other factors. The EW is the weight of polymer that is neutralized by 1 equivalent of alkali.

The mechanical strength of the membrane decreases with an increase in the water content, while the electrical conductance of the membrane increases when the water content increases. The electrical conductance decreases with an increase in the EW. Ideally, the membrane should have the following properties:

● chemical stability;
● mechanical stability;
● high conductivity;
● preferable ion selectivity.

The electrochemical processes in the membrane cell is, of course, the same as for the usual diaphragm cell:

At the anode:	$2Cl^- \rightarrow Cl_2 + 2e$	(8.5)
At the cathode:	$H_2 + 2OH^- \leftarrow 2H_2O + 2e$	(8.6)

Overall reaction: $2Cl^- + 2H_2O \rightarrow Cl_2 + H_2 + 2OH^-$ (8.7)

However, the catholyte is NaOH solution free of Cl^- instead of a mixed solution of NaOH and NaCl as in the diaphragm cell. Since Na^+ ion is hydrated, it is accompanied by about 2–4 molecules of water per Na^+, depending on the NaOH concentration and other factors, when it permeates the membrane by electrolysis. This transport is an important factor for the water balance in the cathode compartment.

It is preferable that the membrane permits passage of Na^+, but not

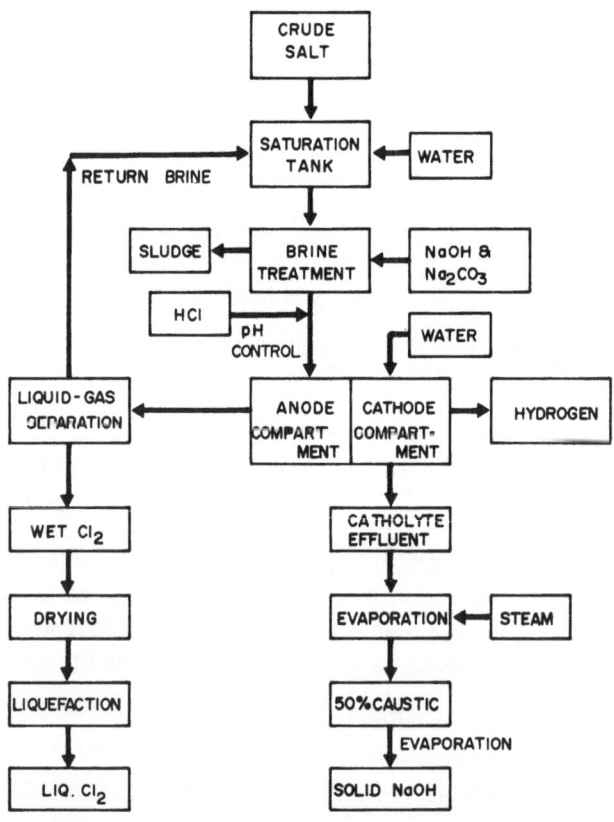

FIGURE 8.14. Flowsheet of the membrane cell chlor-alkali plant.

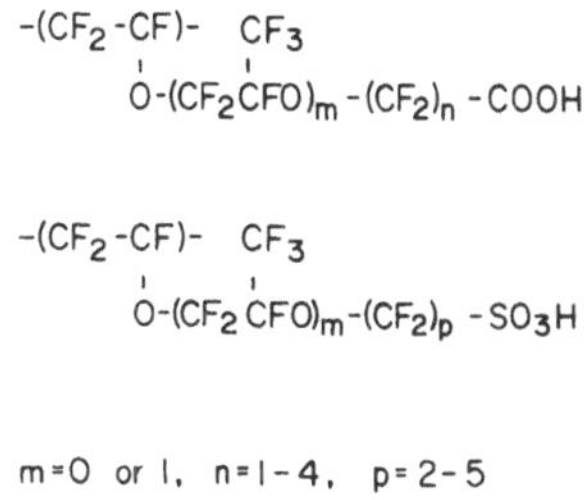

-(CF2-CF)- CF3
O-(CF2CFO)m-(CF2)n-COOH
-(CF2-CF)- CF3
O-(CF2CFO)m-(CF2)p-SO3H
m=0 or 1, n=1-4, p=2-5

FIGURE 8.15. Chemical structure of Asahi Chemical's perfluorocarboxylic acid-type membrane.

OH^- or Cl^-. In practice, however, there is some leakage of such anions through the membrane. Leakage of OH^- is a direct cause of a decrease in current efficiency and an increase of the anolyte pH.

The flowsheet for the membrane cell process is essentially the same as for the amalgam cell plant, as shown in Fig. 8.14. Purified and acidified brine saturated with NaCl is fed to the anode compartment. The spent brine is recycled to the brine treatment yard. Since the anolyte pH is high, due to permeation of OH^-, it is neutralized by HCl. Chlorine is separated from the spent brine, and is treated by the usual method.

Fresh water or dilute caustic soda solution is fed to the cathode compartment. Caustic soda produced is recovered and is evaporated if necessary.

Since caustic soda solution contains a minimum amount of NaCl, design and operation of the evaporator are simple compared with the diaphragm-type chlor-alkali cell plant.

Because some hydroxyl ions permeate through the membrane, the feed brine must be acidified with a considerable amount of muriatic acid, depending on the current efficiency or OH^- leakage rate. This is a problem of membrane cell technology.

Modifications of ion-exchange membranes have been investigated extensively because cell performance parameters such as current efficiency and the cell voltage are closely related to the characteristics of the membrane used.

CURRENT EFFICIENCY(%)
80
60
40
1100
1300
1500
EW

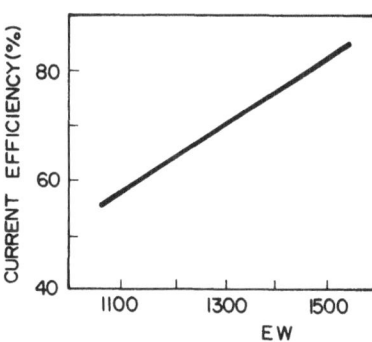

FIGURE 8.16. Effect of equivalent weight on current efficiency.

A perfluoro-carboxylic acid membrane developed by Asahi Chemical Company, as shown in Fig. 8.15, is reported to show better results than the perfluoro-sulfonic acid type of membrane.[16] DuPont's Nafion 315, used typically with chlor-alkali cells, is a composite membrane consisting of 2 mils of 1500-EW resin on 4 mils of 1100-EW resin with Teflon reinforcing cloth. The current efficiency increases with increase of the EW as shown in Fig. 8.16, but unfortunately the electical resistance also increases greatly.[17]

The rate of permeation of OH^- depends on the electrochemical characteristics of the barrier layer in contact with the catholyte. The thin 1500-EW resin film of Nafion 315 is thus designed for this purpose. It is of interest that the current efficiency of the membrane cell is further improved when the barrier resin of the membrane is modified with ethylenediamine.[18]

Membrane cells have already been started up at several installations, although plant capacity is small. Figure 8.17 illustrates Hooker MX cells using Nafion which are being operated at a pulp mill in Canada. The MX cells are of the filter-press bipolar type. In contrast, the MMA cell of the Diamond Shamrock Corporation is of the monopolar type. Table 8.4 summarizes dimensions of some membrane cells now availabe.[19,20]

FIGURE 8.17. Hooker MX-type membrane cell operated at Dryden plant, Reed Paper Ltd., Canada. (Courtesy Hooker Chemicals and Plastics Corporation.)

TABLE 8.4

Dimensions of Various Membrane Cells (1979 Data)

		Hooker[a] MX	Diamond Shamrock[b]		Asahi[c] Glass	Asahi Chemical[d]
			DX-7	MMA-212		
Current	kA	3	3.2	53.5	20	10.8
Current density	A/dm^2	31	31	31		40–50 30–40
Cell voltage	V/cell	3.9	4.0	4.0		3.80–4.15 3.75 @ 40 A/dm^2
DC power consumption	kWh/ton NaOH	2800	2700–3100		3300	3120–3410 2500–2700
Current efficiency	%	89–96	87–95			93–96
Membrane used		Nafion	Nafion		Own	Nafion own
Size	m^2	ca. 1	1.02	1.440		ca. 3
Estimated life	years	2	28		1.5	
NaOH concentration	%	11–12	28		40	21–24 25–40
NaCl conc. in NaOH		35 ppm	50 ppm in 28% NaOH		14 ppm	0.0004–0.0022 %

[a] S. A. Dahl, Chem. Eng., p. 60 (August 18, 1975).
[b] E. J. Peters and D. R. Pulver, paper presented at meeting of Electrochemical Society, Atlanta, GA, October 1977.
[c] Asahi Glass literature.
[d] Asahi Chemical literature.

8.3. AMALGAM PROCESS VS. DIAPHRAGM PROCESS

Comparison of the amalgam process and the diaphragm process has been discussed frequently,[21] and the results have been useful in feasibility studies for plant construction. Today, we need to consider the membrane cell process together with the conventional processes.

Feasibility studies for the process industry should be made carefully from the viewpoints of

- industrial economy, including plant location, regulation, and occupational health and safety;
- technology and experience;
- product quality and marketability; and
- environmental protection.

These factors are, of course, related to each other.

Studies of industrial economy involve two major factors: investment and operating cost. These factors will be affected greatly by social and political situations.

The rise in land cost and the energy crisis are of great concern for the electrochemical industries. It is estimated that the cost of electric power reaches more than 60% of the manufacturing cost of the chlor-alkali industry, and a surprisingly high 70–75% in the aluminum industry.

Reduction of the initial investment and of the power consumption, namely the cell voltage, are the targets for electrochemical engineering at present. Therefore, the performance of electrolyzers to be used should be carefully examined and considered before final plant design.

The current density, and hence the reaction rate, can be varied and controlled over a wide range, and is a unique variable parameter of electrochemical process as compared with nonelectrochemical process. The amalgam-type chlor-alkali cell is a typical example since it can be operated at current densities between 10 and 200 A/dm^2, although the optimum condition may exist in a more narrow range of current density. The fixed charge of either the unit cell or the entire plant is reduced with increase in the current density. On the other hand, the cell voltage or the power consumption is high at high current densities. Therefore, optimum conditions are determined by economic factors such as the first investment and the power price, and the characteristics of the electrolyzer (see Chapter 14).

Generally, several cells, or several hundred cells in some cases, are installed and operated in a plant so as to optimize the conversion efficiency of the rectifier. Typical chlor-alkali plants operate 20–200 cells, and a water electrolysis plant usually has more than 1000 cells. On the other hand, a small cell is comparatively expensive, and the maintenance cost increases with increase in number of cells. Also, the ohmic voltage drops in hardware will increase when many electrolyzers are connected in series. Therefore, the number of cells in a plant should be optimized.[22]

A large evaporator required to concentrate cell liquor of the diaphragm cell is expensive and can be as high as 40–45% of the total investment of the plant. Also, the evaporator consumes 2–5 tons of steam to produce a ton of NaOH as 50% liquor. Therefore, the economy of the diaphragm-type chlor-alkali plant depends greatly on the evaporation of caustic liquor. Likewise, the plant using membrane cells stands in the same situation.

The feasibility of the process to be used must be studied from various points of view. Energy consumption, difficulty of operation, man-hours for operation, maintenance, adaptability of operation, and effects of impurities in electrolytic solution are some examples to be considered. While the metal anode is more expensive than the graphite anode, it expands the operating time for chlorine cells of both amalgam-type and diaphragm-type, so that the maintenance cost is reduced significantly. The cell voltage is smaller because of low chlorine overvoltage on the metal anode and the minimum gap between the anode and cathode which is possible in this case.

Since the electrolytic solution passes through the porous diaphragm from the anode compartment to the cathode compartment in the diaphragm cell, the percent conversion of brine is an important factor for cell operation. It must be kept constant as possible for steady cell performance. With the amalgam cell, on the other hand, the brine conversion rate can be varied over a relatively wide range without serious difficulty, although the solution *IR* drops and the chlorine overvoltage will be affected somewhat. The cathode process of the amalgam cell is affected by impurities in the brine, as shown in Table 8.6, so that purification of the feed brine is most important in this case.[23] The procedure of brine treatment and the effects of impurities on the cathode process have been discussed by several authors.[24–29]

Performance of the diaphragm cell is also affected by impurities, especially the solution hardness cations, Mg and Ca, due to plugging of micropores of the diaphragm with precipitates. It is reported that these hardness ions penetrate into ion-exchange membranes and that membranes fail owing to formation of deposits or clusters.[30]

The estimated manufacturing cost of chlorine and caustic soda in the diaphragm cell plant is slightly higher than that in the amalgam cell plant, as shown in Table 8.5. However, product quality is also an important factor in the selection of a process, as shown in Table 8.6. The amalgam cell caustic contains a minimum amount of impurities, whereas diaphragm cell caustic contains about 1% NaCl, 0.1% $NaClO_3$, and some other impurities, and the market is accordingly limited. Corrosion of equipment owing to chloride and chlorate in caustic soda, especially when made of the austenitic stainless

TABLE 8.5

Comparison of the Production Cost of Amalgam Cell Caustic
and Diaphragm Cell Caustic in Yen per Metric Ton[a]

	Amalgam process	Diaphragm process
Operating cost		
Crude salt	5,900	6,000
Electricity	11,000	8,300
Electrode (graphite)	1,500	1,400
Steam		2,800
Others	2,300	3,700
SUBTOTAL	20,700	22,200
Fixed charge		
Labor	700	900
Maintenance	900	1,100
Depreciation	3,800	4,500
Tax and insurance	600	700
Overhead in plant	1,100	1,200
Others	2,200	500
SUBTOTAL	9,300	8,900
Manufacturing cost	30,000	31,100
General overhead	4,400	4,900
Interest to installation	1,300	1,600
Interest to operation	800	900
Gross manufacturing cost	36,500	38,500
	$146[b]	$154[b]

[a] 97% NaOH basis in a plant of 60,000 ton/year in 1972.
[b] (a 250 yen/$.

steels, is another problem to be considered.[31] The quality of the membrane cell caustic is about the same as of the amalgam cell product.

Quality of chlorine gas is almost independent of the process while the oxygen content increases with increase in the percent conversion of brine. It is reported that the oxygen content in the gas from diaphragm cells using Ru-oxide-coated anodes is high, probably due to relatively low-oxygen overvoltage of these materials.[32] Therefore, the gas requires treatment before sending to hydrocarbon chlorination processes to avoid explosion hazard. Of course, the improvement of the anode materials is being conducted.

Protection of the environment is a most important topic for all industries today. The investment and the operating cost for this purpose in a

TABLE 8.6

Comparison of Impurities in Diaphragm, Amalgam,
and Membrane Cell Caustic Soda

	Diaphragm cell caustic soda[a]		Amalgam cell caustic[a]	Asahi Chemical membrane cell[b]	
	Regular (%)	Purified (%)	Typical (%)	21.6% NaOH	48% NaOH
Na_2CO_3	0.10–0.30	0.15	0.02	0.02%	0.04 %
NaCl	1.00	0.08	0.0004	20 ppm	44.3 ppm
Na_2SO_4	0.013–0.020	0.01	0.000	0.000	0.000
SiO_2	0.018–0.025	0.009	0.0002	2.0 ppm	4.4 ppm
$NaClO_3$	0.05–0.10	0.0002			
CaO	0.0017	0.0010	0.0015	0.4 ppm	0.9 ppm
MgO	0.0010–0.0020	0.0010	0.0005		
Al_2O_3	0.0013–0.0030	0.0015	0.0002	0.4 ppm	0.9 ppm
NH_3		0.00015			
Fe	0.0005	0.00025	0.00006	1.3 ppm	2.9 ppm
Ni	0.00003	0.00001	0.00001		
Cu	0.00003	0.00002	0.00001		
Mn	0.00001–0.00006	0.00003	0.00002	0.01 ppm	0.02 ppm

[a] The figures apply to 50% liquid caustic soda. Olin catalog.

[b] M. Seko, paper presented at 22nd Chlorine Plant Managers Seminar, Atlanta, GA, the Chlorine Institute, February 1977.

chemical process industry is significantly large. Availability of the technology for abatement of waste discharge and prevention of any leakage is a key to success and also to expenditure. Efforts for abatement of mercury discharge in amalgam cell plants is an example.[33] Of course, reduction of consumption and prevention of leakage of mercury have been practiced for many years, but extensive investigations, especially in Japan, have been done since the early 1970s. Today's technology for abatement of mercury discharge has been developed to a very high degree of perfection. As a consequence, the investment for a 300 ton/day plant is as large as 592.3 million yen (ca. 2.4 million dollars), which is 5–7% of the total investment, with an operating cost of 1635 yen/ton NaOH or $7/ton, for example. The mercury consumption in Japan is only 1.469 g/ton NaOH on the average, which is one of the best results in the world.

Waste asbestos of the diaphragm-type chlorine cells should be carefully treated because asbestos is strongly associated with the danger of lung cancer. Abatement of SO_x and NO_x in the flue gas of the steam generator is also expensive. These expenditures will raise the production cost of diaphragm cell caustic and chlorine.

Siting of the process plant may depend greatly on the availability of space and the local population density in nearby residential areas, and hence the feasibility study to determine which process to use must be conducted on a case-by-case basis. In any event, good public relations will become increasingly important in the future for the chemical process industries, including the electrochemical industries.

REFERENCES

1. D. J. Harvey and J. R. Fowler, *Chem. Eng. Prog.*, p. 47 (April 1976).
2. A. B. Nisercola, R. P. Tracy, I. A. Franson, and R. J. Knoth, Abstract No. 37, Meeting of Electrochemical Society, Washington, D.C., May 1976.
3. H. C. Twiehaus and N. J. Ehlers, *Chem. Ind.*, p. 230 (August 1948).
4. P. E. Feathers and J. E. Wyche, *Chem. Eng.*, p. 122, (May 14, 1973).
5. F. Hine, M. Yasuda, I. Sugiura, and T. Noda, *J. Electrochem. Soc.* **121**, 220 (1974).
6. F. Hine and M. Yasuda, *Denki Kagaku* (J. Electrochem. Soc. Jpn.) **39**, 530 (1971).
7. R. B. MacMullin, *Denki Kagaku* (J. Electrochem. Soc. Jpn.) **38**, 570 (1970).
8. F. Hine, M. Yasuda, and K. Fujita, Abstract No. 411, meeting of Electrochemical Society, Minneapolis, MN, May 1981.
9. M. P. Grotheer and C. J. Harke, *Chlorine Bicentennial Symposium*, p. 209, Electrochemical Society (1974).
10. F. Hine, M. Yasuda, and T. Tanaka, *Electrochim. Acta* **22**, 429 (1977).
11. W. H. Koh, Abstract No. 391, meeting of Electrochemical Society, St. Louis, MO, May 1980.
12. E. H. Cook, C. A. Lazarz, and A. C. Schulz, Abstract No. 390, meeting of Electrochemical Society, St. Louis, MO, May 1980.
13. P. J. Kienholz, *Chlorine Bicentennial Symposium*, p. 198, Electrochemical Society (1974).
14. D. J. Vaughan, *DuPont Innovation* **4**(3), 10 (1973).
15. W. G. F. Grot, G. E. Munn, and P. N. Walmsley, Abstract No. 154, meeting of Electrochemical Society, Houston, TX, May 1972.
16. M. Seko, Abstract No. 417, meeting of Electrochemical Society, Minneapolis, MN, May 1981.
17. G. E. Munn, Abstract No. 436, meeting of Electrochemical Society, Atlanta, GA, October 1977.
18. T. Berzins, Abstract No. 437, meeting of Electrochemical Society, Atlanta, GA, October 1977.
19. E. J. Peters, D. R. Pulver, and E. H. Price, Abstract No. 443, meeting of Electrochemical Society, Atlanta, GA, October 1977.
20. S. A. Dahl, *Chem. Eng.*, p. 60 (August 1975).
21. R. B. MacMullin, *Chem. Ind.* (July 1947).
22. F. Hine, *J. Electrochem. Soc.* **117**, 139 (1970).
23. R. B. MacMullin, Electrolysis of brines in mercury cells, in *Chlorine*, edited by J. S. Sconce, ACS Monograph 154, Reinhold, New York (1962).
24. G. Angel, T. Lunden, S. Dahlerus, and R. Brännland, *J. Electrochem. Soc.* **99**, 435 (1952); **102**, 124, 246 (1955).
25. F. Hine, S. Yoshizawa, and S. Okada, *Kogyo Kagaku Zasshi* (J. Chem. Soc. Jpn.,Industrial Chemistry Section) **62**, 769, 773, 778 (1959).
26. F. Hine, T. Sugimori, S. Yoshizawa, and S. Okada, *Kogyo Kagaku Zasshi* (J. Chem. Soc. Jpn., Industrial Chemistry Section) **62**, 955 (1959).

27. F. Hine, S. Yoshizawa, S. Okada, N. Yokota, T. Kadota, and J. Kushiro, *Kogyo Kagaku Zasshi* (J. Chem. Soc. Jpn., Industrial Chemistry Section) **62**, 961 (1959).
28. F. Hine, S. Matsuura, and S. Yoshizawa, *Electrochem. Tech.* **5**, 251 (1967).
29. F. Hine, M. Yasuda, F. Wang, and K. Yamakawa, *Electrochim. Acta* **16**, 1519 (1971).
30. C. J. Molnar and M. M. Dorio, Abstract No. 442, meeting of Electrochemical Society, Atlanta, GA, October 1977.
31. F. Hine and M. Okubo, *Boshoku Gijutsu* (Corrosion Engineering) **25**, 509 (1976).
32. L. I. Krishtalik, *Electrochim. Acta* **26**, 329 (1981).
33. F. Hine, N. Yokota, and T. Takasaki, *Int. Chem. Eng.* **17**(1), 1 (1977).

Chapter 9

FUSED SALT ELECTROLYSIS AND ELECTROTHERMICS

9.1. BACKGROUND OF FUSED SALT ELECTROLYSIS

It is well known that noble metals such as copper can be electrodeposited at high current density from aqueous solution. However, hydrogen evolution may take place as a competitive reaction when cathodic deposition of less-noble metals such as Fe and Ni is carried out. Metals which are even less noble, such as Al and Mg, do not electrodeposit from aqueous solution at all, although these metals can be obtained by electrolysis of nonaqueous solutions, such as electrolytes of organic solvents or ammonia. These methods may be used on a laboratory scale but not on a commercial scale because of difficulties in operation and high cost.

Hydrated ions of Na^+ and Cl^-, for example, may form and migrate freely in aqueous solution of NaCl, but Na^+ and Cl^- do not exist as bare ions. The number of water molecules, which combine with ions such as $Na(H_2O)_x^+$ and $Cl(H_2O)_y^-$, for example, depends on the size of the ionic species of interest, the solution composition, and other factors such as charge. Generally speaking, small ions of the same group in the Periodic Table have a large energy for hydration and a large number of water molecules; thus, the migration velocity of these hydrated ions is slow, as shown in Table 9.1.[1] Hydrogen and hydroxyl ions show separate behaviors because of different mechanisms for their migration.

Sodium chloride is molten at temperatures higher than 803°C, and the Na^+ and Cl^- ions of the NaCl crystal are released. Of course, these ions still interact with their neighbors. In some cases, ionic species in molten salt may form complex ions instead of solvated ions. Formation of $CaCl_3^-$ in the molten salt consisting of $CaCl_2$ and KCl is an example.

Because there is no water molecule to participate in reaction, even very active (less-noble) metals such as K and Na can be reduced on the cathode with high-current efficiencies. Properties of ionic species in molten salt are about the same as in aqueous solutions: the electrical conductivity of molten

TABLE 9.1.

Ion Hydration Energy and Number of Bound Waters of Hydration

	Hydration energy (kcal/g ion)	Number of waters of hydration
H^+	255	4
Li^+	131	6
Na^+	97	4
K^+	77	2–3
Rb^+	73	2
Ag^+		3–4
Mg^{2+}		9–13
Ca^{2+}		8–10
Ba^{2+}		6–8
Zn^{2+}		11–13
Fe^{2+}		11–13
Cu^{2+}		11–13
Cd^{2+}		10–12
Pb^{2+}		5–7
F^-	123	5
Cl^-	83	3
Br^-	73	2
I^-	63	0–1
ClO_4^-		0
NO_3^-		2

salt is of the order of 1 mho/cm, the diffusion coefficient of ions is about 10^{-5} cm^2/sec, and the viscosity of molten salt electrolytes is in the order of 1 centipoise (c.p.), for example.

The electrochemical reaction rate in molten salt electrolyte at high temperatures is large and the overvoltage is generally low. Under some conditions, an unwanted voltage drop, called the "anode effect," appears at the anode. Sparking occurs and the cell voltage fluctuates greatly as the anode surface is covered by gas. Heavy current passes locally at spots on the anode surface and is detrimental because local superheating causes vaporization of molten salt. Local contact between salt and anode is broken and reestablished, giving rise to a cyclic behavior for the anode effect. These phenomena have been studied extensively as a major difficulty in commercial cells for molten salt electrolysis. Insufficient wetting contact of fused salt with anode is thought to be a cause of the anode effect.

In some cases, metallic products dissolve in the molten salt and are oxidized, which causes current inefficiency, especially at high temperatures. The current density can be improved somewhat with an increase of the current density because the loss of product is due mainly to chemical reaction and is independent of current density. For this reason, electrolysis of

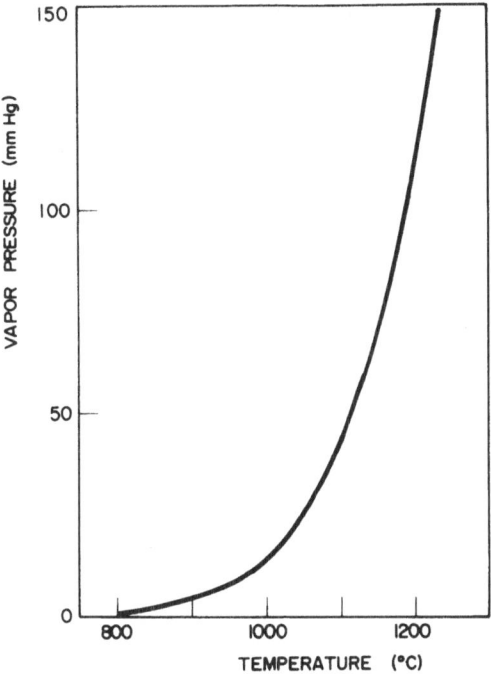

FIGURE 9.1. Vapor pressure of NaCl melt.

molten salt is conducted at relatively high current densities. Since there is temperature variation in the cell, dissolved metal is supersaturated in regions where the temperature is low and is precipitated as very fine mist like particles. This "metal fog" may be oxidized either at the anode or in the bulk of electrolyte, with a lowering of the current efficiency for metal deposition.

Consequently, electrolysis of molten salt should be conducted at a temperature as low as possible so as to keep the current efficiency high. Figure 9.1 shows the vapor pressure of NaCl melt as example.[2] High vapor pressures of electrolyte and products are also problems at high temperatures. The melting point of the molten salt may be lowered by the addition of an appropriate foreign salt. The electrolyte of eutectic composition has the lowest melting temperature in a binary salt mixture, as shown in Fig. 9.2, for the phase diagram of the $CaCl_2$-NaCl system.[3]

It should be noted that the electrical conductivity of the mixed electrolyte is lower than that of the single electrolyte, which is the same behavior as for aqueous solutions as described in Section 4.1.3. Hence, the ohmic voltage drop becomes high when a foreign salt is mixed with the molten electrolyte.

Operation of the molten salt electrolytic cell at a constant temperature is important since otherwise the anode effect and formation of metal fog tend

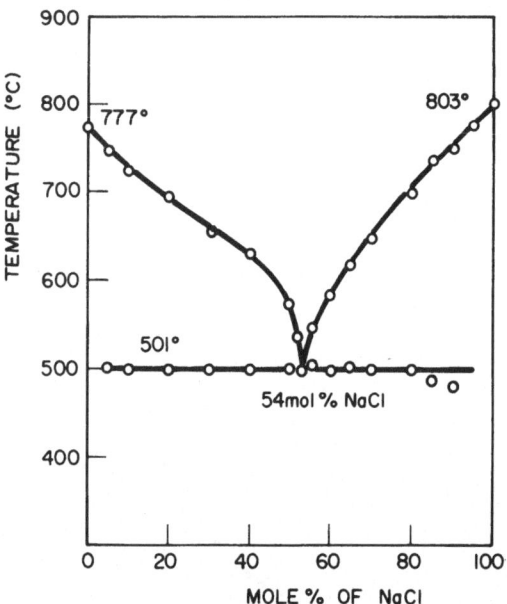

FIGURE 9.2. *Phase diagram of the CaCl$_2$–NaCl system.*

to occur. The thickness of the frozen crust along the cell wall may change with change of the operating temperature, causing serious damage or failure of the cell. Substantial dc electric power is, of course, consumed by the electrochemical reaction, but a major part of the dc power is consumed as a source of heating and is an important factor for evaluating the voltage balance and the energy balance for the molten salt electrolysis under discussion.

Selection of construction materials is difficult because molten salt at high temperatures is strongly corrosive to both metallic and nonmetallic materials. Generally, the inside wall of a steel electrolyzer is lined with carbon plates and/or firebricks, although even those materials are attacked gradually. Since heat is transferred from the wall to the environment, the salt is cooled until it freezes and the frozen crust protects the cell wall from corrosion. The control of an adequate layer of frozen salt is an important technique in molten salt electrolysis.

The selection of a suitable anode material is another important requirement. Carbon and graphite are usually used. These materials resist molten metal halides even at high temperatures, whereas they are strongly attacked by oxygen-bearing compounds such as hydroxides, oxides, chlorates, and sulfates. It is for this reason that metal halide melts are often preferred. Consumption of carbon anodes in the Hall–Heroult process is about 0.5 t/t-Al because alumina feedstock brings oxygen into the cryolite melt.

The Dow cell for magnesium production requires 0.1 ton of carbon anode for 1 ton of Mg compared to only 0.01–0.02 t in the so-called "German cell" because the Dow cell uses hydrated magnesium chloride, whereas anhydrous feedstock is used in the German cell.

9.2. PRODUCTION OF ALUMINUM

Aluminum was first produced by reduction of aluminum chloride with sodium. In 1886, C. M. Hall in America and P. L. Heroult in France independently invented electrolysis of fused cryolite to obtain aluminum, and the process is now known as the "Hall–Heroult process." Soon after, the Bayer process for manufacturing pure alumina was developed, permitting commercial production of aluminum via the electrochemical route.

Aluminum compounds are distributed widely in the world as oxides, silicates, and other forms, although only bauxite, consisting of hydrated aluminum oxide with some impurities (mostly iron oxide), is used as a raw material for the Bayer process for reasons of both technology and economy.

The consumption of electric energy per unit production of aluminum is as large as about 14,000 kWh/t-Al even with the modern technology of the Hall–Heroult process. Of course, reduction of the energy consumption has been investigated extensively. Also, new processes that will require a smaller amount of electricity than the present process are being studied. Electrolysis of aluminum chloride instead of cryolite is an example, although the process has not been commercialized as yet.

9.2.1. The Bayer Process for Production of Alumina

The chemical composition of bauxite is 55–60% Al_2O_3, 5–10% Fe_2O_3, 3–7% SiO_2, and a small amount of TiO_2 with water, depending on the source. Of these, SiO_2 ties up NaOH and Al_2O_3 in the extraction stage to form the insoluble compound $Na_2O \cdot xAl_2O_3 \cdot ySiO_2$ as a form of kaolin, causing losses of alkali and aluminum, so that ore with low SiO_2 content is preferable.

Figure 9.3 illustrates the flowsheet of the Bayer process. Bauxite is calcined at 350–500°C to remove organic impurities and water, and is ground to 20 mesh size or smaller by ball mill or rod mill. Alumina in bauxite ore is extracted by caustic soda in a digester operated at 150–170°C by the reaction:

$$Al_2O_3 + 2NaOH = 2NaAlO_2 + H_2O \tag{9.1}$$

The NaOH concentration in the effluent is 40–50% by weight, and the

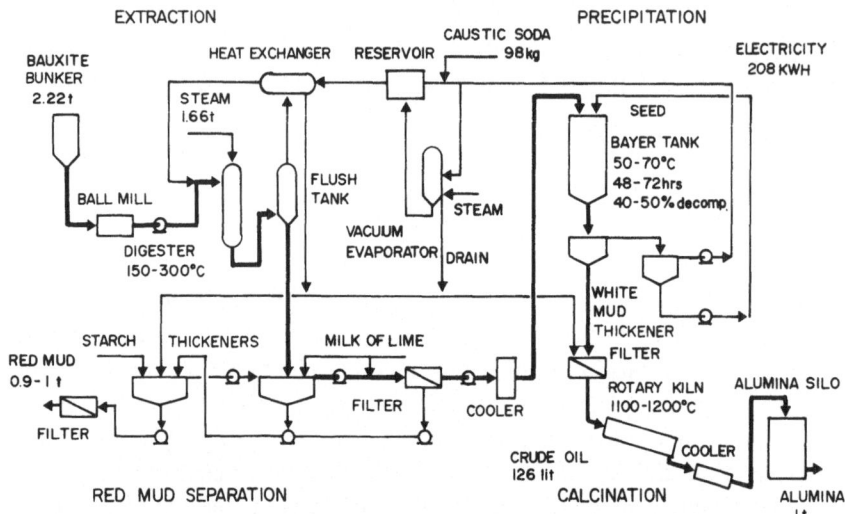

FIGURE 9.3. Flowsheet for producing alumina by the Bayer process.

concentration ratio of Al_2O_3 to NaOH is 0.55–0.65 in general. The extract is diluted to an adequate concentration. A precipitate, the so-called "red mud", which is mostly oxyhydroxide of iron, is separated by means of thickening and filtration, and the filtrate is sent to hydrolysis to obtain aluminum hydroxide. A small amount of seed material is added to the hydrolysis tank containing supersaturated solution at a controlled temperature of from 50 to 70°C, depending on plant procedure. The reaction requires 2–5 days. The aluminum hydroxide precipitate is filtered and washed with fresh water. The mother liquor is recycled to the digester after evaporation to control the NaOH concentration. The filtrate is also returned to the hydrolysis tank. Aluminum hydroxide is calcined in a rotary kiln at 1200–1300°C to prepare alpha alumina.

A certain amount of red mud is discharged from the thickener and is washed and filtered to recover caustic liquor prior to purge. Treatment of red mud is a problem from the viewpoint of environmental protection.

9.2.2. The Hall–Heroult Process for Aluminum Production

The electrolyte is essentially a cryolite melt, $3NaF \cdot AlF_3$. Fluorides such as NaF, KF, and LiF are added if necessary. Addition of LiF effectively improves the electrical conductivity of molten cryolite, and hence contributes to reduction of the cell voltage or the energy consumption. There are several phases of the mixture of NaF and AlF_3, depending on composition and temperature, as shown in Fig. 9.4. Cryolite consists of $3NaF \cdot AlF_3$ with a melting point of 1009°C.

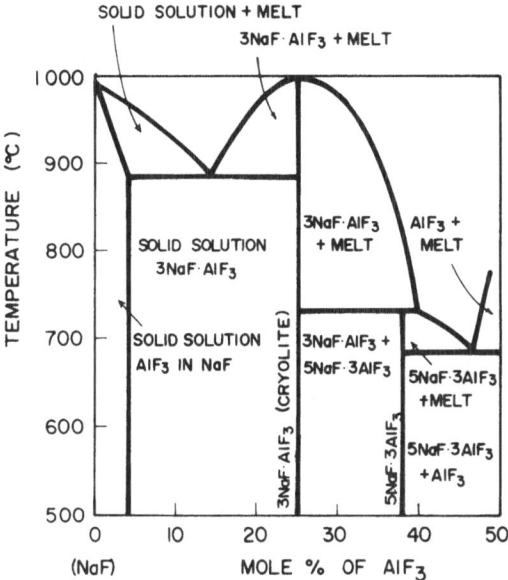

FIGURE 9.4. Phase diagram of the NaF–AlF₃ system.

The phase diagram of the cryolite–alumina system has been well studied, and the eutectic point is estimated to be 20 mol % Al_2O_3 at $962 \pm 1°C$, as shown in Fig. 9.5. The electrical conductivity decreases with an increase of the AlF_3 content in NaF, as shown in Fig. 9.6. The conductivity of molten cryolite at 1000°C is 2.9 mho/cm and decreases when Al_2O_3 is added. For example, the conductivity of the eutectic melt containing 20 mol % Al_2O_3 is about 2 mho/cm (see Fig. 9.7).

The liquid aluminum produced (melting point = 660°C) accumulates on the carbon cathode which serves as the cell bottom and itself functions as cathode. The aluminum has a density of 2.3 g/cm³ at 925°C and 2.35 g/cm³ at 1000°C, compared to about 2.15 g/cm³ for the less dense electrolyte at the operating temperatures of 950–970°C.[4]

Various mechanisms for electrochemical processes in the aluminum cell have been proposed by a number of authors. Table 9.2 summarizes these theories,[5] which can be classified into five groups:

(1) decomposition of NaF;

(2) decomposition of Na_2O;

(3) decomposition of AlF_3;

(4) decomposition of Na_3AlF_6; and

(5) decomposition of Al_2O_3,

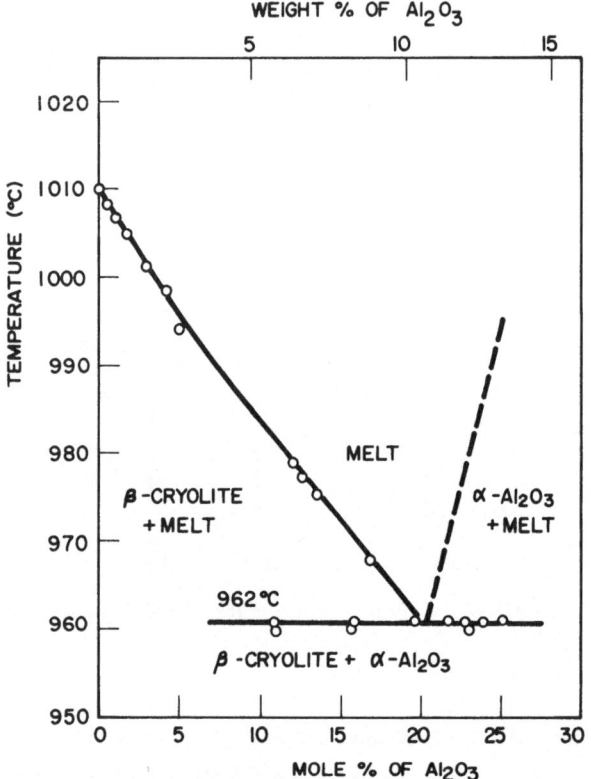

FIGURE 9.5. Phase diagram of the Na_3AlF_6–Al_2O_3 system.

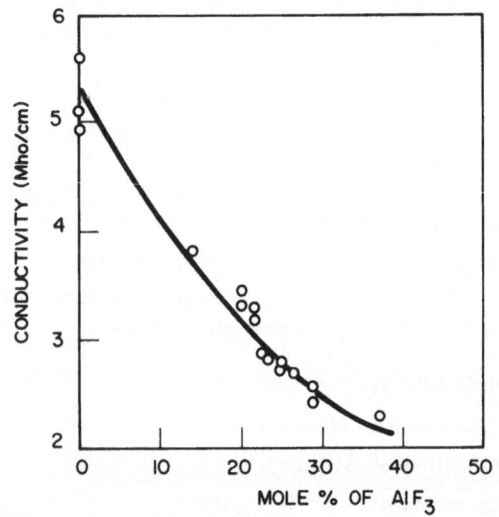

FIGURE 9.6. Conductivity of NaF–AlF_3 melts at 1000°C.

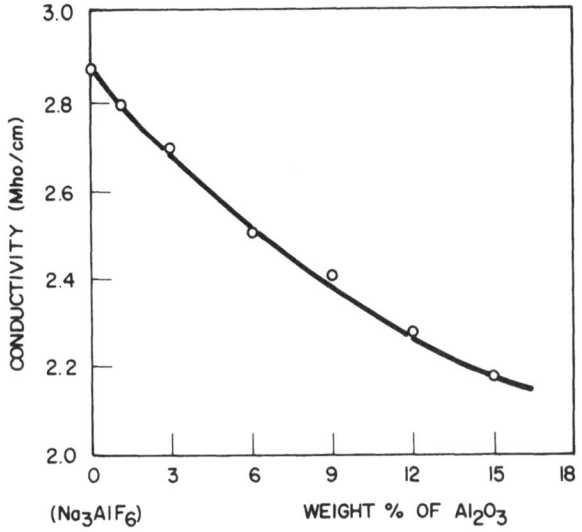

FIGURE 9.7. Conductivity of the $Na_3AlF_6–Al_2O_3$ melts at 1000°C.

whereas the overall reaction is the same as shown in Eq. (9.2):

$$\tfrac{1}{2}Al_2O_3(s) + \tfrac{3}{4}C(s) = Al(l) + \tfrac{3}{4}CO_2(g) \qquad (9.2)$$

The enthalpy and free energy for the reaction at 1250 K or 977°C are as follows[6]

$$-\Delta H^0 = 131 \text{ kcal/mol}$$

$$-\Delta G^0 = 81 \text{ kcal/mol}$$

and

$$-\Delta S^0 = 40 \text{ cal/mol K}$$

The stoichiometric consumption of carbon based on reaction (9.2) is 0.33 ton/ton Al compared to 0.4–0.5 ton/ton Al of commercial cells. It is evident that the carbon anode in aluminum cells is of the category of consumable anodes. Thus, the Söderberg anode was invented to supply carbon automatically, and is designed to prepare a solid carbon electrode beginning with a paste consisting of coke and pitch. A part of the heat generated in the cell is used to convert the paste to carbon. The carbon anode is continuously lowered to maintain the anode-to-cathode gap at about 40–50 mm, while the carbon is consumed by electrolysis. Figure 9.8 illustrates a cell with a vertical-type Söderberg anode. In another version, the spikes of the Söderberg anodes are inserted from the horizontal direction. The casing is made of steel.

TABLE 9.2

Electrode Processes in Aluminum Reduction Cell

	Anodic reaction	In electrolyte	Cathodic reaction
1. Decomposition of NaF			
Drossbach	$6F^- \rightarrow 3F_2 + 6e$ $3F_2 + Al_2O_3 \rightarrow 2AlF_3 + \frac{3}{2}O_2$ $\frac{3}{2}O_2 + \frac{3}{2}C \rightarrow \frac{3}{2}CO_2$ (or CO)	$2Na_3AlF_6 \rightarrow 6Na^+ + 6F^- + 2AlF_3$ $2AlF_3 + 3Na_2O \rightarrow 6NaF + Al_2O_3$	$6Na^+ + 6e \rightarrow 6Na$ $6Na + Al_2O_3 \rightarrow 3Na_2O + 2Al$
Andrie	$6F^- + Al_2O_3 \rightarrow 2AlF_3 + \frac{3}{2}O_2 + 6e$ $\frac{3}{2}O_2 + 3C \rightarrow CO$	$2AlF_3 \cdot NaF \rightarrow 6NaF + 2AlF_3$ $6NaF \rightarrow 6Na^+ + 6F^-$	$6Na^+ + 2AlF_3 + 6e \rightarrow 6NaF + 2Al$
2. Decomposition of Na$_2$O			
Gadeau	$3O^{2-} + \frac{3}{2}C \rightarrow \frac{3}{2}CO_2 + 6e$	$6NaF + 3AlF_3 \rightarrow 2Na_3AlF_6$ $Al_2O_3 + 6NaF \rightarrow AlF_3 + 3Na_2O$ $3Na_2O \rightarrow 6Na^+ + 3O^{2-}$	$6Na + 2AlF_3 + 6e \rightarrow 6NaF + 2Al$
3. Decomposition of AlF$_3$			
Guskow and Terebesi	$6F^- \rightarrow 3F_2 + 6e$ $3F_2 + Al_2O_3 \rightarrow AlF_3 + \frac{3}{2}O_2$ $\frac{3}{2}O_2 + \frac{3}{2}C \rightarrow \frac{3}{2}CO_2$	$2AlF_3 \rightarrow 2Al^{3+} + 6F^-$	$2Al^{3+} + 6e \rightarrow 2Al$
4. Decomposition of Na$_3$AlF$_6$			
Cuthbertson and Waddington	$2AlF_6^{3-} + Al_2O_3$ $\rightarrow 4AlF_3 + \frac{3}{2}O_2 + 6e$	$Na_3AlF_6 \rightarrow 3Na^+ + AlF_6^{3-}$ $3Na_2O + 4AlF_3 \rightarrow 2Na_3AlF_6 + Al_2O_3$	$6Na^+ + Al_2O_3 + 6e \rightarrow 3Na_2O + 2Al$
5. Decomposition of Al$_2$O$_3$			
Fedotieff	$2AlF_6^{3-} + Al_2O_3$ $\rightarrow 4AlF_3 + \frac{3}{2}O_2 + 6e$	$Al_2O_3 \rightarrow Al^{3+} + AlO_3^{3-}$ $Na_3AlF_6 \rightarrow 3Na^+ + AlF_6^{3-}$ $AlO_3^{3-} + 3Na^+ \rightarrow Na_3AlO_3$ $Na_3AlO_3 + 2AlF_3 \rightarrow Na_3AlF_6 + Al_2O_3$	$Al^{3+} + 3e \rightarrow Al$
Almand	$O^{2-} \rightarrow O_2 + 2e$	$Al_2O_3 \rightarrow 2Al^{3+} + 3O^{2-}$	$Al^{3+} + 3e \rightarrow Al$

Source: P. Drossbach, *Elektrochemie geschmolzener Salze*, Berlin (1938); E. Grübert, *Z. Elektrochem.* **48**, 393 (1942); S. I. Beljajew, *Metallurgie des Aluminiums*, Vol. 1 (1956).

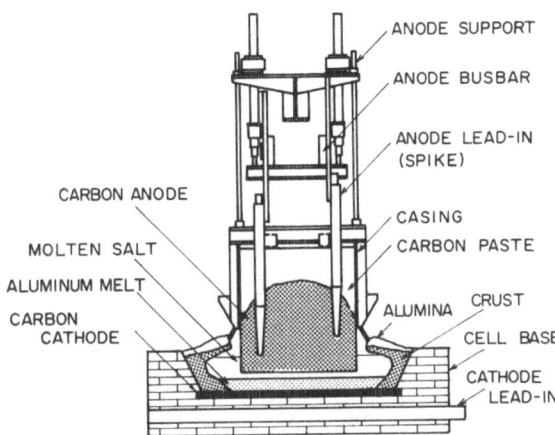

FIGURE 9.8. Aluminum electrolyzer using a vertical-type Söderberg anode. (Quoted from: Trattato Di Ingegneria Elettrochimica, by P. Gallone, by permission of Tamburini Editore, Milano.)

The estimated decomposition voltage based on the Gibbs free energy change for reaction (9.2) is 1.17 V, and hence the energy consumption without any other voltage loss should be only 3490 kWh/ton Al. By comparison, the 15,000–16,000 kWh/ton Al required in practical cells, as shown in Table 9.3,[7] means that voltage drops other than the decomposition voltage are very large and also that the current efficiency is not 100%, say in the range 85–90%. Reduction of the energy consumption is a most important subject for the aluminum production industry. Extensive efforts have been conducted to improve the power consumption with a best result of about 13,500 kWh/ton Al.[8] The ad hoc committee on new processes of aluminum production sponsored by the Electrochemical Society of Japan in association with the aluminum industries has reported that

TABLE 9.3

Consumption of Raw Materials and Energy for Producing 1 Ton of Aluminum

Raw materials (kg)		Electricity (kWh)	
Alumina	1,948	dc power for electrolysis	15,338
Cryolite	29	Losses in bus bar and rectifier	413
Recovered cryolite	21	For melt	68
Aluminum fluoride	32		
Anode paste	585	Labor (man hr)	
Coke	420	Direct	2.03
Pitch	188	Indirect	1.07
Cathode carbon	15		

TABLE 9.4

Breakdown of the Cell Voltage of Aluminum Electrolyzers

	Prebaked anode (V)	Söderberg anode (V)
Decomposition voltage	1.60	1.60
IR drop in electrolyte	1.45	1.45
Anode overvoltage	0.30	0.50
Cathode overvoltage	0.40	0.40
IR drop in hardware	0.10	0.10
Anode effect	0.15	0.15
Terminal voltage	4.00	4.20

12,300 kWh/ton Al is the limit for the Hall–Heroult process.[7] Today's technology has reached its utmost level and there is little margin for further improvement.

Table 9.4 shows a comparison of the voltage balance in a cell equipped with prebaked anodes with that of a cell using the Söderberg anodes. In this table, the decomposition voltage is listed as 1.60 V instead of 1.17 V as described above and has been obtained by extrapolation of the volt–ampere curve of a practical cell to zero current. It is therefore a superficial rather than a theoretical decomposition voltage. According to the data in Table 9.4, the terminal voltage of the cell using the prebaked anodes is lower than that of the cell using the Söderberg anode. It has been reported that the investment cost of a plant using Söderberg anodes is lower than that of a plant using prebaked anodes when the production capacity is small, as shown in Fig. 9.9.[9] The break-even point is estimated to be about 100,000 ton Al/year. Therefore, the prebaked anode system is paid a great

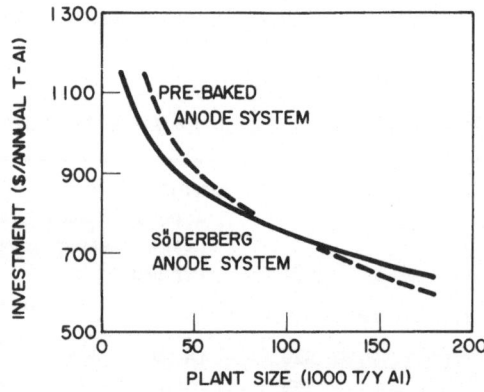

FIGURE 9.9. Average fixed-capital investments for alumina reduction plants (1963 cost).

deal of attention today, and several large-scale plants using this system have been constructed in last few years.

The consumption of the carbon cathode is small, as shown in Table 9.3, and the cathode can be used for three years or more. However, the cathode material is being investigated extensively to improve the energy consumption for aluminum production.

The anode effect occurs during operation of aluminum cells, about once a day, if the alumina content in the molten salt decreases to about 1%. Although its behavior serves to indicate the need for alumina feeding during operation, the anode effect should be minimized to reduce energy consumption.

Since the Hall–Heroult process requires a lot of electric energy and carbon for unit production of aluminum, new processes which may be able to substitute for the Hall–Heroult process are being investigated. Alcoa's electrochemical process using aluminum chloride instead of cryolite is an example.[10] The molten electrolyte is an eutectic consisting of $AlCl_3$, NaCl, KCl, and LiCl, and consumption of the carbon anode is considerably low. Chlorine generated at the anode is utilized for chlorination of alumina. Consumption of electricity for electrolysis is said to be about 9000 kWh/t-Al.

Nonelectrochemical processes have also been investigated. Alcan has attempted the disproportionation reaction of AlCl to obtain Al and $AlCl_3$.[11] Bauxite is fired with carbon, and is brought to reaction with $AlCl_3$ at 1800°F to convert it into AlCl. These new processes are still under development, but commercialization is anticipated in the near future.

9.2.3. Production of High-Purity Aluminum by Means of Electrorefining

Aluminum produced by the Hall–Heroult process contains very small amounts of impurities such as 0.05% Si and 0.05% Fe, and is of 99.9% purity. This purity is sufficient to cover almost all requirements for pure aluminum. However, there are some markets which require extra-pure aluminum because the physical and chemical properties of the high-purity metal are more desirable than those of the regular product. Demand is found in the electrical industry, decoration, and production of special alloys, as examples. Such demands for high-purity aluminum are increasing at a considerable rate.

Refining of aluminum is carried out by a specially designed electrolytic cell, the "three-phase electrolyzer" shown in Fig. 9.10. The anode consists of a copper–aluminum alloy containing 30–35% Cu. The density of 30% Cu–70% Al alloy is 3.05–3.10 g/cm^3 at 700°C. The molten salt electrolyte is composed of AlF_3, NaF, BaF_2, $BaCl_2$, CaF_2, and others. The density of the electrolyte consisting of 23% AlF_3, 17% NaF, and 60%

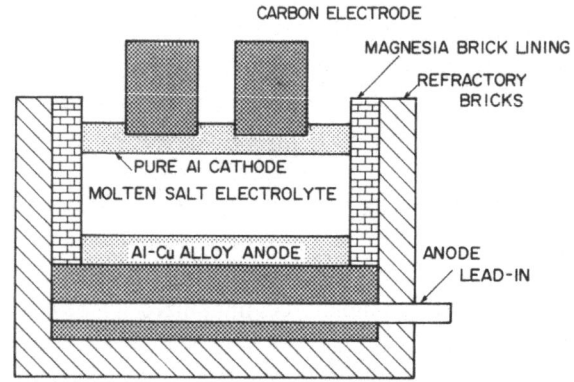

FIGURE 9.10. Three-phase-type aluminum refining cell.

$BaCl_2$ is about 2.7 g/cm^3 at 740°C. Therefore, pure aluminum, whose specific gravity is 2.3, rises to the cell top, while the alloy anode settles at the bottom. Thus, three phases are established in the cell. Aluminum in the alloy dissolves anodically to form Al^{3+}, whereas Cu does not dissolve. The quality of aluminum deposited at the cathode is 99.995% or higher.

Figure 9.10 only illustrates the principle of the aluminum refining cell. In practice, the composition of the alloy anode varies with electrolysis and the molten salt electrolyte is contaminated gradually because crude aluminum is fed to the cell. Part of the alloy is oxidized by air and forms scale, which must be removed from the cell since otherwise the product is contaminated and the cell performance is decreased. It causes some consumption of both alloy and electrolyte, as shown in Table 9.5.

TABLE 9.5
Operating Conditions of Three-Phase-Type Aluminum Refining Cell

Current (kA)	10–40
Current density (A/dm^2)	30–40
Terminal voltage (V)	5–8
Current efficiency (%)	93–95
Temperature (°C)	740–760
Anode alloy composition (%)	
Al	65–75
Cu	35–25
DC power consumption (kWh/t-Al)	18,000–20,000
Consumption (kg/t-Al)	
Al anode	1,030–1,050
Electrolyte	30–50
Copper	5–10
Graphite	30–50

In the molten salt electrolyzer, the electrolyte, electrodes, raw materials, and products, or all of them, are liquid. Therefore, physical properties such as viscosity, density, and surface tension, as well as the conductivity and the melting point, are very important factors for electrolysis. The three-phase refining cell for pure aluminum is a good example of technology that is based on small differences in physical properties of ingredients in the cell. Although the decomposition voltage for the refining cell is small, the terminal voltage of commercial cells is higher than that of the Hall–Heroult cells because of a large ohmic drop through the electrolyte, and hence the consumption of electrical energy is very large, as shown in Table 9.5.

9.3. ELECTROLYTIC PRODUCTION OF MAGNESIUM

More than 40% of magnesium is used for making light weight alloys with aluminum. Magnesium has some other markets such as chemical industries, die casting, nodular iron, sacrificial anodes for cathodic protection, and extrusion. Magnesium is also used as reducing agent for making titanium by the Kroll process:

$$2Mg + TiCl_4 = Ti(sponge) + 2MgCl_2 \tag{9.3}$$

Byproduct magnesium chloride in this process is recycled for electrolysis to recover magnesium and chlorine. Magnesium production in the United States in the last ten years has grown at the rate of about 8% per year.

Two types of electrolytic cells are used for magnesium production: the Dow cell and the German I. G. cell. In the U.S., the Dow cells are common, while I. G. cells are mostly used in Europe, and in the U. S. to a smaller extent.

The feedstock of the I. G. cell is anhydrous magnesium chloride, but the Dow cell uses magnesium salt containing some molecular water. As a result, there are differences in current efficiency, consumption of graphite anodes, and the chlorine concentration in the cell gas, as shown in Table 9.6. The consumption of electric power is about the same.

Although it is clear that anhydrous salt is desirable as a feedstock to a magnesium electrolyzer, dehydration of magnesium chloride is troublesome and expensive. Partial dehydration up to 1.5 mole of water for 1 mole $MgCl_2$ seems to be an optimum for the Dow process, where magnesium salt is recovered from ocean water in the Gulf of Mexico. The economics are influenced greatly by the plant site, the market or the end-use of products, and the availability of raw materials.

The molten salt for the Dow cell consists of $MgCl_2$, NaCl, and $CaCl_2$, but the I. G. cell uses a more complicated composition. The phase diagrams

TABLE 9.6
Operating Conditions of Magnesium Cells

	U.S. cell	I.G. cell
Composition of molten salt		
$MgCl_2$ (%)	25	13
NaCl (%)	60	35
$CaCl_2$ (%)	15	40
KCl (%)		12
Conductivity (mho/cm)	1.5–2.0	
Raw material	$MgCl_2 \cdot 1.5H_2O$	Anhydrous $MgCl_2$
Current (kA)	55	32
Cell voltage (V)	6–9	7
Temperature (°C)	700	720–780
Current efficiency (%)	75	90
Power consumption (kWh/kg Mg)	17–22	17–18
Anode consumption (kg/kg Mg)	0.1	0.015
Chlorine concentration (%)	5–10	95

of the $CaCl_2$-NaCl system and the $MgCl_2$-NaCl system are shown in Figs. 9.2 and 9.11, respectively.[12]

Figure 9.12 illustrates the I. G. cell.[13,14] The inside of the steel vessel is lined with firebrick. Parallel graphite anode blades and steel cathode plates are positioned vertically. Chlorine gas generated at the anode is collected by the anode cover, located at the top. The cell is sealed completely, and the magnesium salt is anhydrous. The anode reaction is thus only chlorine

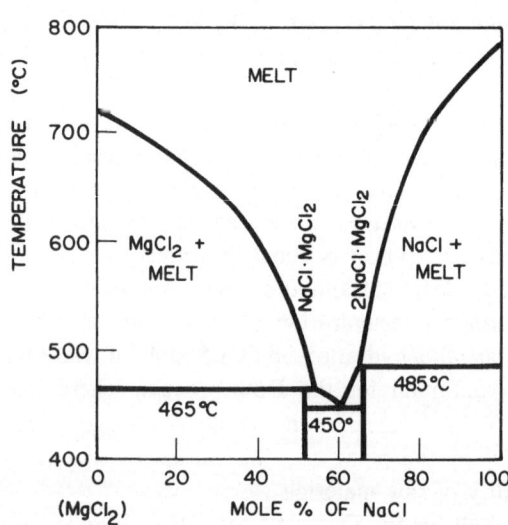

FIGURE 9.11. Phase diagram of the $MgCl_2$–NaCl system.

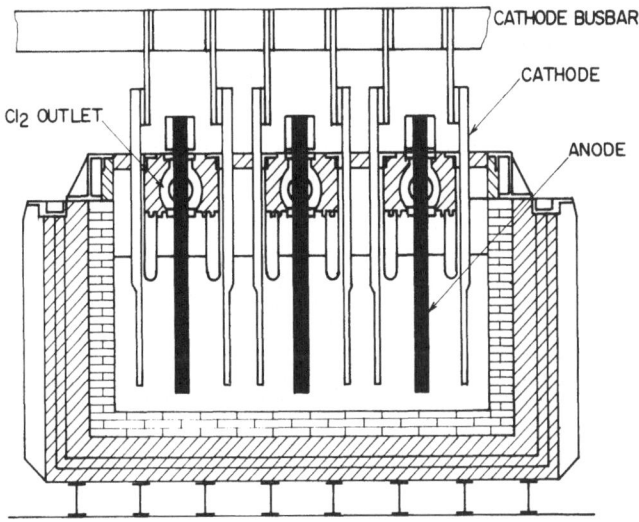

FIGURE 9.12. I.G. magnesium electrolyzer. (From: Angewandte Elektrochemie, by A. Schmidt, by permission of Verlag Chemie.)

evolution with very little consumption of the graphite anode (see Table 9.6). A simple cell design is thus possible.

Since the feedstock is the hydrated salt in the Dow cell, water electrolysis occurs to some extent with the chlorine evolution, which consumes the graphite anode extensively. The Dow cell is designed so as to permit the graphite anode rods to be changed easily. The concentration of chlorine in the cell gas is low at 5–10% Cl_2 with the rest consisting of CO_2, HCl, and some air.[15] In the Dow plant, the cell gas is washed with water to absorb HCl, and is then burned with hydrocarbons to recover HCl, which is absorbed with dilute acid to prepare 20% muriatic acid. The acid is used for neutralizing magnesium hydroxide precipitate recovered from ocean water.[16]

The specific gravity of magnesium is 1.580 at 700°C, and is less than that of the molten salt. The metal formed at the cathode rises to the cell top and is collected. The cathode screen is designed to allow liquid metal to move easily.

The terminal voltage of the I. G. cell at 32 kA is 6.8 V compared to the 2.59 V decomposition voltage at 700°C. The average current densities at the anode and the cathode are 60 A/dm^2 and 68 A/dm^2, respectively. A major part of the difference between the terminal voltage and the decomposition voltage is thought to appear in the *IR* drop and the anode overvoltage, both of which convert into Joule heat which maintains the operating temperature of the cell.

The anode-to-cathode space of a new cell operated in a Stassfurt,

Germany, plant is about 10 cm.[16] With an assumpion of 2.2 mho/cm for the conductivity of the molten salt, the *IR* drop is estimated to be about 3 V, and hence the *IR* drop between the two electrodes should be minimized to further reduce energy consumption, although the current efficiency is greatly decreased as the electrode gap narrows. Addition of LiCl to the molten salt to increase conductivity might be beneficial.[14]

Carnallite, $MgCl_2 \cdot KCl \cdot 6H_2O$, and dolomite, $MgCO_3 \cdot CaCO_3$, are the minerals used as raw material for magnesium production. Magnesium chloride recovered from ocean water and brines is also used. Dehydration of the minerals containing 6 moles of water per one mole of $MgCl_2$ is somewhat complicated. Four molecules of water can be removed easily by heating, but the dehydrated salt is decomposed into magnesia (see Fig. 9.13):

$$MgCl_2 \cdot 6H_2O \xrightarrow{116.7°C} MgCl_2 \cdot 4H_2O + 2H_2O \qquad (9.4)$$

$$MgCl_2 \cdot 4H_2O \xrightarrow{181.5°C} MgCl_2 \cdot 2H_2O + 2H_2O \qquad (9.5)$$

$$MgCl_2 \cdot 2H_2O \longrightarrow MgO + H_2O + 2HCl \qquad (9.6)$$

Magnesium hydroxide is precipitated and recovered from solution containing Mg^{2+}, such as sea water, by addition of alkali. The hydroxide is thickened and filtered. The filter cake is acidified or dissolved by HCl to prepare $MgCl_2 \cdot 6H_2O$. A part of the hydroxide cake is also dried and

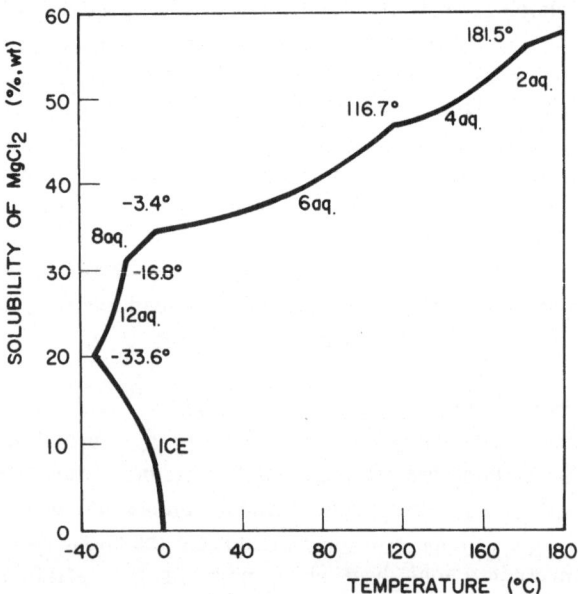

FIGURE 9.13. Phase diagram of the $MgCl_2$–H_2O system.

calcined at 900°C to form MgO. The $MgCl_2 \cdot 6H_2O$ and MgO products together with coke and chlorine gas are reacted in a fluidized bed furnace to produce anhydrous magnesium chloride.

Magnesium can also be produced in a nonelectrochemical process by the reduction of magnesia with ferrosilicon containing 75% Si:

$$4MgO + Si = 2Mg + Mg_2SiO_4 \qquad (9.7)$$

Magnesium vapor is collected and condensed in a separate reservoir. Calcium carbide, aluminum and its alloys, carbon, and other materials have been studied as possible agents for producing magnesium via a chemical route.[17]

9.4. ELECTROCHEMICAL PRODUCTION OF SODIUM

About 100 years ago, in order to produce metallic sodium as a cheap reducing agent for making aluminum, Castner invented an electrochemical process using a caustic soda melt. The Castner process has long been operated, but almost all plants have been converted to the Downs process because of the large consumption of electrical energy.

The Castner cell is made with a steel shell of about 50-cm-diam and about 60-cm-deep, and is positioned in a furnace. The steel cathode is located at the center of the cell and is surrounded by a separator made of steel mesh. The iron anode, which is cast together with the cell cover, is located around the outside of the separator.

Although the melting point of pure caustic soda is 318°C, commercially pure caustic soda melts at about 300°C because it contains a trace of impurities. The operating temperature of the Castner cell is thus kept in the range 310–320°C. At high temperatures, sodium produced at the cathode dissolves into the molten salt electrolyte and causes decreased current efficiency.

Highly pure caustic soda should be supplied to the cell, since otherwise the cell is corroded seriously by impurities, especially sodium chloride.

The chemical reaction in the Castner cell for making metallic sodium is as follows:

At the anode:	$2OH^- \rightarrow H_2O + \frac{1}{2}O_2 + 2e$	(9.8)
At the cathode:	$2Na \leftarrow 2Na^+ + 2e$	(9.9)
Side reaction:	$Na + H_2O \rightarrow NaOH + \frac{1}{2}H_2$	(9.10)
Overall reaction:	$NaOH \rightarrow Na + \frac{1}{2}H_2 + \frac{1}{2}O_2$	(9.11)

$$\underbrace{Na}_{\substack{\text{cathode}\\\text{product}}} \quad \underbrace{\frac{1}{2}H_2 + \frac{1}{2}O_2}_{\substack{\text{anode}\\\text{product}}}$$

cathode　　anode
product　product

With Eq. (9.11), it is evident that the maximum current efficiency for sodium is only 50%. The current efficiency of commercial cells is somewhat less at about 45% for several reasons such as dissolution and oxidation of metallic sodium produced:

$$2Na + 2H_2O = Na_2O_2 + 2H_2 \qquad (9.12)$$

$$Na_2O_2 + 2Na = 2Na_2O \qquad (9.13)$$

$$Na_2O + H_2O = 2NaOH \qquad (9.14)$$

The consumption of electric power is 10 kWh/kg-Na or more, as shown in Table 9.7.

Although electrolysis of NaCl melt seems to be simple and suitable from a theoretical point of view in comparison with the Castner process, the technology is more difficult because of the high 803°C melting point of NaCl:

At the anode: $2Cl^- \rightarrow Cl_2 + 2e$ (9.15)

At the cathode: $2Na \leftarrow 2Na^+ + 2e$ (9.16)

Overall reaction: $2NaCl \rightarrow 2Na + Cl_2$ (9.17)

The melting point of NaCl is lowered by the addition of $CaCl_2$, and the eutectic mixture, 54 mol% NaCl + 46 mol% $CaCl_2$, melts at 501°C (see Fig. 9.2). The Downs process uses this composition of molten electrolyte, and is operated at 600–650°C.

TABLE 9.7
Operating Conditions of Sodium Cells

	Castner cell	Downs cell
Current (kA)	5–10	35
Cathode current density (A/dm^2)	80–110	100
Anode current density (A/dm^2)	90–100	100
Cell voltage (V)	4.0–4.2	5.7–6.0
Current efficiency (%)	44–46	85
Power consumption (kWh/kg Na)	10.8	9.6
Temperature (°C)	310–350	600–650
Anode material	Steel or copper	Graphite
Cathode material	Steel or nickel	Steel or copper
Composition of molten salt (%)		
NaOH	96	
Na_2CO_3	2	
NaCl	2	40
$CaCl_2$		60

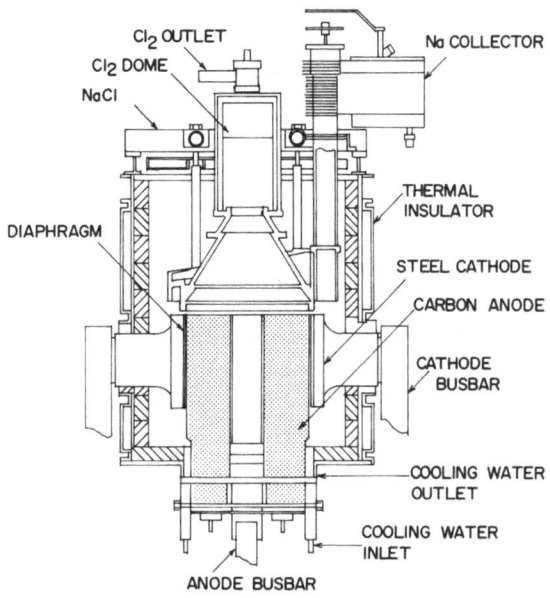

FIGURE 9.14. Downs sodium cell. (From: Electrochemical Engineering, by C. L. Mantell, by permission of McGraw-Hill.)

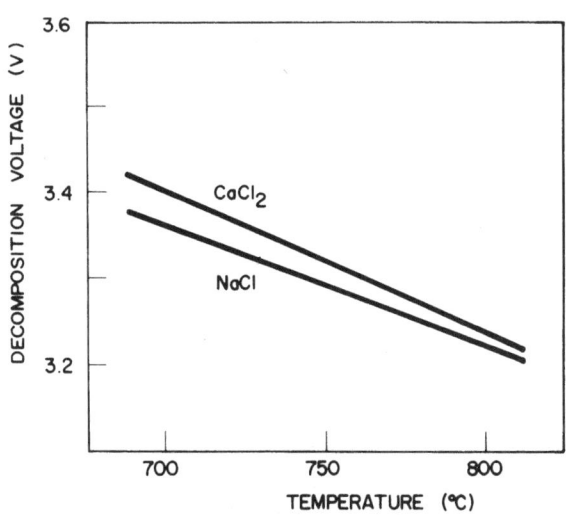

FIGURE 9.15. Decomposition voltage of NaCl and CaCl$_2$ melts.

Figure 9.14 is an example of the Downs cell.[18] The anode and the cathode are made of carbon and cast iron, respectively, and the separator is made of steel. The anode of a 32 kA cell is 2.029 m long and 0.41 m wide. There are four anodes in a cell, so that the total area is 13.32 m². The terminal voltage is 5.7–6.0 V at 100 A/dm² at the anode, as shown in Table 9.7. The decomposition voltage of NaCl at 650°C is 3.4 V (see Fig. 9.15). The balance, 2.3–2.6 V, is converted to Joule heat. Chlorine gas generated at the anode is recovered at the cell top while molten sodium is collected and brought to the receiver, where it is cooled slightly and is recovered every two hours.

Some calcium deposits with sodium at the cathode because the decomposition voltage for $CaCl_2$ is close to that for NaCl, as shown in Fig. 9.15. The melting point of Ca is high (842°C) compared to 97.5°C for Na. Therefore, calcium powder, which disperses in molten sodium, is filtered out to obtain pure sodium. The recovered calcium is brought to a separate reactor, where it reacts with NaCl to recover sodium. As a result, the electrochemical deposition of calcium does not contribute to current inefficiency for sodium production.

9.5. PRODUCTION OF CALCIUM CARBIDE

The process for making calcium carbide from limestone and coke is not an electrochemical reaction, but an endothermic chemical reaction ($\Delta H = 108$ kcal/mol) carried out at very high temperatures, close to 2000°C:

$$CaO + 3C = CaC_2 + CO \tag{9.18}$$

For such chemical reactions, use of the electric furnace is convenient and economical. Since design and operation of the carbide furnace closely resemble that of electrolyzers for molten salt electrolysis such as an aluminum cell, electrothermics is classified as a part of electrochemical processes for the sake of convenience.

Electrothermics involves chemical processes operated at high temperatures with electric furnace of various kinds. The electrometallurgical production of metals and alloys is also included. Since electrothermics does not involve charge transfer reactions and instead uses electricity as a heat source, ac electrical power is generally used although dc power is also used in some cases.

Electric furnaces are classified into several types such as resistance, arc, inductance, consumable electrode, and others. The carbide furnace is of the resistance type with consumable electrodes, and the design is relatively simple, as shown in Fig. 9.16. The general size is in the range

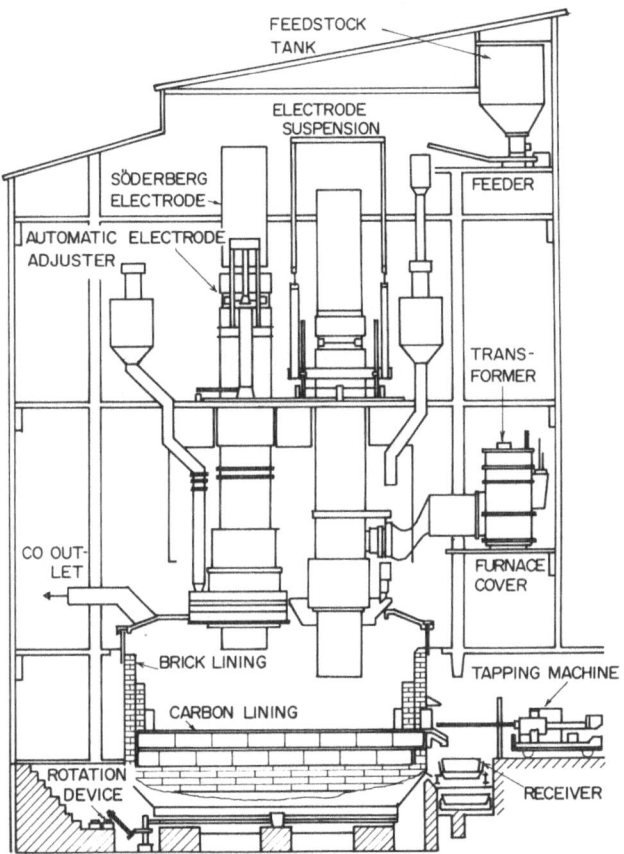

FIGURE 9.16. Electric furnace for producing calcium carbide. (Courtesy of Denki Kagaku Kogyo Co.)

5000–15,000 kW, but recently a large furnace of 50,000 kW is also being operated.[19]

Limestone of about 98% purity is calcined to make lump calcium oxide. Screenings of calcium oxide are mixed with high quality coke to make paste for the Söderberg electrode. These raw materials should be dried sufficiently to avoid introduction of water to the furnace.

An open furnace has been used previously, but the closed-type furnace is common today with a view toward environmental protection. The energy consumption of modern furnaces is also improved by reduction of heat loss. The steel casing of the furnace and the bottom are lined with firebricks and carbon plates, respectively.

Carbide produced is tapped intermittently, which is a difficult and

FIGURE 9.17. Tapping machine for calcium carbide production. (Courtesy of Denki Kagaku Kogyo Co.)

labor-intensive operation. Recently an automatic machine for tapping has been developed by Denki Kagaku Kogyo Co., as shown in Fig. 9.17, which has improved the economics of carbide production.

The carbide furnace is generally operated with ac electric power, with some exceptions. The terminal voltage of the secondary side of the transformer can be changed over a wide range, 130–300 V, which is valid for use of off-peak electricity. The current density on the electrode is in the range 650–900 A/dm². Table 9.8 shows the consumption of raw materials and utilities for unit production of calcium carbide.

In Japan, there are suitable deposits of limestone and relatively cheap electricity is available in some locations. Hence, the Japanese carbide

TABLE 9.8

Consumption of Raw Materials and Utilities
for Production of Calcium Carbide (Basis: 1 ton)

Calcium oxide	ton	0.920
Carbon	ton	0.550
Electrode	ton	0.015
Electric power	kWh	3030
Labor	man hr	2.50

industry has the largest capacity at 2 million tons per year, or about one-fifth of the world production, followed by Germany and the United States.

The largest market for calcium carbide is, of course, acetylene for organic synthesis. Cheap acetylene from petrochemical sources provided considerable competition for the carbide industry. However, the carbide industry and the products from carbide have been given renewed attention today because of the sky-rocketing increase of price and shortage of crude oil.

REFERENCES

1. N. Kameyama, *Denki Kagaku no Riron oyobi Oyo* (Theory and Application of Electrochemistry), Vol. 1, p. 68, Maruzen, Tokyo (1946).
2. *Yoyuen Bussei Hyo* (Data of Molten Salts), edited by the committee, p. 387, Kagaku Dojin, Kyoto (1963).
3. T. Sato and T. Amano, *Kinzoku no Kenkyu* (Researches of Metals) 11, 305 (1934).
4. P. Pascal and A. Jouniaux, *Z. Elektrochem.* 22, 71 (1916).
5. Supplied by Sumitomo Chemical Industry Co. (1968).
6. *Aruminiumu no Shin Seiren Gijutsu Chosa Hokokusho* (Report on the New Technologies of Aluminum Production), p. 6, edited by the committee, Electrochemical Society of Japan (1976).
7. *Kagaku Purosesu Shusei* (Collection of Chemical Processes), p. 268, edited by the committee, Tokyo Kagaku Dojin (1970).
8. Anon. *Chem. Eng.*, p. 33 (June 9, 1975).
9. Anon. *Chem. Eng.*, p. 120 (September 2, 1963).
10. Anon. *Chem. Eng.*, p. 45 (January 22, 1973).
11. Anon. *Chem. Eng.*, p. 71 (July 22, 1963).
12. W. Klemm and P. Weiss, *Z. Anorg. Allgem. Chem.* 245, 281 (1940).
13. A. Schmidt, *Chem. Ing. Tech.*, 37, 596 (1965).
14. A. Schmidt, *Angewandte Elektrochemie*, p. 232, Verlage Chemie, Weinheim (1976).
15. R. M. Hunter, *Trans. Electrochem. Soc.* 86, 21 (1944).
16. C. L. Mantell, *Electrochemical Engineering*, 4th ed. p. 407, McGraw–Hill, New York (1960).
17. *Denki Kagaku Binran* (Handbook of Electrochemistry), p. 1026, Maruzen, Tokyo (1964).
18. W. C. Gardiner, FIAT Final Report 820 (PB 44,671) (1946).
19. *Kagaku Purosesu Shusei* (Collection of Chemical Processes), p. 171, edited by the committee, Tokyo Kagaku Dojin (1970).

ELECTROREFINING AND
ELECTRODEPOSITION OF METALS

10.1. ELECTROCHEMICAL PRODUCTION OF COPPER

Copper is found in nature as the metal and in various types of ore such as oxide, hydroxide, carbonate, sulfide, and in mixtures and double salts of these compounds. Of these, the sulfide ore, chalcopyrite $Cu_2S \cdot FeS \cdot FeS_2$, is commonly used in Japan. Figure 10.1 illustrates an example of the flowsheet for copper concentrate processing. The ore is treated to separate rock and to concentrate minerals. The powdered mineral is sintered and lumps are calcined with limestone to concentrate the copper as matte. Impurities, mostly as silicates, are removed in the slag. Chemical compositions of concentrate, matte, and slag are shown in Table 10.1.[1] Gas from the calcination blast furnace contains SO_2 of high concentration and is sent to the sulfuric acid plant. Matte is refined in the converter in which iron sulfide reacts with silica and excess air to form silicate, while the copper compound in the matte is reduced to crude copper in a period of about 3 hr.

Slag formation: $\qquad 2FeS + 3O_2 + 2SiO_2 = 2FeO \cdot SiO_2 + 2SO_2 \qquad$ (10.1)

Copper production: $\quad Cu_2S + 2Cu_2O = 6Cu + SO_2 \qquad\qquad\qquad$ (10.2)

Since the slag of the converter contains some copper, it is recycled to the blast furnace to recover metal. The crude copper produced is cast to make slabs weighing 100–500 kg, which are sent to the electrorefining cell as anodes.

Either wood or concrete tanks, both lined with lead sheet, were once widely used as the electrolytic cell. Today, the cell is fabricated with polyvinyl chloride (PVC) and/or fiberglas-reinforced plastics (FRP) as corrosion-resistant material, and is 4–5 m long, 1–2 m wide, and 1.0–1.5 m deep. The typical copper electrorefining plant operates 300–500 cells working at 5–20 kA of current. Recently, very large cells, 4 m wide, 28 m long, and 1.5 m deep, have been operated in a modern plant in Japan. These

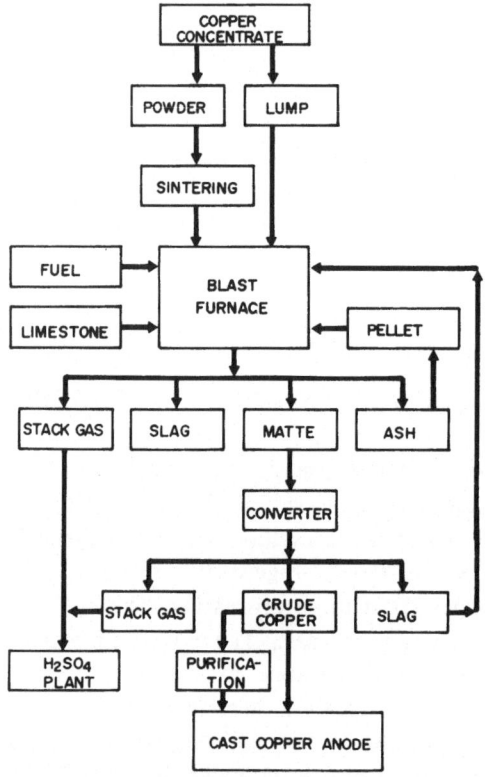

FIGURE 10.1. Flowsheet for copper concentrate processing.

TABLE 10.1

Examples of Chemical Analysis of Copper Concentrate, Matte, and Slag

	Concentrate	Matte	Slag
Au (g/ton)	3.2	13.5	0.07
Ag (g/ton)	149.0	537.6	3.74
Cu (%)	14.90	40.52	0.34
S (%)	31.92	23.90	0.75
Pb (%)	0.37	2.87	0.06
Zn (%)	1.38	2.27	3.21
Fe (%)	30.84	26.76	33.94
SiO$_2$ (%)	9.84		36.06
Al$_2$O$_3$ (%)			5.15
CaO (%)			7.52
MgO (%)			1.92
BaO (%)			0.24

TABLE 10.2

Analysis of the Anode Copper and the
Electrodeposited Copper in Percent by Weight

	Anode copper	Electrodeposited copper
Cu	98.4	99.99
Pb	0.55	0.0000
Bi	0.005	0.0000
As	0.025	0.0002
Sb	0.026	0.0004
Se	0.035	0.0000
S	0.035	0.0006
O	0.3–0.4	
Fe		0.0006
Ni		0.0000

cells are equipped with more than 1000 pairs of the anode and cathode in comparison with 50 pairs in conventional cells. The electrolyte solution consists of H_2SO_4-acidified cupric sulfate, and is circulated between the cell and a reservoir to control the solution composition and the temperature. The solution is also filtered continuously to remove anode slime.

Almost all impurities in the anode precipitate as slime, although some impurities such as Fe and Ni go into solution. The basic principle of electrochemical refining of copper is that these less-noble metal ions are not reduced, and only copper is obtained at the cathode. Table 10.2 shows an example of the chemical analysis of the anode copper and of the electrorefined copper.[1]

Since some silver, gold, and traces of platinum in the copper ore are contained in the anode slime, recovery of the noble metals is an important process for the copper industry. More than 70% of silver and almost 100% of gold in Japan are useful byproducts of electrochemical refining of copper.

10.1.1. Electrorefining of Copper

The anode consisting of crude copper and the starting cathode of pure copper sheet are positioned in the electrolytic solution of acidic copper sulfate. The anodic and the cathodic reactions in the cell are simple:

$$\text{At the anode} \qquad \text{Cu (crude)} \rightarrow \text{Cu}^{2+} + 2e \qquad (10.3)$$

$$\text{At the cathode} \qquad \text{Cu (pure)} \leftarrow \text{Cu}^{2+} + 2e \qquad (10.4)$$

$$\text{Overall reaction} \qquad \text{Cu (crude)} \rightarrow \text{Cu (pure)} \qquad (10.5)$$

TABLE 10.3

An Example of the Breakdown of the Terminal Voltage
of Copper Electrorefining Cell

Electrolyte composition	
H_2SO_4 (g/liter)	185.5
Cu (g/liter)	42.0
Fe (g/liter)	0.5
Ni (g/liter)	14.8
As (g/liter)	3.8
Operating temperature (°C)	50
Current density (A/dm^2)	1.9
Electrode gap (cm)	4
Anode overvoltage (V)	0.003
Cathode overvoltage (V)	0.022
Solution IR drop (V)	0.140
Terminal voltage (V)	0.165

It is clear that the anodic reaction is the same as the cathodic reaction, but in a reverse direction. Consequently, the Gibbs free energy change, ΔG, and the decomposition voltage, E_d, are zero so that the terminal voltage, and thus the energy consumption, of the copper electrorefining cell is low. Since the overvoltages, both anodic and cathodic, are relatively small, about 85 % of the terminal voltage is the IR drop in the electrolytic solution, as shown in Table 10.3, which represents an example of the voltage balance of the copper refining cell. The electrical conductivity of the mixed solution of $CuSO_4$ and H_2SO_4 increases with increase of H_2SO_4 and decrease of $CuSO_4$ in the concentration range of the electrolyte for the copper refining cell.

The anode copper is dissolved to supply Cu^{2+} in the electrolyte and the cathode grows by electrolysis, as illustrated in Fig. 10.2A and 10.2B. The electrode gap must be minimized to reduce the solution IR drop because of the largest factor of the terminal voltage, but short-circuiting due to formation of dendrites at the cathode during electrolysis is troublesome, as illustrated in Fig. 10.2C. A small amount of surface active agent such as natural glue and/or synthetic high polymers is added to the electrolyte to prevent dendrite formation. Organic surfactant is adsorbed on the cathode surface, and the overvoltage increases somewhat. A fine grain size of deposited metal results, and dendrite formation does not take place. Decrease of the anode-to-cathode gap spacing is a direct procedure for reducing the terminal voltage. Table 10.4 shows typical conditions for operation of the copper refining cell, representing the result of such efforts in a modern cell in comparison with conventional cell parameters.[1-3]

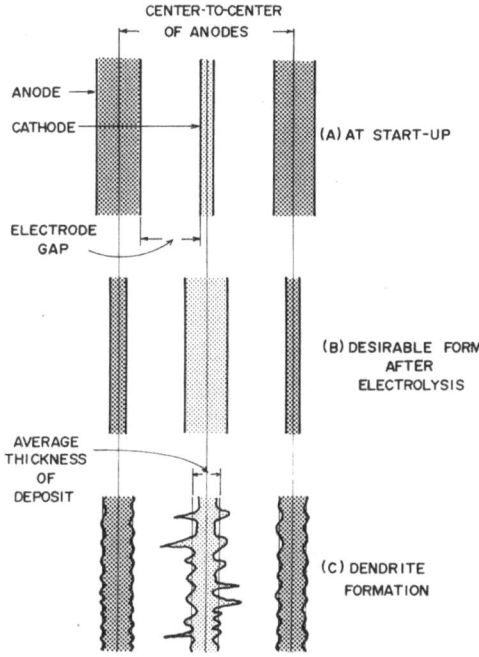

FIGURE 10.2. Changes of anode and cathode during electrolysis and morphology of electrodeposition.

TABLE 10.4
Operating Conditions of Copper Refining Cell

	Conventional cell	Modern cell
Electrolyte composition		
Cu content (g/liter)	35–49	40–50
Free acid (g/liter)	150–200	180–200
Temperature (°C)	50–55	60–65
Circuation rate of electrolyte (liter/min)	15–20	40–50
Amperage (kA)	5–6	20–30
Current density (A/dm^2)	1.5–2.0	2.0–3.0
Terminal voltage (V)	0.3–0.4	0.20–0.25
Current efficiency (%)	90–95	93–98
Energy consumption (ac kWh/ton)	420–500	280–300
Final weight of cathode (kg)	30–50	50–150
Initial weight of anode (kg)	100–200	300–450
Center-to-center distance of anodes (mm)	110–130	50–110
Anode life (days)	15–25	20–25
Percentage of anode scrap (%)	15–30	10–20
Slime formation (%)	1.0–1.5	0.3–0.5

The mechanism of the copper electrode process has been investigated extensively.[4,5] An outline of the reaction sequence is as follows:

$$Cu^{2+} + e \rightarrow Cu^{+} \tag{10.6}$$

$$Cu^{+} + e \rightleftarrows Cu_{ad} \tag{10.7}$$

$$Cu_{ad} \rightleftarrows Cu \text{ (crystal)} \tag{10.8}$$

$$2.3RT/a_a F = 40mV, \qquad 2.3RT/a_c F = 120mV$$

Both the charge transfer step (10.7) of Cu^{+} to surface-adsorbed, discharged Cu_{ad} and the crystallization step (10.8) on the cathode are estimated to be fast or in partial equilibrium so that the charge transfer step (10.6) of Cu^{2+} to Cu^{+} is rate-determining. Many physicochemical studies of the crystal growth of copper have also been made because of its importance as a model system for the investigation of electrocrystallization (see Ref. 6 in detail).

The cathode potential rapidly becomes more negative when the current density becomes very high. This polarization is caused by limitation of the mass transport of Cu^{2+} from the bulk of solution to the cathode surface due to slow diffusion and/or migration. Some electrode processes are controlled by a slow step of mass transfer, while others are controlled by the charge transfer step (see Section 3.3).

10.1.2. Electrochemical Winning of Copper

In general, copper ore is treated by a pyrometallurgical reduction process to obtain crude copper, which is further purified by electrolysis, as described in the preceding section. In some cases, however, hydrometallurgy is used to recover copper from low-grade ore, mostly in the U.S. Copper oxide containing ore is treated with sulfuric acid to leach copper. Pressurized ammonia solution is also used. In the case of sulfide mineral, the ore is roasted to convert into oxide prior to leaching. The solution is purified sufficiently to remove iron and other impurities which may have an effect on electrolysis, and is sent to the cell.

Calcined pyrite, a byproduct of SO_2 gas production in sulfuric acid manufacture, is used in iron smelting. Since a small amount of copper in the calcined ore is undesirable for steel making, it is leached, and the dilute solution of copper sulfate is recovered and is sent to electrolysis.[7,8]

Copper in the solution can be recovered by cathodic reduction, but a suitably insoluble anode must be used. Anodes of lead and its alloys are used in commercial cells containing sulfate solution. Pure lead is corrosion resistant, but its mechanical strength is insufficient. Antimony additions of 2–10% are used to improve the mechanical strength. Other metals such as Ag, Te, Sn, Si, Cu, and Ca are also alloyed with lead for this purpose.[9]

The anode surface is covered with $PbSO_4$ in sulfuric acid solution,

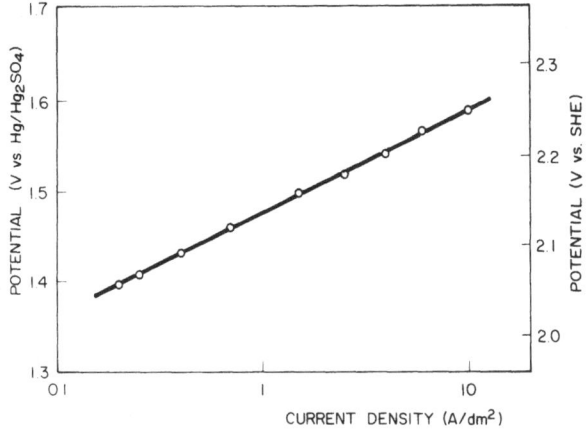

FIGURE 10.3. Anodic polarization curve of pure lead in 4 M H_2SO_4 at 30°C.

which is oxidized further to a thick brown layer of PbO_2 when the electrode is polarized at higher potentials. It is interesting that the $PbSO_4$ layer has a quite high resistivity, whereas the PbO_2 layer is a good conductor. The oxygen overvoltage of the lead anode, or most likely PbO_2, is very high, as shown in Fig. 10.3.

The solution composition and the operating conditions such as the current density and the solution temperature of the electrowinning cell are about the same as for the electrorefining cell. The cathodic process is, of course, the same as in the electrorefining cell, whereas the anodic reaction is oxygen formation instead of copper dissolution:

At the anode: $\quad H_2O \rightarrow 2H^+ + \frac{1}{2}O_2 + 2e \qquad E_A = 1.26 \text{ V} \quad (10.9)$

At the cathode: $\quad Cu \leftarrow Cu^{2+} + 2e \qquad\qquad E_C = 0.30 \text{ V} \quad (10.10)$

Overall reaction: $\quad Cu^{2+} + H_2O \rightarrow Cu + 2H^+ + \frac{1}{2}O_2 \quad E_d = 0.96 \text{ V} \quad (10.11)$

where E_A and E_C are the reversible potentials for the anode and the cathode reactions, respectively. The values have been corrected by the Nernst equation for the solution composition and temperature. It is clear that the Cu^{2+} concentration decreases and the H^+ concentration increases when electrolysis is carried out. Therefore, the depleted solution is recycled to leaching.

The voltage balance is as follows:

Decomposition voltage	0.96 V
Anode overvoltage	0.91 V
Cathode overvoltage	0.05 V
Solution IR drop	0.20 V
Terminal voltage	2.12 V

The terminal voltage of the electrowinning cell is very high compared to that of the electrorefining cell shown in Table 10.3, and hence the consumption of electrical energy for electrowinning is as large as 2,500–4,500 kWh/t-Cu depending on the flowsheet and the operating conditions. The largest factor is the decomposition voltage followed by the anode overvoltage. The sum of these factors is more than 88% of the total voltage.

It is well known that electrochemical plants are among the industries that heavily consume energy, and that energy costs sometimes exceed more than one-half of the manufacturing cost. The energy consumption, and thus the terminal voltage, for electrochemical processes, should be minimized for both industrial economy and energy conservation. Electrochemical winning of copper as well as other metals from raw ores is thus doubtful because of these factors. However, the process enables the conservation of resources, which today is also an important consideration. Therefore, feasibility studies for each process must be conducted in detail before planning and operation.

10.2. ELECTRODEPOSITION OF LESS NOBLE METALS

Hydrogen evolution must be considered as a competitive reaction when a less noble metal whose reversible potential is more negative than that of the hydrogen electrode process is cathodically deposited from an aqueous solution containing its ions:

$$2H^+ + 2e \rightarrow H_2 \tag{10.12}$$

$$M^{n+} + ne \rightarrow M \tag{10.13}$$

Assume that the reactions (10.12) and (10.13) may occur independently of each other for simplicity. The polarization curves for the reactions are illustrated in Fig. 10.4, where E_H and E_M are the reversible potentials of the hydrogen evolution reaction and of the metal deposition, respectively. The cathodic polarization for the hydrogen electrode process is shown by curve 1, which does not intersect with the polarization curve for the metal deposition, curve 2, if the hydrogen overvoltage on the metal under discussion is low. In this case, it is clear that only hydrogen evolution takes place at the cathode.

Curve 3, on the other hand, illustrates the polarization curve for hydrogen evolution when the hydrogen overvoltage is significantly high. Assume that i_H and i_M are the partial current densities for hydrogen evolution and for metal deposition, respectively, at the cathode potential E'. The current efficiency ξ for metal deposition is thus shown by the equation:

$$\xi = \frac{i_M}{i_H + i_M} = \frac{i_M}{i_T} \tag{10.14}$$

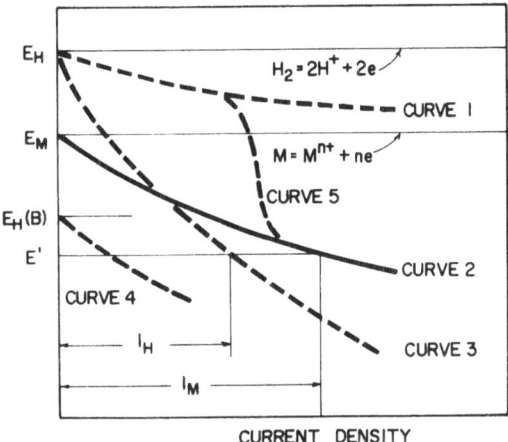

FIGURE 10.4. Cathodic polarization curves related to metal deposition.

where i_T is the overall current density, assumed to be the sum of i_H and i_M for simplicity.

There are two ways to increase the current efficiency: one is use of specific additives for increasing the hydrogen overvoltage, and the other is the use of high-pH solution as the electroplating bath.

Some ions and/or molecules, both inorganic and organic, adsorb and occupy the sites on the cathode surface that are otherwise active for the hydrogen electrode process, so that hydrogen evolution is retarded. Addition of surfactant to the electrolytic solution for copper electrorefining to obtain smooth deposits has been described in the preceding section. Although surfactants are used for the same purposes in some cases of electrodeposition of less noble metals, other surface-active agents are used to increase the hydrogen overvoltage and improve the current efficiency for metal deposition.

It is clear from the Nernst equation that the reversible potential for hydrogen evolution becomes less noble when the solution pH increases, that is, the reversible potential tends to the more negative direction by about 60 mV for each unit increase of the solution pH. Also, the hydrogen overvoltage increases with an increase of the pH due to an insufficient concentration of H^+ on the cathode surface, illustrated by curve 4 starting at $E_H(B)$ in Fig. 10.4. At high-current densities and in high-pH solutions, the diffusion-limited current density for hydrogen discharge appears, and the cathode potential tends sharply to the less noble direction. Of course, hydrogen evolution can take place even in alkaline solution via the different route:

$$H_2O + e = \tfrac{1}{2}H_2 + OH^- \tag{10.15}$$

However, the reversible potential for this reaction is very negative compared to that for the cathodic deposition of various metals. Since many metal ions may precipitate as oxide and/or hydroxide, there are limitations to the application of alkaline solution as an electroplating bath. Control of solution pH is thus an important procedure for eletroplating. Electrolytic solutions containing complex salts of the metal of interest are useful for increasing the current effciency, because the complex salt is soluble even in high-pH solutions.

Curve 5 is of interest since as the cathode potential tends to the negative direction away from curve 1, even in acidified solution, it intersects curve 2, and hence metal deposition may take place.

The polarization curves in Fig. 10.4 are illustrated with a simple assumption of no interference between the hydrogen electrode process and the zinc electrode process, for example. In highly pure acid solution, the polarization curve for hydrogen evolution can start from the thermodynamic potential E_H at zero current as shown by either curve 1 or curve 3. However, if zinc ions are in the solution, the situation is one described best as a mixed potential electrode. Because the Zn/Zn^{2+} electrode is more negative and more reversible than the hydrogen electrode, the mixed potential is close to the Zn/Zn^{2+} electrode potential by the concepts of corrosion theory. Thus the hydrogen electrode is already polarized significantly even at zero current in the outside circuit. The zinc metal surface should be covered by hydrogen, poised for vigorous hydrogen evolution. If the cathode is polarized more negative than the Zn/Zn^{2+} electrode potential, zinc ions in the electrolyte may be reduced on the cathode through the intermediate such as the absorbed metallic species, Zn_{ad}. These intermediates also cover the active sites on the cathode and retard hydrogen evolution. Consequently, zinc can deposit on the cathode with a very small amount of hydrogen evolution—thus, at high current efficiencies.

It is well known that the hydrogen electrode process is affected by the solution impurities, additives, and the intermediates of the reaction including metallic species in the electrochemical cells for metal processing, as shown by curve 5 in Fig. 10.4.

10.3. ELECTROLYTIC PRODUCTION OF PURE ZINC

Pure zinc is produced by two routes: an electrochemical process and a distillation process. Of these, electrolytic zinc accounts for more than 60% of the total production, and is one of the most advanced processes in the field of electrochemical metallurgy, especially in Japan.

The largest market for zinc is the production of brass, which requires high purity metals. Iron and lead are the most harmful impurities. Elec-

TABLE 10.5

Solution Composition and Current Density in Three Processes for Zinc Electrowinning

	Slightly acidified	Moderately acidified	Highly acidified
Composition (g/liter)			
Zn	20–50	55–60	80
H_2SO_4	45–120	150–160	220
Current density (A/dm^2)	2.7–4.0	5.5–7.3	10–11

trolytic zinc is free of these impurities, and is thus well suited for production of alloys. Some copper, nickel, cobalt, and cadmium can be recovered from the electrolyte and the anode slime of the electrolytic zinc cell, which is an added advantage of the electrochemical process. Cadmium is associated in nature with zinc and lead and there is no single route for its recovery alone, so the electrochemical process for zinc is a major source for making cadmium.

The most important zinc ore is sphalerite or zinc sulfide. Silicate and carbonate ores are also used. Sulfide ore is roasted to convert it to oxide, and zinc is leached with electrolytic solution recycled from the electrolyzer. The solution composition is 10% H_2SO_4 and 2.5% Zn before leaching. After leaching, the concentrated solution is purified prior to electrolysis. The electrolyte composition depends on local practice but it is classified into three types based on whether it is slightly, moderately, or highly acidified, as shown in Table 10.5.[10]

Although zinc is a considerably non-noble metal, the current efficiency for electrodeposition is high enough, say 90–95%, because of a very high hydrogen overvoltage on the zinc cathode, even in acidified solution, as described in the preceding section. For example, the logarithm of the exchange current density in A/cm^2 is -10.8 in $1N\,H_2SO_4$ at 20°C, and the Tafel slope is 120 mV/decade.[11] It is clear that the partial current density for hydrogen evolution, i_H, is much smaller than that for electrodeposition of zinc, i_{Zn}, as illustrated by curves 2 and 3 in Fig. 10.4. Consequently, the current efficiency for metal deposition might be affected seriously by any contamination in the solution which might lead to a decrease in the hydrogen overvoltage on the zinc cathode. Table 10.6 represents the allowable concentrations of impurities in the zinc electrowinning cell. The same concept applies to the cathode process of the amalgam-type chlorine cell, where the amalgam cathode is also strongly influenced by impurities in the electrolyte (see Section 7.2).

Because of hindrance to the exchange of electrons between the zinc surface and solvated protons in solution, hydrogen evolution is retarded.

TABLE 10.6

*Allowable Concentration of Impurities
in Zinc Electrowinning Cell*

Ge	0.01 mg/liter
Sb	0.02 mg/liter
Ni	0.5 mg/liter
As	1 mg/liter
Co	1 mg/liter
Cd	1 mg/liter
Cu	10 mg/liter
Fe	30 mg/liter
Mn	350 mg/liter
Cl	50 mg/liter

Simultaneous electroposition of zinc becomes possible at sufficiently negative potentials, as illustrated by curve 5 in Fig. 10.4. It is clear that operation at high current densities is important to keep the current efficiency high, although the terminal voltage increases. The electrode gap should be minimized, and the concentration of sulfuric acid must be increased as much as possible so as to reduce the solution *IR* drop. Accordingly, the process uses highly acidified solution. Since hydrogen evolution is stimulated by impurities in the solution, rigorous solution purification must be emphasized, although the allowable concentration of impurities depends on operating conditions and plant economics.

Lead or its alloys, say Pb-Ag alloy containing 0.5–1.0% Ag, are used as the anode, and the cathode is aluminum sheet. Deposited zinc is usually recovered from the cathode every two days or so. The distance between the anode and the next anode is as large as 50–100 mm in order to avoid short-circuiting caused by dendrite formation.

The reversible potential for the oxygen formation reaction at the anode is positive and relatively large and the oxygen overvoltage is also high. On the other hand, the reversible potential for electrodeposition of zinc is quite negative. as follows:

At the anode:	$H_2O = 2H^+ + \frac{1}{2}O_2 + 2e$	1.23 V vs. SHE
At the cathode:	$Zn = Zn^{2+} + 2e$	-0.77 V vs. SHE

Decomposition voltage:	2.00 V
Anode overvoltage:	1.0 V
Cathode overvoltage:	0.1 V
Solution *IR* drop:	0.5 V

Terminal voltage:	3.6 V

The terminal voltage of the zinc electrowinning cell is high, and hence the consumption of electric energy is 3,000–3,700 kWh/t-Zn.

10.4. INITIATION, GROWTH, AND MORPHOLOGY OF ELECTRODEPOSITED METAL

Mechanisms of electrodeposition of metal by cathodic reduction of metal ions followed by crystallization into a growing lattice have been investigated and proposed by many researchers (see Refs. 6 and 12). Hydrated metal ion, $M(H_2O)_x^{n+}$, is transported from the bulk of solution to the outside of the double layer through the diffusion layer on the cathode surface. Water molecules may be removed sequentially, and the ion discharges to deposit an adatom, M_{ad}, at the active site on the cathode surface.

The morphology of electrodeposited metal depends greatly on the operating conditions, and can be classified into various types. Some types of deposit differ greatly from those obtained by nonelectrochemical process, due to different mechanisms for initiation and growth of the metal crystal. The adatom moves on the cathode surface or step toward special sites such as edges and kinks, as shown in Fig. 10.5, according to Bockris.[6] It is considered that the formation of a nucleus on the step is somewhat difficult until most of the edges and/or kinks are filled up. Spiral growth may appear when the crystal growth proceeds along a screw dislocation. This type of crystal growth occurs frequently.

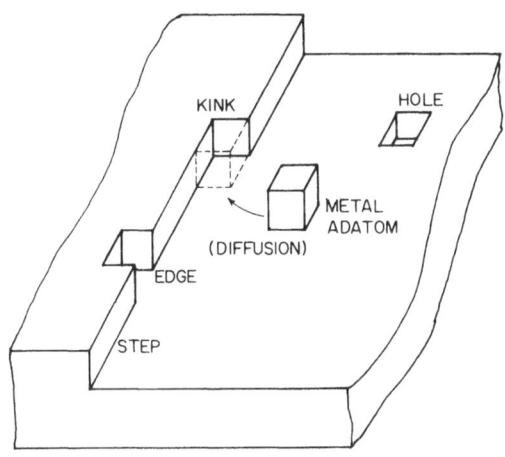

FIGURE 10.5. A model for electrodeposition of metal.

There are many types of crystal growth of electrodeposition of metals:

(1) pyramids
(2) layers
(3) blocks
(4) ridges
(5) cubic layers
(6) spiral
(7) whiskers
(8) dendrites
(9) texture

According to Bockris, types 3 through 8 are considered to be modifications of types 1 and 2.[6] If a polycrystal grows with systematic orientation, a macroscopic repetition of pattern, or texture, may appear on the cathode surface.

Morphology of electrodeposition is affected by several factors such as the solution composition including the pH and impurities, the operating temperature, the flow rate of the electrolyte, the cathode potential and over-voltage, the current density, and the substrate. In the case of copper deposition, for example, pyramids are obtained at relatively low current densities, while at high current densities, layers or polycrystalline deposits are obtained as shown in Fig. 10.6.[6]

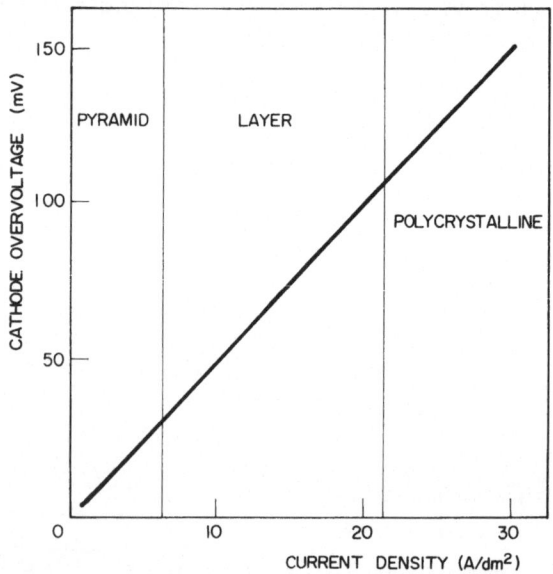

FIGURE 10.6. Correlation between polarization curve and growth form for copper electrocrystallization.

Since the local current density at the peaks of a cathode with a rough surface is larger than the average current density, the rate of electrodeposition is large compared to the average rate. By the same token, the deposition rate in the valleys between peaks on the rough surface is minimum. Consequently, the surface roughness is increased by electroplating.

It is believed that a surface active agent added to the solution becomes adsorbed on the cathode surface preferentially on the peaks but not in the valleys. By retarding charge transfer to metallic ions on the peaks, the agent causes leveling of the electrodeposited surface. The smooth plating characteristic is called the "throwing power" and is an important factor for electroplating.

Wagner obtained the current distribution on a cathode with a sine wave profile surface, and showed the throwing power to be a dimensionless function of parameters of the polarization curve, solution conductivity, and electrode dimensions as follows[13]:

$$\text{Throwing power} = h_c/\rho a \qquad (10.16)$$

where h_c is the slope of the polarization curve in V cm^2/A, ρ is the resistivity of the electrolyte in ohm cm, and $2a$ is the peak-to-peak distance of the sine wave profile in centimeters. In general, current distribution tends to be uniform when $h_c/\rho a > 1$.

REFERENCES

1. *Kagaku Binran* (Handbook of Chemistry), New edition, p. 1502, Maruzen, Tokyo (1958).
2. Panel discussion at the first meeting of Symposia on Electrochemical Industries, sponsored by Japan Society of Chemical Engineers, Osaka, June 12, 1979.
3. C. L. Mantell, *Electrochemical Engineering*, 4th ed., p. 149, McGraw-Hill, New York (1960).
4. E. Mattson and J. O'M. Bockris, *Trans. Faraday Soc.* **55**, 1586 (1959).
5. J. O'M. Bockris and H. Kita, *J. Electrochem. Soc.* **109**, 928 (1962).
6. J. O'M. Bockris and G. A. Razumney, *Fundamental Aspects of Electrocrystallization*, Plenum, New York (1967).
7. C. L. Mantell, *Electrochemical Engineering*, 4th ed., p. 198, McGraw-Hill, New York (1960).
8. R. Remirez. *Chem. Eng.* p. 96 (September 8, 1969).
9. A. T. Kuhn, ed. *Industrial Electrochemical Processes*, p. 535, Elsevier, Amsterdam (1971).
10. *Denki Kagaku Binran* (Handbook of Electrochemistry), 2nd ed., p. 971, Maruzen, Tokyo (1964).
11. H. Kita, *J. Electrochem. Soc.* **113**, 1095 (1966).
12. B. E. Conway and J. O'M. Bockris, *Electrochim. Acta* **3**, 340 (1961).
13. C. Wagner. *J. Electrochem. Soc.* **98**, 116 (1951).

Chapter 11

BATTERIES

Electrochemical operations consist essentially of two basic types: electrolysis and conversion of chemical energy to electricity. Electrolysis is a kind of heterogeneous catalytic reaction accompanied by the transfer of electric energy. On the other hand, a battery supplies electricity to an external load by means of conversion of internal chemical energy into electric energy. Certain batteries are also able to convert electric energy into chemical energy when electricity is supplied and can store it until an external load is applied.

Volta first discovered the phenomenon of generation of electricity accompanying chemical reaction in 1800. In 1836, the first practical battery, the Daniell cell, was invented and became a convenient dc power source. The battery is presumably one of the oldest types of equipment based on electrochemical operation.

In the earliest days of electrochemical operations, the battery was often used as a dc power source, e.g., for electroplating and water electrolysis, and it contributed greatly to the development of electrochemical science and technology.

Since the invention of the dynamo in 1875, the electrochemical process industries have been able to receive abundant electric energy at low cost, so that batteries are no longer used for this purpose. However, research and development of various types of batteries, either conventional or based on new concepts, have continued in response to the demands of a wide variety of markets. Today, batteries are utilized in various field, mostly as small-scale power sources. As a new development, the scale-up and industrial use of batteries is being reconsidered as a possible means for energy conservation through the storage of off-peak nighttime electric utility generation capacity.

There are two types of batteries: the primary battery, and the secondary battery, also called the storage battery. The primary cell is utilized one time only as an electric power source by the electrochemical conversion of chemical to electric energy, after which the cell ends its duty and life. On the other hand, the secondary battery can store electricity supplied by an outside dc power source, and can discharge the dc power on demand.

The continuous supply of reactants to the anode and cathode of an electrochemical cell identifies the fuel cell. The fuel cell may be considered as an energy converter or generator rather than a battery. In a similar fashion, the solar cell is also an energy converter that utilizes the sun's spectral energy as a source for conversion to direct electrical current.

While early work on fuel cells was done in the 19th century, full development took place in the 1960s as part of the U.S. space effort. Fuel such as hydrogen and hydrocarbons is brought to the anode compartment, where the fuel is oxidized electrochemically while an oxidation agent such as oxygen is reduced at the cathode. The chemical energy is thus converted into electric energy by electrochemical processes without a large amount of heat generation. In principle, the dc electric power can be obtained continuously if a sufficient amount of fuel is sent to the cell and if the reaction products are removed continuously. Hybrids between the fuel cell and the battery are possible. For example, an air-consuming cathode is used in the zinc–air battery. Also, periodic replacement of consumable negative electrodes may be considered as a mechanically recharged battery or an intermittently fed fuel cell when used with an air cathode.

Commercial batteries must have various characteristics and meet requirements such as

- large emf;
- low overvoltages, both anodic and cathodic;
- low IR drop in the electrolyte;
- minimum deviation of the terminal voltage during operation;
- large amount of electric energy for unit weight or volume of cell;
- minimum self-discharging;
- long life;
- low cost; and
- ease of operation.

The Daniell cell was one of the earliest to use zinc as an anode material, and zinc is still the best anode material in modern batteries such as the dry cell. The advantageous properties of zinc as a battery anode are that the reversible potential of zinc is sufficiently negative, and that the anodic dissolution reaction is relatively fast (low overvoltage). Also, a most important property is the very large overvoltage for the hydrogen evolution reaction, which helps to retard self-discharging of batteries, thus improving the energy efficiency. It is difficult to meet all of the requirements in practice, so that there is a limited selection of electrode processes suitable for battery electrodes.

11.1. LECLANCHÉ-TYPE BATTERIES

The Leclanché cell is called a dry cell since the electrolyte is a gelatinous paste containing $ZnCl_2$, NH_4Cl, and some starch. The anode and the cathode are zinc and manganese dioxide, respectively. The composition and the electrode processes are as follows:

(Negative pole) $Zn \mid ZnCl_2$, $NH_4Cl \mid MnO_2$–carbon (positive pole)   (11.1)

At the cathode (positive): $2MnOOH \leftarrow 2MnO_2 + 2H^+ + 2e$ (11.2)

At the anode (negative): $Zn \rightarrow Zn^{2+} + 2e$ (11.3)

$$Overall\ reaction: \quad Zn + 2MnO_2 + 2H^+ \rightarrow Zn^{2+} + 2MnOOH \quad (11.4)$$

or

$$Zn + 2NH_4Cl + 2MnO_2 \rightarrow Zn(NH_3)_2Cl_2 + Mn_2O_3 \cdot H_2O \quad (11.5)$$

Zinc sheet is stamped to fabricate the anode can, or container. Since zinc is somewhat brittle, it is heat-treated to recover ductility after cold work. The ductility and cold working properties are improved by the addition of a small amount of lead, say 0.3%, to the zinc. Impurities such as Fe, Cu, As, Sb, Bi, Sn, and Mg are harmful by accelerating localized corrosion or self-discharge so that zinc of greater than 99.9% purity must be used.

The surface of the zinc anode is frequently amalgamated by addition of a small amount of $HgCl_2$ into the electrolyte so as to eliminate localized corrosion. The amalgamation of the zinc surface increases the hydrogen overvoltage and produces an associated decrease of self-discharge current.

Carbon powder is mixed with manganese dioxide particles so as to increase surface area and the electric conductivity. Electrolytically produced manganese dioxide is more conductive and more suitable as active mass than is naturally occurring material because it has many lattice defects.

The electrolytic production of manganese dioxide begins with the treatment of rhodochrosite, $MnCO_3$, with sulfuric acid to obtain manganese sulfate solution. This electrolyte is purified enough to remove iron, zinc, and other impurities. Electrolysis is conducted with lead or graphite anodes at 80°C to obtain a manganese dioxide deposit on the anode surface with a current density in the range 0.5–1.5 A/dm^2. The anode product is removed from the substrate and is crushed to a powder size of about 200 mesh. The product is complex in that manganese dioxide has several types of crystal structure such as α, β, and γ, and various oxidation states. The reversible potential of the β-MnO_2/Mn^{2+} system is stable, whereas the potential of the γ-MnO_2 fluctuates because γ-MnO_2 can form solid solutions with lower oxides in various types. The overvoltage of the γ-MnO_2 electrode is lower than that of α-MnO_2.[1,2] In an electrolytic solution of pH = 7.5, α-MnO_2 is reduced to $MnO_{1.66}$ by discharge reaction, but γ-MnO_2 is changed to $MnO_{1.57}$. By x-ray analysis, the discharge reaction may take place below the surface of the oxide, causing an expansion of the lattice. The expansion was less in α-MnO_2 than in γ-MnO_2.[2] Kozawa stated that γ-MnO_2 prepared by electrodeposition from $MnSO_4$ solution had a large capacity for ion exchange reaction compared to β-MnO_2 prepared by thermal decomposition of $Mn(NO_3)_2$.[3] In conclusion, γ-MnO_2 is the best material or structure as the cathode for the Leclanché-type primary cell.

According to Korver et al.[4] and Vosburgh et al.,[5] the product of the cathodic process is MnOOH [see reaction (11.6)], and the oxyhydroxide converts into Mn^{2+} and MnO_2 in acidic solution by means of the disproportionation reaction (11.7). In neutral solutions, further reduction may take place to form $Mn(OH)_2$ and Mn^{2+} [reactions (11.8) and (11.9)].

$$MnO_2 + H^+ + e \rightarrow MnOOH \qquad (11.6)$$

$$2MnOOH + 2H^+ \rightarrow MnO_2 + Mn^{2+} + 2H_2O \qquad (11.7)$$

$$MnOOH + H^+ + e \rightarrow Mn(OH)_2 \qquad (11.8)$$

$$Mn(OH)_2 + 2H^+ \rightarrow Mn^{2+} + 2H_2O \qquad (11.9)$$

On the other hand, MnO_2 is reduced to $MnO_{1.5}$ in alkaline solutions by means of a solid-phase reaction, and the intermediate is further reduced to $MnO_{1.0}$ by means of the heterogeneous reaction as shown in Fig. 11.1, suggested by Kozawa.[6] Mn^{4+} in MnO_2 is reduced to Mn^{3+}, and Mn^{3+} transfers to the site of adjacent Mn^{4+}. H_2O is decomposed to produce a proton and OH^- simultaneously at the interface between solid and liquid.

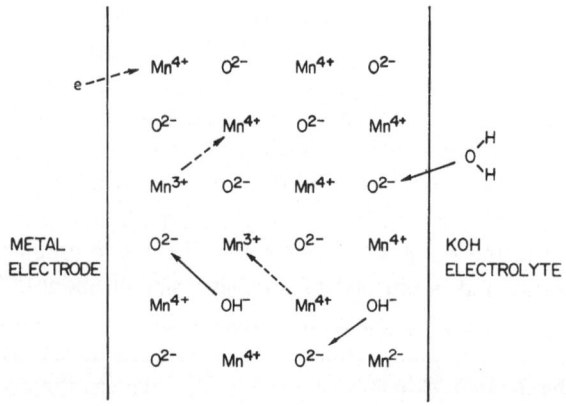

FIGURE 11.1. Kozawa's model of electrical conduction in MnO_2.

OH^- transfers to the sites of O^{2-} in the lattice. That is, a sequence of reactions and steps as

$$MnO_2 + H_2O + e \rightarrow MnOOH + OH^- \tag{11.10}$$

$$MnOOH(s) \rightarrow Mn(III) \text{ (in the solid)} \tag{11.11}$$

$$Mn(III) + e \rightarrow Mn(II) \text{ (cathodic reduction)} \tag{11.12}$$

$$Mn(II) \text{ (in the solid)} \rightarrow Mn(OH)_2(s) \tag{11.13}$$

NH_4^+ in the electrolyte of the manganese dioxide battery is thought to be a proton-donor against MnOOH. NH_3, which is produced by charge transfer of NH_4^+, is combined with Zn^{2+} to form a complex ion.[7]

$$2MnOOH + 2NH_4^+ \rightarrow Mn^{2+} + MnO_2 + 2H_2O + 2NH_3 \tag{11.14}$$

$$Zn^{2+} + nNH_3 \rightarrow Zn(NH_3)_n^{2+} \tag{11.15}$$

11.2. LEAD–ACID BATTERIES

Lead–acid batteries are widely utilized in various fields such as automobiles, trains, and communication systems. The construction of the cell and the electrode processes are as follows:

After charge: (negative pole) $Pb \mid H_2SO_4 \, aq \mid PbO_2(Pb)$ (positive pole)

$$\tag{11.16}$$

After discharge: $(Pb) \, PbSO_4 \mid H_2SO_4 \, aq \mid PbSO_4(Pb) \tag{11.17}$

Positive pole: $PbSO_4 + 2H_2O \underset{\text{discharging}}{\overset{\text{charging}}{\rightleftharpoons}} PbO_2 + 4H^+ + SO_4^{2-} + 2e$

$$\tag{11.18}$$

Negative pole: $Pb + SO_4^{2-} \underset{\text{discharging}}{\overset{\text{charging}}{\rightleftharpoons}} PbSO_4 + 2e \tag{11.19}$

Overall reaction $2PbSO_4 + 2H_2O \underset{\text{discharging}}{\overset{\text{charging}}{\rightleftharpoons}} PbO_2 + Pb + 2H_2SO_4$

$$\tag{11.20}$$

Thus, the emf, E_f, is shown by the equation below in which pure solid phases are considered to have unit activity:

$$E_{PbO_2} = E^0_{PbO_2} + \frac{RT}{2F} \ln \frac{(a_{H^+})^4 (a^{2-}_{SO_4})}{(a_{H_2O})^2}$$

$$E_{Pb} = E^0_{Pb} - \frac{RT}{2F} \ln (a_{SO_4^{2-}})$$

$$E_{PbO_2} - E_{Pb} = (E^0_{PbO_2} - E^0_{Pb}) + \frac{RT}{2F} \ln \frac{(a_{H^+})^4 (a_{SO_4^{2-}})^2}{(a_{H_2O})^2}$$

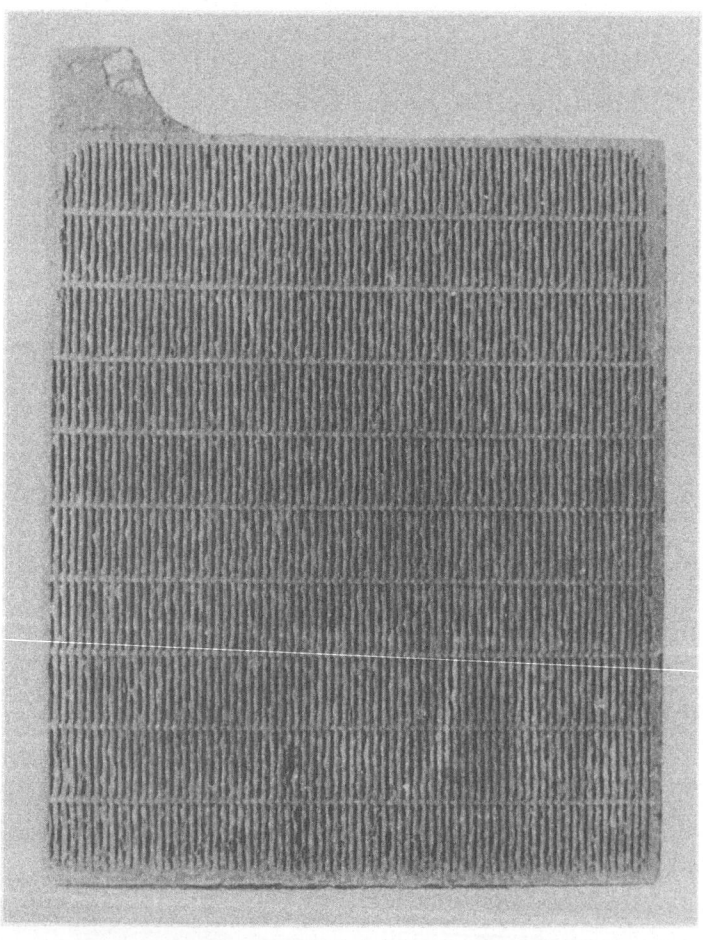

FIGURE 11.2. Tudor plate electrode for lead–acid battery. (Courtesy of Nippon Battery Co., Kyoto.)

The standard emf may be designated E_f^0:

$$E_f^0 = E_{PbO_2}^0 - E_{Pb}^0$$

Also, the mean activity is composed of ion activities, as follows:

$$a_{\pm} = [(a_{H^+})^2 \cdot (a_{SO_4^{2-}})]^{1/3}$$

Therefore,

$$E_f = E_{PbO_2} - E_{Pb} = E_f^0 + \frac{RT}{F} \ln \left[\frac{(a_{\pm})^3}{a_{H_2O}} \right] \qquad (11.21)$$

$$E_f^0 = 2.0184 \text{ V at } 25°C$$

In order to obtain a large area of electrode surface, and hence a large capacity, per unit volume, two types of electrode configuration have been developed: the Plante-type and the paste-type. The former is a solid plate having an uneven surface shown in Fig. 11.2, and is called the Tudor plate. Another style of the Plante-type electrode is the chloride-type plate which consists of a perforated plate into which spiral ribbons are plugged to obtain a large surface area. The paste-type electrode consists of a grid made of Pb-Sb alloy, filled initially with a paste made of lead oxide powder mixed with sulfuric acid. Since the paste-type electrode has a very large area compared with the Plante electrode, the paste plate design is used almost exclusively.

One of the problems with the paste electrode, especially with the positive plate, is degradation caused by deformation due to charging and discharging, as shown in Fig. 11.3. Stress due to expansion of PbO_2 produced by repetition of charging and discharging is thought to be a main reason for failure. The phenomenon is caused by the formation of clusters consisting of PbO_2 and SbO_x from corrosion of the alloy grid.[8] To avoid such problems, Pb-Ca alloy containing less than 0.1% of Ca has been investigated and applied.

The paste for the positive electrode consists of a mixture of lead powder and litharge, PbO. The electrode is formed by anodic oxidation to PbO_2 in sulfuric acid solution. The crystal structure and the chemical composition depend on the conditions of anodic oxidation, that is, there are two types of crystal structure of PbO_2: the orthorhombic α-PbO_2 and the tetragonal or rutile β-PbO_2.[9] The proportion of these structure is considered to be an important factor for the characteristics of the battery positive, although the picture is not complete. The PbO_2 crystallites grow with repetition of charging and discharging, with associated degradation of cell performance.[10]

Formation of $PbSO_4$ of low electrical conductivity is a problem of the negative electrode which causes reduction of battery capacity. Carbon black,

FIGURE 11.3. A positive electrode after failure. (Courtesy of Nippon Battery Co., Kyoto.)

barium sulfate, and lignins are mixed in the paste to prevent growth of the PbSO$_4$ crystals. Lignins might adsorb selectively on the electrode surface, and retard initiation of nuclei and growth of undesirable PbSO$_4$ forms.

The anodic oxidation of Pb to PbSO$_4$ takes place at potentials more positive than ca. −0.5 V vs. SHE in sulfuric acid, and further oxidation from PbSO$_4$ to PbO$_2$ occurs at +1.5 V, as shown in Fig. 11.4. There is measurable solubility of PbSO$_4$ in sulfuric acid, say 10^{-4} mol/liter, whereas PbO$_2$ is almost insoluble. Although the charge transfer reaction at the negative takes place on the lead substrate, the possibility exists for the slightly soluble lead sulfate to participate during charging. Soluble lead sulfate may also participate in the oxidation of PbSO$_4$ to PbO$_2$ at the positive electrode during charging. It is well known that there are various kinds of oxide, hydroxide, and basic salts of lead depending on the solution composition, pH, temperature, and the electrode potential or the degree of oxidation of

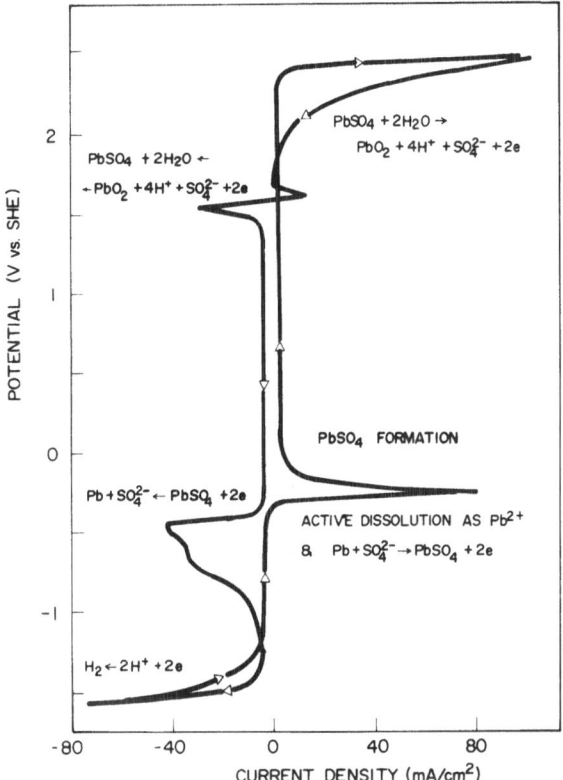

FIGURE 11.4. Polarization curve of Pb in deaerated 1 M H₂SO₄ at 40°C.

environment. The profile of these oxides on the Pb electrode in sulfuric acid
has been illustrated by Burbank, as shown in Fig. 11.5.[11]

The lead electrode is covered by the $PbSO_4$ layer in a wide range of
operating conditions, and the layer rejects SO_4^{2-} from permeation whereas
H^+ and H_2O can pass through. At positive potentials, H^+ and H_2O
penetrate the $PbSO_4$ layer to the lead substrate, and the deposited layer is
oxidized to oxides. Therefore, the thickness of the $PbSO_4$ layer decreases
with an increase of the potential and the layer becomes more porous. The
oxide layer at positive potentials consists of both orthorhombic and
tetragonal phases and is a nonstoichiometric compound, shown as PbO_t. At
more positive potentials, PbO_2 appears in the oxide layer, and the
composition changes to $PbO_t \cdot xPbO_2$, which represents various oxides such
as Pb_2O_3, Pb_3O_4, Pb_5O_8, Pb_7O_{11}, and others. These oxides are considered
to be mixtures of PbO and PbO_2. The oxide may tend to PbO_2 when the
electrode is polarized at more positive potentials. It is interesting that the
inner layer adjacent to the Pb substrate is α-PbO_2, whereas β-PbO_2 forms at

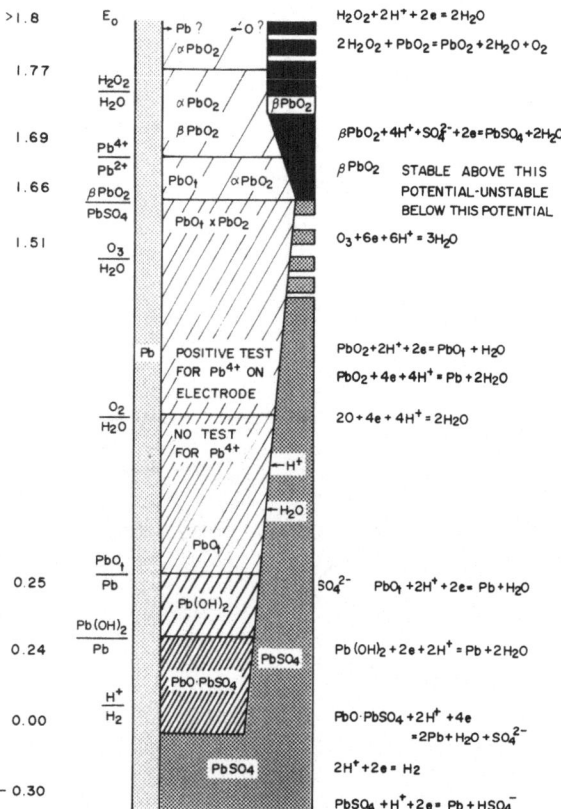

FIGURE 11.5. The structures formed on lead in H_2SO_4 as functions of the electrode potential vs. SHE. (By permission of The Electrochemical Society, Inc.)

the outside of the oxide layer next to the solution. The specific resistance of β-PbO_2, probably a semiconductor, is in the order of 10^{-4} ohm cm compared to 10^{-3} ohm cm for α-PbO_2.

11.3. ALKALI STORAGE BATTERIES

The electrolyte in alkali storage batteries is 20–25% KOH, and the solution sometimes contains a small amount of LiOH to improve cell performance. The positive electrode consists of nickel oxide. The Edison cell uses iron powder as the negative electrode material, while the Jungner cell uses cadmium. Because of the large hydrogen overvoltage of cadmium, self-discharge is limited. Cell performance is also improved and almost all alkali storage batteries today use cadmium electrodes.

The construction and the electrode processes of the Jungner-type alkali storage battery are as follows:

After charging:

$$\text{(negative pole)} \; Cd \,|\, KOH \; aq. \,|\, NiOOH(Ni) \; \text{(positive pole)} \quad (11.22)$$

After discharging:

$$(Cd) \; Cd(OH)_2 \,|\, KOH \; aq. \,|\, Ni(OH)_2(Ni) \quad (11.23)$$

At the positive pole:

$$2Ni(OH)_2 + 2OH^- \underset{\text{discharging}}{\overset{\text{charging}}{\rightleftarrows}} 2Ni(OH)_3 + 2e \quad (11.24)$$

or

$$2Ni(OH)_2 + 2OH^- \underset{\text{discharging}}{\overset{\text{charging}}{\rightleftarrows}} 2NiOOH + 2H_2O + 2e \quad (11.25)$$

At the negative pole:

$$Cd + 2OH^- \underset{\text{discharging}}{\overset{\text{charging}}{\rightleftarrows}} Cd(OH)_2 + 2e \quad (11.26)$$

Overall reaction:

$$2Ni(OH)_2 + Cd(OH)_2 \underset{\text{discharging}}{\overset{\text{charging}}{\rightleftarrows}} Cd + 2Ni(OH)_3 \quad (11.27)$$

or

$$2Ni(OH)_2 + Cd(OH)_2 \underset{\text{discharging}}{\overset{\text{charging}}{\rightleftarrows}} Cd + 2NiOOH + 2H_2O \quad (11.28)$$

For the Edison cell, the structure and the reactions are the same as for the Jungner cell, if Fe substitutes for Cd.

The efficiency for charge transfer for the positive electrode may substantially increase if the active mass consisting of $Ni(OH)_2$ can be oxidized further into $Ni(IV)$ rather than $Ni(III)$ during charging, although the electrode is oxidized only to $Ni(III)$ in practical cells. The active mass consists of Ni(III), Ni(IV), OH⁻, and O^{2-}, and has a structure of a nonstoichiometric semiconductive material.

A conventional alkali cell uses the pocket-type electrode in which the active mass, consisting of metal or metal oxide powder, is packed into a pocket made of a perforated-iron (or nickel-electroplated) sheet, shown in Fig. 11.6. The effective surface area of this electrode is relatively small, and the internal resistance is large. The sintered electrode illustrated in Fig. 11.7 has a large surface area because of microporosity, and hence the utilization efficiency of the active mass is high. Nickel or nickel oxide powder is deposited on a perforated substrate of nickel electrodeposited on iron sheet, and is sintered at high temperatures.

FIGURE 11.6. A pocket-type electrode for an alkali battery. (Courtesy of Nippon Battery Co., Kyoto.)

FIGURE 11.7. A sintered plate electrode for alkali battery. (Courtesy of Nippon Battery Co., Kyoto.)

11.4. FUEL CELLS AND APPLICATION OF FUEL CELL CONCEPTS TO CHEMICAL PROCESSES

According to Liebhafsky and Douglas, "A fuel cell is an electrochemical device in which the chemical energy of a conventional fuel is converted directly and usefully into low-voltage direct-current electrical energy."[12] Since research and development of fuel cells have proceeded over the last 20 years, the hardware, the performance, and the applications of the fuel cell have changed extensively. However, the definition of the fuel cell shown above remains valid.

There is worldwide awareness and concern that natural resources for energy such as fossil fuels may be nearly depleted by the end of this century. Thus, energy conservation and its effective use are worldwide tasks. Fuel cell concepts offer several possibilities as an important procedure for this purpose.

Some fuel cells have already been used in special applications such as space flight and to power beacon lamps. There is long-range interest in the feasibility of the fuel cell as a commercial-scale energy source, requiring the development of a very large fuel cell which can be operated efficiently with cheap fuel. The so-called "TARGET" project in the United States is an example. There are also strong incentives for the operation of large-scale fuel cells in the chemical process industries, especially in electrochemical plants where a large amount of electricity is utilized for chemical processing. A variety of byproducts of chemical processing could be used as the feedstock of the fuel cell. Hydrogen from chlor-alkali cells, for example, is a candidate. Thus, it would be of great advantage if a sufficient amount of the dc electric power requirement could be generated and supplied by the on-site fuel cell station.

The enthalpy of H_2O from H_2 and O_2 is large:

$$H_2 + \tfrac{1}{2}O_2 = H_2O + \Delta H$$
$$\Delta H = -68.3 \text{ kcal/mol} \tag{11.29}$$

It is impossible to recover all of the enthalpy shown by ΔH as electric energy because there is a loss based on the entropy change according to thermodynamic principles. However, the electric energy equivalent to the Gibbs free energy, ΔG, can be utilized:

$$E_f = -\frac{\Delta G}{nF} = -\frac{\Delta H - T\Delta S}{nF} \tag{11.30}$$

At 25°C or 298 K, $\Delta S = -39.0$ cal/mol K. Substituting ΔS into Eq. (11.30),

$$E_f = \frac{68.3 - 298 \times 0.039}{2 \times 23.06} = 1.23 \text{ V} \tag{11.31}$$

That is, the emf, E_f, of the hydrogen–oxygen fuel cell based on reaction (11.29) is calculated to be 1.23 V.

The maximum efficiency for energy conversion, ζ_{max}, is

$$\zeta_{max} = \frac{\Delta G}{\Delta H} \tag{11.32}$$

For the hydrogen–oxygen fuel cell, for example,

$$\zeta_{max} = \frac{56.7}{68.3} = 0.83$$

The theoretical conversion efficiency of 83% is large compared to other systems, and is a basic reason for interest in fuel cell concepts. The conversion efficiency from thermal energy to electric energy in a modern electric power station of, say 10^6-kW capacity, is only about 40%, for example.

Of course, there are some losses of energy in the fuel cell under operation, so that we cannot recover ΔG completely. The terminal voltage lowers from E_f to V due to the internal resistance, which is the sum of the overvoltages and the IR drops in the cell, when the fuel cell is operated at the total current I. Also, some current inefficiencies caused by the side reactions and the physical losses of fuel must be considered. Therefore, the actual efficiency for energy conversion, ζ_a, is

$$\zeta_a = \frac{[\Delta G - nF(E_f - V)]\xi}{\Delta H} = \frac{nFV\xi}{\Delta H} \qquad (11.33)$$

where ξ is the current efficiency.

There are many innovations for stimulating the gas electrode reactions because the overvoltage of the gas electrode is a major factor of the inefficiency for energy conversion in this system.

Use of the porous electrode having a three-phase zone, shown as a model in Fig. 11.8, is an example. The porous electrode is treated with a waterproofing agent to prevent permeation of aqueous electrolyte through micropores. Gases, either hydrogen or oxygen, are able to come from the backside of the electrode (gas phase) through the holes to contact the electrolyte. Thus, solid carbon, aqueous electrolytic solution, and fuel gas meet close to the front surface of electrode next to the solution, at a region called the three-phase zone.

An appropriate electrode catalyst is applied at the three-phase zone to stimulate the electrochemical reaction. The elements of the platinum group

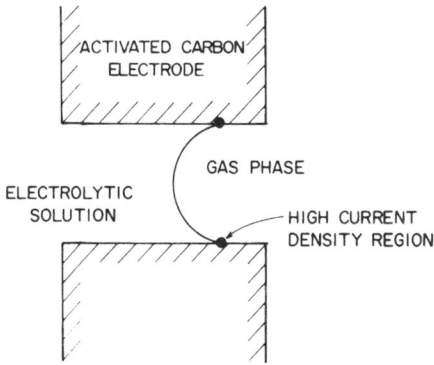

FIGURE 11.8. The three-phase zone of a fuel cell electrode.

metals such as Pt, Ir, and Os are used as the catalyst for the hydrogen electrode, whereas the Ib group elements such as Cu and Ag are thought to be
useful for the oxygen electrode.

Extensive studies on catalysts have been conducted. Marko and
Kordesch, for example, has developed unique materials consisting of Co-Al
or Fe-Mn-Ag as low-cost and efficient catalysts for the oxygen cathode
process.[13]

It is interesting and important that fuel cell technology also has some
possibilities for recovery of waste heat in the chemical process industries.
Recovery of the waste heat generated as the formation energy of HCl from
H_2 and Cl_2 is an example. Generally, hydrogen is burned with chlorine in the
reactor to produce hydrochloric acid. A large amount of heat from the
process must be cooled by water and is wasted. Useful electric energy of a
great extent could be recovered if hydrogen and chlorine could be sent to a
fuel-cell-type reactor, although there are some difficulties for fuel cell
technology with acid solutions. Application of the oxygen cathode in chlor-
alkali cells, diaphragm-type or membrane-type, is another example.

REFERENCES

1. C. C. Liang, Manganese, in *Encyclopedia of Electrochemistry of the Elements*, edited by
   A. J. Bard, Vol. 1, p. 377, Marcel Dekker, New York (1973).
2. W. C. Vosburgh and P. S. Lau, *J. Electrochem. Soc.* **108**, 485 (1961).
3. A. Kozawa, *J. Electrochem. Soc.* **106**, 552 (1959).
4. M. P. Korver, R. S. Johnson, and N. C. Cahoon, *J. Electrochem. Soc.* **107**, 587 (1960).
5. W. C. Vosburgh, M. J. Pribble, A. Kozawa, and A. Sam, *J. Electrochem. Soc.* **105**, 1
   (1958).
6. A. Kozawa and R. A. Powers, *J. Electrochem. Soc.* **113**, 870 (1966).
7. C. C. Liang, Manganese, in *Encyclopedia of Electrochemistry of the Elements*, edited by
   A. I. Bard, Vo. 1, p. 385, Marcel Dekker, New York (1973).
8. A. C. Simon, *J. Electrochem. Soc.* **114**, 1 (1967).
9. W. Mindt, *J. Electrochem. Soc.* **116**, 1076 (1969).
10. T. F. Sharpe, *J. Electrochem. Soc.* **116**, 1639 (1969).
11. J. Burbank, *J. Electrochem. Soc.* **106**, 369 (1959).
12. H. A. Liebhafsky and D. L. Douglas, in *Fuel Cells*, edited by G. J. Young, p. 1, Reinhold,
    New York (1960).
13. A. Marko and K. Kordesch, U. S. Patents 2,615,932 and 2,669,598.

III ELECTROCHEMICAL
ENGINEERING

Chapter 12

CONFIGURATION OF ELECTROLYZERS

12.1. CONFIGURATION AND TYPE OF ELECTROLYZERS

12.1.1. Case Study of Configuration of Chlor-Alkali Cells

As an example of cell configuration, we may consider chlor-alkali cells and the amalgam cell in particular. The chlor-alkali industry has steadily improved the design of this type of cell to have a large capacity and its technology is well developed.

The older style amalgam cells operated in Japan more than 40 years ago were made of concrete. The side wall of the cell was lined with ceramic tile to prevent corrosion by acidic brine. The cell cover was also made of concrete, with its inside surface coated with asphalt. The Krebs amalgam cell was composed of paired long channels in parallel rows, one being be electrolytic cell and the other the amalgam decomposition cell. In an alternate design, the Osaka soda cell consisted of a single box divided by a partition into two compartments to provide both electrolysis and amalgam decomposition zones.

Improvement by the use of rubber-lined steel cells was introduced in Germany in the early 1940s and the amalgam cell technology has been since studied and improved extensively in the United States, Europe, and Japan.

Although the cell bottom was at one time rubber-lined, the cathode of the modern cell is composed of bare steel plate with considerably low carbon content. The surface of the steel plate becomes uniformly covered by a thin layer of sodium amalgam if the sodium content in the amalgam is higher than 0.005%. Incidentally, wettability of iron or steel by pure mercury is negligible, especially for high-carbon steel.

The thickness of the amalgam layer on the steel-plate cathode is from about 2 to 6 mm. Cell operation becomes easier with an increase of the amalgam thickness, although the mercury inventory increases. The flow velocity of the amalgam ranges from 10 to 30 cm/sec, depending on the slope of the cell bottom. The velocity of mercury and/or sodium amalgam is almost zero at the cathode plate because of wetting of the steel by the

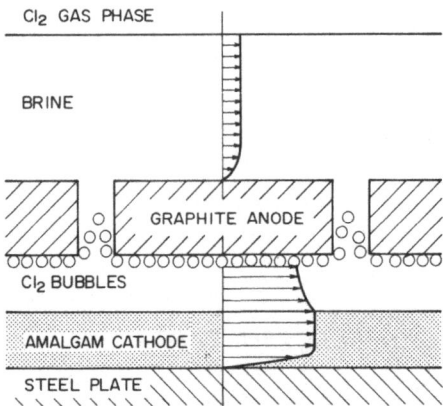

FIGURE 12.1. Flow pattern of brine and amalgam cathode.

amalgam. Therefore, there is a rapid variation of flow velocity in the mercury pependicular to the steel plate, as shown in Fig. 12.1. This causes vigorous agitation of the amalgam and enhances diffusion of sodium in the amalgam.

Since the purified brine is fed to the cell at a relatively small rate, the flow velocity of brine is very small well into the bulk of the solution, especially on the top of the graphite anode, where the solution is agitated by chlorine bubbles.[1] The amalgam flow accompanies the brine in the interelectrode gap, however. This helps to remove chlorine bubbles from the downward facing anode, and the *IR* drop due to the bubble effect is reduced.

The device for adjusting the interelectrode gap is of importance for the amalgam cell. The thickness of amalgam layer is affected by various factors such as the slope of cell, the mercury inventory, the sodium content, the current density, the cathode potential, and other factors. So-called "amalgam butter", which is voluminous, forms when impure brine is fed to the cell, affects flow distribution, and also causes unwanted short-circuiting with the anode. The mercury flow ripples occasionally, thus changing the gap locally. The graphite anode is consumed by electrolysis with a consumption rate of 1.5–2.0 kg/ton NaOH, depending on the material and the operating conditions.[2] These many factors may cause changes of the anode-to-cathode gap in the amalgam cell during operation.

The electrode gap should be minimized to keep the cell voltage low, but troubles due to short-circuiting must be eliminated. Metal anodes such as the DSA do not tolerate short-circuiting as well compared to graphite anodes. On the other hand, the current efficiency decreases somewhat when the interelectrode gap becomes too small, probably due to enhancement of the unwanted recombination reaction between sodium amalgam and chlorine. Consequently, devices and methods for anode adjustment to maintain an

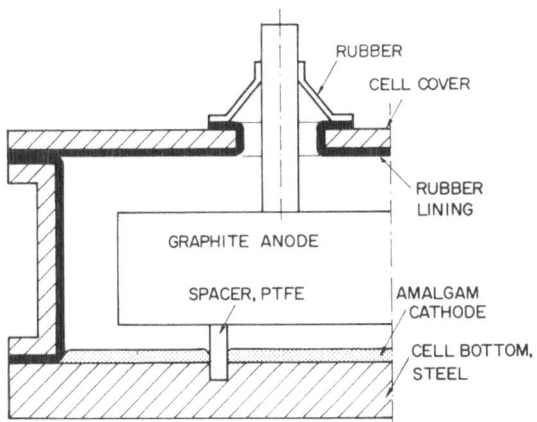

FIGURE 12.2. Device for anode adjustment.

adequate gap distance have been investigated extensively. A simple method, the first tried in chlorine cells, is illustrated in Fig. 12.2. Glass or PTFE plates are positioned on the steel cathode bed as spacers, and the graphite anodes are placed upon them. The anodes move down gradually by gravity when the working surface is consumed by electrolysis, so that the anode-to-cathode gap is automatically kept constant. Such a procedure was widely applied in practice in the 1960s. The spacer also enables the better regulation of any uneven flow of mercury, especially in a wide cell. On the other hand, the mercury flow stagnates in the vicinity of the spacer, and the flow of graphite particles is also retarded. This local build-up of graphite particles lost from the anode causes unwanted decomposition of sodium amalgam in the chlorine cell. As the amalgam cell was scaled up, the cell cover was also modernized. A number of graphite anode plates were suspended from the supporter located above the cell, as shown in Fig. 12.3. The supporter itself is moved up and down by a computer-controlled electric motor driven system which thus maintains the electrode gap constant or within a suitable range.

The diaphragm-type chlorine cell has been improved extensively, beginning in the 1970s with the introduction of DSA type metal anodes. The use of lightweight cell covers made of fiberglas-reinforced plastics (FRP) has also been a factor in the modernization of the diaphragm cell. Figure 12.4(a) shows an old-style cell in which graphite anodes were positioned on the concrete cell bottom with lead shielded by pitch to prevent corrosion. Cell design was changed greatly when metal anodes were introduced, as shown in Fig. 12.4(b). The anode, made of expanded titanium mesh coated with noble metal oxide, is supported by a titanium-clad copper bar, and the copper lead-

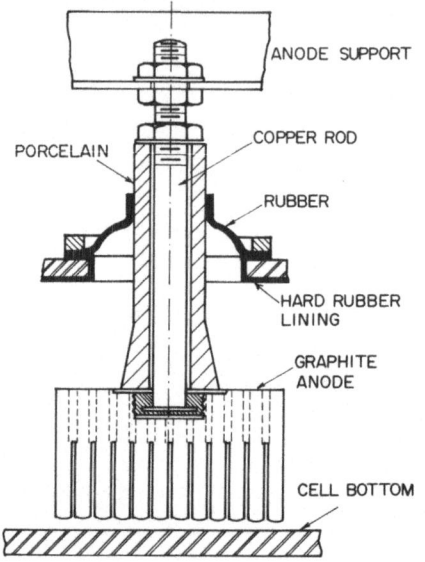

FIGURE 12.3. Device for anode adjustment of Olin E-510 cell.

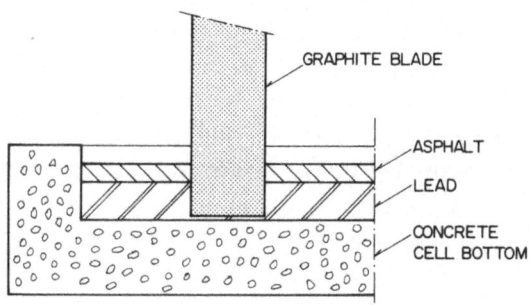

(A) GRAPHITE ANODE WITH CONCRETE BOTTOM

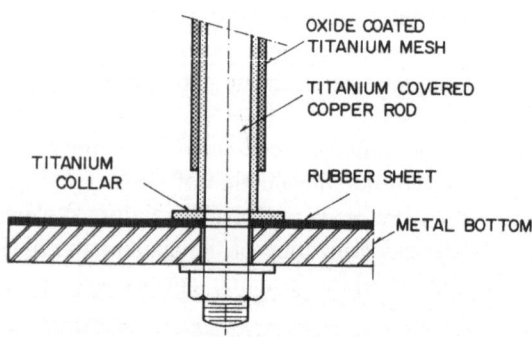

(B) METAL ANODE WITH METAL BOTTOM

FIGURE 12.4. Configurations of anode arrangement of diaphragm-type chlorine cell.

in bar is fixed to the cell bottom which is made of heavy copper or steel plate. Since the anode box structure serves as the brine downcomer during electrolysis, the space between the cathode fingers needed in the older-style cell is no longer necessary, permitting a simple cathode configuration to be used, as described in Chapter 8.

The vertical-type rotating-disk cathode cell is a unique amalgam cell. Designed by I. G. Farbenindustrie AG in Germany during World War II, it was operated by Chemische Werke Hüls AG until 1964. This type of cell, no longer in use, is described in various articles.[3,4] Table 12.1 summarizes dimensions and operating conditions of the German cell.[5]

In 1945, Professor Okada and associates at Kyoto University began the development of a rotating-cathode cell, similar to the I. G. Farben cell, with the support of the Japan Soda Industry Association. Three laboratory cells of 100, 200, and 3000 A capacity and a 10-kA pilot cell were designed and operated. Finally, a 20 kA prototype cell shown in Figs. 12.5 and 12.6 was fabricated at the Osaka Works of the Asahi Glass Company.[6] A large consumption of graphite anodes, and hence an increase of the cell voltage, was a serious problem with these cells because of difficulty with the anode adjustment. In a 20-kA cell, pairs of graphite anode plates were positioned between two cathode plates, as illustrated in Fig. 12.6. The graphite plates were moved toward the counter cathodes by a screw mechanism to minimize the anode-to-cathode gap as the anode was being consumed. This device worked well although it was somewhat complicated. Such difficulties could be readily avoided if metal anodes were used. Low current efficiency was another problem with these cells. The I. G. Farben–Hüls cell was operated at only 86% current efficiency, as shown in Table 12.1, probably because of vigorous agitation of the chlorine containing brine and unwanted contact with sodium amalgam. Difficulties with the vertical rotating cathode cell

TABLE 12.1

Operating Conditions and Performance of the
Rotating-Cathode-Type Amalgam Cell Operated by Chemische Werke Hüls AG

Number of cells	190	Graphite consumption (kg/ton Cl_2)	4
Amperage (kA)	49	Operating temperature (°C)	80–85
Current density (A/dm^2)	45	Amalgam concentration (% Na)	0.2
Cell voltage including bus bar (V)	4.7–5.0	Current efficiency (%)	86
IR drop in bus bar, estimated (V)	0.3	Hydrogen content in cell gas (%)	1.2–2.0
Diameter of cathode disk (m)	1.845	Brine concentration (g/liter)	
Number of disks per cell	4	Feed	300
Rotating speed (RPM)	7	Return	275
Number of graphite plates	40	pH of feed brine	5–7

FIGURE 12.5. A 20-kA vertical-type rotating-disk cathode prototype cell.

have prevented its widespread adoption, and hence there is no practical cell
of this type presently in use in either Germany or Japan.

Another unique cell is the horizontal-type rotating-cathode cell
developed by Asahi Glass Company, as shown in Figs. 12.7 and 12.8.[7]
Mercury is fed to the center of the rotating-steel-disk cathode, where it flows
to the circumference of the disk by centrifugal force. Sodium amalgam is
collected and sent to the vertical decomposer by gravity. Mercury from the
decomposer is pumped to the cell top for recirculation. The anode
adjustment is made in the same manner as for the conventional horizontal
mercury cell described previously in Fig. 12.3. Since the brine containing

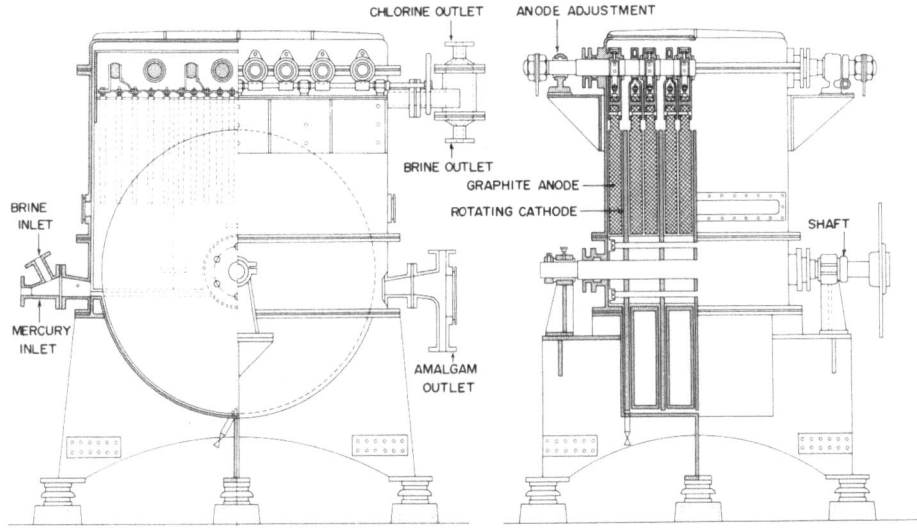

FIGURE 12.6. A 20-kA vertical-type rotating-disk cathode cell.

FIGURE 12.7. The horizontal-type rotating-cathode cells in a plant. (Courtesy of Asahi Glass Company.)

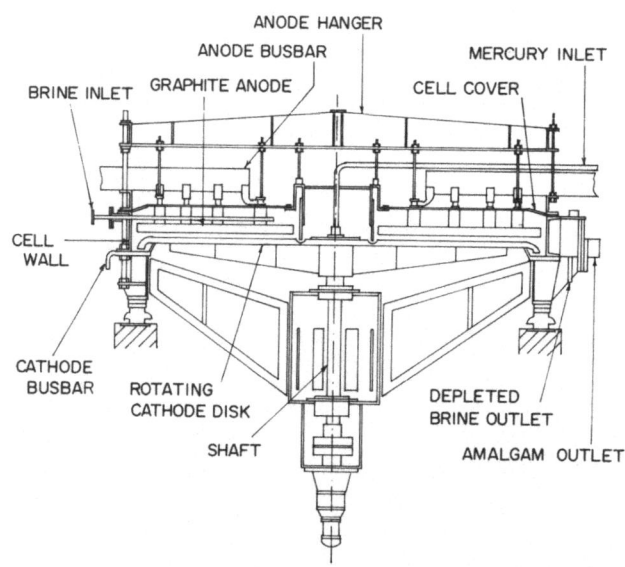

FIGURE 12.8. The horizontal-type rotating-disk cathode chlorine cell.

chlorine bubbles flows quickly along with the mercury flow on the rotating cathode, the bubble effect is greatly reduced. Therefore, the cell can be operated at low voltage even at relatively high current densities. A typical value of cell voltage is 4.5 V at 72.7 A/dm².

12.1.2. Monopolar Cells vs. Bipolar Cells

How to enlarge the electrode surface per unit cell volume or unit floor space is a question of great importance for cell design since the electrochemical process is a heterogeneous reaction occurring on the electrode surface, but not in the bulk of solution. The traditional horizontal mercury-type chlorine cell is an inadequate style from this view point, and probably motivated the innovation of the vertical-type rotating-cathode cell developed in Germany (Fig. 12.5). Some horizontal-type diaphragm cells have been operated previously, but almost all such cells were shut down. Modern diaphragm-type chlorine cells are vertical and many anode–cathode pairs are fabricated in a stack.

There are two methods for connecting the electric circuit: one is the monopolar and the other the bipolar manner, as shown in Fig. 12.9. In the monopolar cell, each side of a cathode plate functions as a cathode, and the electric current flows to the cathode plates in parallel. The same applies to the anode plates, i.e., both sides operate as anodes and anode plates are in

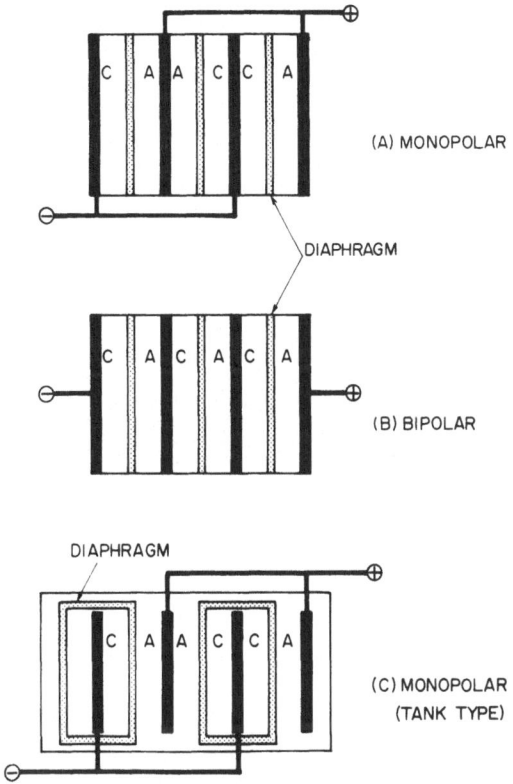

FIGURE 12.9. Monopolar cell and bipolar cell arrangements.

parallel. In the bipolar cell, one side of an electrode plate works as the anode, and the other side as the cathode. The electric current flows in series. As shown in Fig. 12.9, the surface areas of the working electrodes in both cases are the same. The total current flowing in the monopolar cell is nI_1 and the terminal voltage is V since a stack consists of n elementary cells in parallel, where I_1 is the current passing through an element. On the other hand, the current passing through the bipolar cell is I_1, but the terminal voltage is nV. As a slight variation of the monopolar arrangement, pairs of anodes and cathodes are positioned in a tank as shown in Fig. 12.9(c), with the current supply carried out in parallel. The Hooker and Diamond Shamrock chlorine cells are examples of this type, called the "tank-type cell". The lead–acid automobile battery is a monopolar tank-type cell with the separators to prevent short-circuiting of two electrodes. The diaphragm is not necessary in this case.

Table 12.2 shows a simple comparison between the monopolar and the bipolar arrangements from the viewpoints of electrical connection and of

TABLE 12.2
Comparison between Monopolar Cell and Bipolar Cell

	Monopolar cell	Bipolar cell
Current	nI_1; heavy bus bar is required	I_1; leakage between compartments must be avoided
Cell voltage	V plus some IR drops in conductor	nV; IR drop in hardware is negligible
Rectifier	Large amperage at low voltage	Small current at high voltage
	Expensive	Relatively cheap
Safety problems	Preferable due to low voltage	Dangerous due to high voltage
Design and fabrication	Simple	Relatively simple
Handling of feedstock and products	Simple	Complicated
Maintenance	Simple	Complicated

KEY: n = number of cells per circuit; I_1 = current through the electrode surface; V = cell voltage.

operation. There is little difference to choose between the two styles. The bipolar cell has merit from the viewpoint of electrical connection, whereas operation of the monopolar cell is easier. Selection is a matter of the designer's choice. The Pintsch Bamag water electrolysis cell is bipolar-type, but the Hitachi cell is monopolar. The Hooker and Diamond Shamrock chlorine cells are monopolar, while the Glanor cell jointly developed by DeNora and PPG is bipolar (see Fig. 8.9). The Asahi Glass membrane cell is monopolar, whereas many other cells, such as the Asahi Chemical, Hooker, and Tokuyama Soda cells, are bipolar, with a few exceptions.

Modern diaphragm-type chlorine cells, for example the Hooker H-type, use the deposited asbestos diaphragm. but it is difficult to arrange sheet diaphragm material for use in these cells. Old-style chlorine cells such as LeSueur, Townsend, and Allen–Moore were equipped with asbestos sheet or fabric as diaphragm, but have been abandoned because assembly of the fabric diaphragms on electrodes is extremely labor intensive. Water electrolysis cells and HCl cells are usually equipped with fiber fabric diaphragms so as to separate gaseous products at the anode and the cathode, hence the filter-press type cell configuration is preferable for fabrication and maintenance. Chlor-alkali cells using ion-exchange membranes are of the filter-press type for the same reason.

The cell configuration is related closely to supply of feedstock and recovery of products. The diaphragm-type chlorine cells, such as the Hooker cell, are the tank monopolar type in which a number of anodes are positioned. Accordingly, the feed brine is fed to the cell from a single inlet, and the cell gas or chlorine is carried out from a single outlet. The cathode fingers are welded to the hollow shell, in which hydrogen and the catholyte are collected. Handling of feedstock and products for these cells thus is simple. In the bipolar cell, on the other hand, supply of feedstock and recovery of products must be provided individually to each element, and hence the design becomes somewhat more complicated and introduces pathways for leakage of electric current. The Glanor cell has a large box or header on the top so as to distribute the brine to each cell element through nozzles with minimum current leakage. Chlorine gas is collected in the header, and is separated from the brine.

Generally, a number of electrolytic cells are electrically connected in series in a plant, whereas the feedstock is supplied and the products are recovered in parallel. Therefore, it is important to minimize leakage of electricity through the electrolyte and/or products from the viewpoints of safety and economy. A number of innovations and improvements have been introduced for this purpose. Designs of the caustic outlets of the amalgam decomposer and the diaphragm cell are examples. Caustic soda solution, some 50% in concentration, produced in the amalgam decomposer is brought to the overflow cylinder and falls into the receiver drop by drop.

Generally, the caustic receiver is rubber-lined and is connected to the main pipe with a rubber hose to minimize leakage current. In the case of the diaphragm cell, the catholyte effluent is taken out through the caustic percolation pipe to the current interrupter which is located at the top of the funnel, as shown in Fig. 8.4.

12.1.3. Some Cell Configurations of Interest

It is believed that organic electrochmistry has a bright future both as a fundamental field of science and in practical applications. In general, the electrical conductivity of solutions is low and reaction rates are small. These problems may cause difficulty in cell design and operation and have prompted investigations for minimizing the electrode gap and increasing the electrode surface. The pumped slurry cell, the packed bed cell, and the fluidized bed cell are examples (see Fig. 12.10).[8]

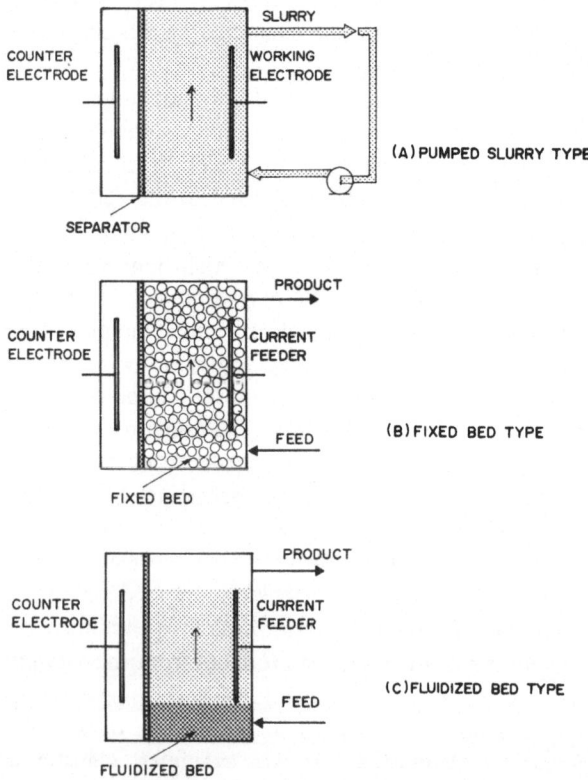

FIGURE 12.10. Various configurations of electrolyzer having a large electrode area.

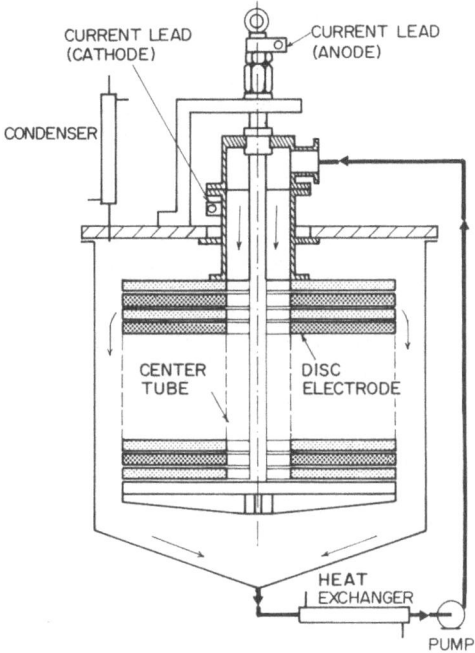

FIGURE 12.11. BASF capillary gap cell.

A unique design, named the "capillary gap cell", has been developed by BASF for electrosynthesis of adiponitrile (ADN) from acrylonitrile (AN) and is shown in Fig. 12.11.[9,10] Monsanto had developed the electrochemical synthetic process for making ADN with a filter-press type cell equipped with ion-exchange membranes.[11,12]

Beck, at BASF, developed a simple cell in which the membrane can be omitted and solution removed quickly to prevent anodic oxidation of the ADN cathode product as a result of good solution flow. Beck initially electrolyzed a mixed solution of 40% AN, 34% tetraethylammonium-p-toluene sulfonate, and 26% H_2O, but the ammonium salt was decomposed and the anode was seriously corroded. When the concentration of tetra ammonium salt in the electrolyte was decreased, the electrical conductance became very low: 10^{-3} ohm^{-1} cm^{-1}. The novel cell design which evolved minimized the eletrode gap as much as possible. The BASF capillary gap cell illustrated in Fig. 12.11 has 100 pairs of graphite plates of 40-cm diam with an electrode gap of only 0.2 mm. The cathode is bare graphite, while the anode plate is coated with a thin layer of electroplated lead dioxide. The mixed solution composed of 55% AN, 28.5% isopropanol, 16% H_2O, and 0.5% tetraethyl ammoniummethyl sulfate (pH = 3.5) was electrolyzed at 30–35°C and at

7–10 A/dm^2. The yield of ADN was about 90% with an energy consumption of 3 kWh/kg.

Bennion and co-workers studied porous, fixed, flow-through carbon electrodes to recover metals such as copper from dilute waste solutions.[13–15] A pair of packed beds of porous carbon or graphite are located in the electrolytic cell. An ion-exchange membrane is provided between the two-bed electrodes if necessary. The solution to be treated is fed to the cell with about 99% of the feed solution passing through the cathode bed, and the remainder through the anode bed. Metallic ions in the solution can be reduced even deep within the cathode bed. Polarity of the anode and cathode may be reversed and electrolysis continued when the cathode bed becomes plugged by metal deposit. Deposited metal is then dissolved anodically, and the concentration of metallic ions in the anolyte effluent is high enough for electrowinning of the metal.

Electrolysis with packed-bed electrodes has been investigated by several authors.[16–26] Coeuret has proposed a parameter K for evaluating the packed bed electrode, using several simplifying assumptions, with the result[21]:

$$K^2 = \frac{\text{electrochemical reaction rate at } L = 0}{\text{flux of electrical conduction in the electrolyte}}$$

$$= i_0 \left(\frac{nF}{RT}\right)(1 - \varepsilon)\, a_g \left(\frac{L^2}{\kappa}\right) \tag{12.1}$$

where i_0 is the exchange current density, ε is the void fraction, a_g is the surface area to volume ratio of particles ($=6/d_g$), d_g is the diameter of particle, L is the height of the packed bed, and κ is the conductivity of the electrolyte.

The effectiveness of the electrode, ξ, is shown by the equation:

$$\xi = \frac{\tanh K}{K} \tag{12.2}$$

When K is small, the overvoltage is practically constant, and the entire surface within the volume of the packed bed is active to work as an electrode: that is, the packed bed becomes a three-dimensional electrode. On the contrary, only the outer zone ($L = 0$) of the bed works as electrode when K is large. Consequently, the dimensionless factor K should be minimized by appropriate methods such as the use of large-size particles, shallow bed depth L in the direction parallel to electric current, and electrolyte having good conductance, and a decrease of the exchange current density.

When the electrolytic solution is pumped into the bottom of the packed bed at a high rate, the so-called "fluidized bed" is established.[24] The

TABLE 12.3

Specific Electrode Areas

Cell configuration	Area per unit volume (m^2/m^3)
Filter press	0.97
Filter press with diaphragm	0.35
Hooker chlorine cell	0.037
Capillary gap cell	4
Packed bed electrode	75
Fluidized bed electrode	75
Bipolar packed bed cells	20

particles may work as an electrode if an adequate current feeder is provided in the bed. The surface area of the working electrode in the fluidized bed cell is established to be very large and therefore suitable for electroorganic processes for which the current density is usually small.[25] In other words, the chief virtue of the fluidized bed lies in the mass transport being large. The fluidized bed cell has also been studied to recover metals from dilute solutions.[26-28] The mass transfer problems in these electrolytic systems have also been investigated.[29,30]

According to Fleischmann, a very large area of working electrode is provided in packed bed cells of various types in comparison to that in conventional cell types, as listed in Table 12.3.[25] King has designed a rod cell consisting of a number of vertical rows of electrically conducting rods, each one being separated from those above and below by a small gap. The electrolyte is fed to the top rods, flows downward over the vertical rows, and is collected from the bottom rows for recirculation. Electrical current is supplied to and removed from the cell through the top and bottom rods, respectively, with the intervening rods taking up a bipolar mode as the current passes alternately through the rods and the electrolyte in the gap.[31] Houghton and Kuhn have reviewed the configuration of various types of electrolytic cells.[32]

12.2. DIAPHRAGMS AND SEPARATORS

An electrolytic cell consists essentially of the anode compartment and the cathode compartment separated from each other by the diaphragm. In some cases, it is desirable to eliminate the diaphragm for the purpose of reducing the cell voltage. For example, the amalgam-type chlorine cell has no diaphragm. The cell design in this case is simple, and is considered to be an important reason behind modern advancement of the amalgam cell.

The anode bag in the metal electrorefining cell and also in the electroplating cell prevents dispersion of the anode slime into the electrolytic solution. The separator in the lead–acid storage battery prevents short-circuiting between electrodes. However, the most important role of the electrolytic diaphragm is physical separation of products at both anode and cathode, with a minimum voltage drop through itself.

Asbestos is used in several forms such as sheet or "asbestos paper" and fabric. Of these, a vacuum-deposited layer about 3 mm thick on the steel cathode screen in the chlorine cell is most popular, as described in Chapter 8. The anolyte consists of concentrated NaCl saturated with active chlorine at one side of the diaphragm. The catholyte is a caustic solution containing NaCl at the other side. Since there is a head of solution between the two compartments, the anolyte passes through the diaphragm to the cathode compartment. Dissolved chlorine which accompanies the anolyte flow is reduced to Cl^- at the cathode. Back migration of a part of the OH^- from the cathode compartment to the anode compartment may take place due to a large difference of the concentration between the two compartments. This results not only in current loss, but also causes unsuitable formation of hypochlorite and/or chlorate at the anode compartment, as follows:

$$H_2O + Cl_2 = HClO + H^+ + Cl^- \tag{12.3}$$

$$2NaOH + Cl_2 = NaClO + NaCl + H_2O \tag{12.4}$$

$$2HClO + ClO^- = ClO_3^- + 2Cl^- + 2H^+ \tag{12.5}$$

$$6ClO^- + 3H_2O = 2ClO_3^- + 4Cl^- + 6H^+ + \tfrac{3}{2}O_2 + 6e \tag{12.6}$$

$$6HClO + 3H_2O = 2ClO_3^- + 4Cl^- + \tfrac{3}{2}O_2 + 6e \tag{12.7}$$

According to reactions (12.3) and (12.4), chlorine and caustic soda will convert into hypochlorous acid and/or hypochlorite when concentrated NaCl solution is electrolyzed without a diaphragm. Such cells were once operated as on-site bleach generators in fiber industries.

In chlor-alkali cells, a suitable diaphragm should be provided to separate chlorine generated at the anode from caustic soda in the cathode compartment. The dissolved chlorine concentration in the anode compartment is low due to cell operation at high temperatures of 90°C or higher, and formation of hypochlorite is thus minimized. To minimize back migration of hydroxyl ions in the cathode compartment of the chlor-alkali cell, the anolyte is flowed to the cathode side through the diaphragm by gravity. Therefore, the asbestos diaphragm in this cell behaves as a filter cloth.

Persulfate can be produced by electrolysis of acidified ammonium sulfate solution with a platinum sheet anode, which has a very high oxygen overvoltage. A microporous porcelain diaphragm almost completely retards the permeation of anolyte through to the cathode compartment. Of course, the electrolytic current can flow through although the IR drop is very large. Subsequently, the anolyte containing persulfate is distilled under a vacuum to obtain hydrogen peroxide. The effluent is recycled to the electrolytic cell. The process is as follows:

$$2SO_4^{2-} = S_2O_8^{2-} + 2e$$

$$2HSO_4^- = 2H^+ + S_2O_8^{2-} + 2e$$ $\left.\begin{array}{l}\\ \\ \end{array}\right\}$ anodic oxidation

$$S_2O_8^{2-} + 2H_2O = 2H^+ + 2SO_4^{2-} + H_2O_2 \qquad \text{vacuum distillation}$$

Electrolysis of sodium sulfate solution to obtain caustic soda and sulfuric acid has been investigated by many authors, but is a difficult technique in practice.[33,34] Hydrogen evolution or OH^- formation is the main reaction at the cathode, the same as in the chlor-alkali cell. Oxygen evolves from the anode, and the sulfuric acid concentration in the anode compartment increases. Caustic soda and sulfuric acid are separated by a thin layer of diaphragm, which requires pores of minimum size, and hence a large IR drop through it.

The use of a diaphragm is best avoided if possible because design, fabrication, and operation of the cell are thus complicated, and consumption of electric power increases because of the IR drop.

The electrolytic separators and diaphragms may be classified into three groups:

(1) porous diaphragms of chlor-alkali cells, HCl cells, etc.;

(2) ion exchange membranes for dialysis, chlorine cells, waste water treatment, etc.; and

(3) battery separators to prevent short-circuiting.

Table 12.4 lists a variety of electrolytic separators and applications.

The requirements for ideal electrolytic separators and diaphragms are somewhat in conflict as pointed out by Jackson et al.[35] Some of these are:

● It should permit easy passage of electricity, but not permeation of molecular species.

● The void fraction should be large to minimize the IR drop.

● The pore size should be small to reject permeation of bubbles.

● Uniform properties of material, both physical and chemical, are required so as to maintain uniform current distribution and high current efficiency.

● Nonconducting material is recommended, otherwise a separator located between two electrodes will behave as a bipolar electrode.

● Diaphragms and separators should resist oxidation when in contact with the positive electrode or its oxidizing products.

● It should not be expensive for practical application, and should be easily available in the market.

Generally, there are several materials which are corrosion resistant to acid solutions, although it is rather difficult to find suitable materials for alkaline solutions. One of the best materials in alkali is chrysotile asbestos, which consists of magnesium silicate in fiber form. It is a useful material as diaphragm for water electrolyzer and chlorine cells.

Of the various types of asbestos minerals such as serpentines and amphiboles, chrysotile (of the serpentine group) is recommended as diaphragm materials in chlor-alkali cells. It is not resistant to acid, however.

12.2.1. Electrochemical Systems Having a Liquid Junction Potential

In diaphragm-type chlor-alkali cells, the anolyte and the catholyte are separated from each other by a thin layer of diaphragm. The composition and the concentrations of these electrolytic solutions are quite different. Liquid junctions are frequently encountered in electrochemical systems.

We may consider that the concentration and the activity of the ionic species j under discussion are C_j and a_j, respectively, on the left side of the

TABLE 12.4
Electrolytic Separators

	Examples of application
Functions	
Separator	Lead–acid batteries
	Anode bag in electroplating bath
Diaphragm	Chlor-alkali cells
	Electroorganic synthesis cells
	HCl cells
Ion-exchange membrane	Chlor-alkali cells
	Electroorganic synthesis cells
	Electrodialysis cells
Materials and forms	
Asbestos	Fabric, sheet, and vacuum-deposited mat
Ceramics	Cylinder, box, and sheet
Fiber fabric	Natural and synthetic
Metallic porous sheet and mesh	Mostly used for molten salt electrolysis
Nonmetallic porous sheet and	Microporous sheet for chlorine cell
mesh	Rubber and synthetic resin

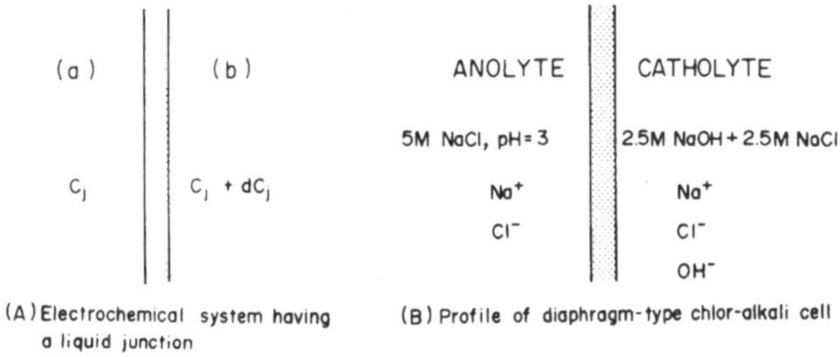

FIGURE 12.12. Model of electrochemical system having a liquid junction.

separator, while $C_j \pm dC_j$ and $a_j \pm da_j$ are on the right side, as illustrated in Fig. 12.12(A). The Gibbs free energy change $d \Delta G$ due to passing a charge of $1F$ is represented as follows

$$d \Delta G = \sum \frac{t_j}{n_j} d\mu_j \tag{12.8}$$

where t_j is the transference number, n_j the number of charges, and μ_j is the chemical potential of species j. Therefore, the potential change dE_L based on $d \Delta G$ is

$$dE_L = -\frac{d \Delta G}{F} = -\frac{RT}{F} \sum \frac{t_j}{n_j} d \ln a_j \tag{12.9}$$

and by integrating, we have

$$E_L = -\frac{RT}{F} \int_A^B \sum \frac{t_j}{n_j} d \ln a_j \tag{12.10}$$

and E_L is called the liquid junction potential.
    Consider a simple case:

$$HCl(C_A) \,|\, HCl(C_B)$$

Since $n_{H^+} = 1$, $n_{Cl^-} = 1$, $t_{H^+} + t_{Cl^-} = 1$, and $a_{H^+} = a_{Cl^-} = a_\pm$, we have

$$E_L = \frac{RT}{F} \left[t_{H^+}\{\ln a_{H^+}(A) - \ln a_{H^+}(B)\} \right.$$

$$\left. - t_{Cl^-}\{\ln a_{Cl^-}(A) - \ln a_{Cl^-}(B)\} \right]$$

$$= \frac{RT}{F} (1 - 2t_{Cl^-}) \ln \frac{a_\pm(A)}{a_\pm(B)} \tag{12.11}$$

Assume that $t_{H^+} = 0.85$ and $t_{Cl^-} = 0.15$. E_L is calculated to be about 41 mV for a decade of difference of the HCl concentration at both sides of the separator for 25°C.

In the case of NaCl solution, E_L is small, about 13 mV/decade, since t_{Na^+} is close to t_{Cl^-}: $t_{Na^+} = 0.39$ and $t_{Cl^-} = 0.61$. The liquid junction potential for the system composed of KCl solution is almost zero over a wide range of concentration because t_{K^+} is almost equal to t_{Cl^-} and is the reason that KCl solution is widely used in electrochemical experiments as a salt bridge.

A more complicated system has been described by Henderson with the result:[36]

$$E_L = \frac{RT}{F} \frac{\sum (u_j/n_j)[C_j(B) - C_j(A)]}{\sum u_j [C_j(B) - C_j(A)]} \ln \frac{\sum C_j(A) u_j}{\sum C_j(B) u_j} \qquad (12.12)$$

where u_j is the mobility of species j. For a $1-1$ valence electrolyte, Eq. (12.12) becomes

$$E_L = (2t_+ - 1) \frac{RT}{F} \ln \frac{C(A)}{C(B)} \qquad (12.13)$$

which agrees with Eq. (12.11).

If the system consists of different electrolytes at the same concentration such as

$$HCl(C) \mid KCl(C)$$

we obtain E_L as follows where $HCl = A$ and $KCl = B$:

$$E_L = \frac{RT}{F} \ln \frac{u_+(A) + u_-(A)}{u_+(B) + u_-(B)} \qquad (12.14)$$

With a well-known relationship:

$$\Lambda = \alpha F(u_+ + u_-) \qquad (12.15)$$

we have

$$E_L = \frac{RT}{F} \ln \frac{\Lambda(A)}{\Lambda(B)} \qquad (12.16)$$

Figure 12.12(B) illustrates the profile of the liquid junction in the diaphragm cell as an example of a complicated electrochemical system. Calculation of E_L in this case is somewhat difficult because t_j is not known exactly. However, E_L is roughly estimated to be about -0.065 V with t_j values assumed as:

In the anode compartment: $t_{Na^+} = 0.42$ and $t_{Cl^-} = 0.58$

In the cathode compartment: $t_{Na^+} = 0.21$, $t_{Cl^-} = 0.15$, and $t_{OH^-} = 0.64$

where the negative sign for E_L shows that the anode side is negative. It is of interest to study in detail these phenomena because of their relationship to electrolysis with diaphragms.

12.2.2. Mass Transfer through a Diaphragm

The role of the electrolytic diaphragm is to separate products in the anode and the cathode compartments from each other, but it must also permit passage of electricity with minimum resistance. It is thus important to consider

- mass transfer and material balance through the diaphragm,
- flow of electrolyte and ionic species through the micropores of the diaphragm, and
- electrical conductance through the diaphragm.

These phenomena are related to several factors of the electrolyte–diaphragm system such as[37–40]

- pore size,
- porosity,
- tortuosity of diaphragm capillary paths,
- permeability, and
- superficial conductivity.

MacMullin has described the material balance of electrolyte and ionic species in a Hooker S-3 diaphragm-type chlor-alkali cell.[41] Figure 12.13 shows the balance sheet per 73.3 A hr in a cell operated at 100°C with an assumption of 50% decomposition of salt and 100% current efficiency. The solid line and the dotted line show the transfer of molecular species and ionic species, respectively. Table 12.5 describes the material balance of ionic species, which is obtained from the relationship between the conductance and the mobilities and concentrations of ionic species, i.e., Na^+, Cl^-, and OH^- [see Eq. (4.1)–(4.3)]. In the anode compartment, $1.156F$ of $2.735F$ ($=73.3$A hr) is carried by Na^+ and $1.579F$ by Cl^-. In the cathode compartment, on the other hand, $0.588F$, $0.402F$, and $1.745F$ of $2.735F$ will be carried by Na^+, Cl^-, and OH^-, respectively.

Transfer of NaCl from the anode compartment to the cathode compartment through the diaphragm is as follows:

NaCl of saturated brine:	5.470 eq.
Transfer as Na^+:	1.156 eq.
Balance:	4.314 eq.

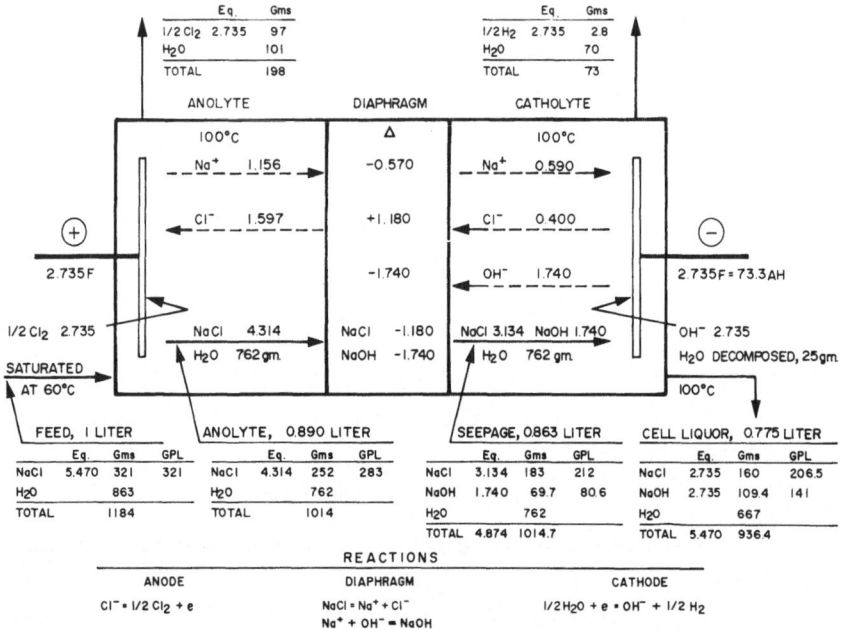

FIGURE 12.13. *Material balance in Hooker S-3 cell (30 kA at 100°C, 50% salt decomposition and 100% current efficiency). (By permission of R. B. MacMullin.)*

TABLE 12.5
Ion Balance at 100°C Calculated by MacMullin

	Ion	λ	Conc. C	$10^3 \kappa^a$	Percent current	Faradays
Anolyte	Na^+	152.0	2.02	308	42.2	1.156
	Cl^-	208.0	2.02	422	57.8	1.579
	Sum	360.0		730 (obs.)	100.0	2.735
Catholyte	Na^+	152.0	2.430	370	21.5	0.588
	Cl^-	208.0	1.215	253	14.7	0.402
	OH^-	901.0	1.215	1098	63.8	1.745
	Sum			1721 (obs.)	100.0	2.735

NOTE: $\lambda \cdot C = 10^3 \kappa$ = equivalent conductance, also called ion mobility; κ = conductivity in $ohm^{-1}\, cm^{-1}$.
[a] Note that 730 and 1721 are the observed values. Others are the fractions calculated by $\lambda \cdot C$.

The concentrations of NaOH and NaCl in the cathode compartment are equal to each other because of the assumption of 50% decomposition, i.e.,

$$C_{NaCl} = C_{NaOH} = \frac{5.470}{2} = 2.735 \text{ eq.}$$

Chlorine of 2.735 eq. evolves from the anode, and hydrogen of an equivalent amount from the cathode. This corresponds to the balance of NaCl concentration in the feed brine or the anolyte (5.735 eq.) and the effluent or the cell liquor (2.735 eq.), and also to the amount of NaOH in the cell liquor. Water consumption by the cathode reaction of hydrogen evolution is 2.735 eq. or about 25 g. Moreover, about 101 g and 70 g of water evaporate with chlorine gas and hydrogen gas leaving the cell, respectively. Thus, the overall water consumption from the cathode compartment would be $70 + 25 = 95$ g per 73.3 A hr.

Mukaibo has theoretically analyzed the problem of mass transfer through the porous diaphragm in chlorine cells which should serve as a basis for experimental verification.[42] H^+ and Cl_2 in the anolyte move to the cathode compartment through the diaphragm of thickness, d, and OH^- back migrates from cathode to anode, as shown in Fig. 12.14.

The differential equations governing mass transfer, r, through a porous, one-dimensional diaphragm under steady-state conditions are as follows:

For H^+:

$$r_H = -K_H \frac{dc_H(x)}{dx} + vc_H(x) + t_H c_H i \qquad (12.17)$$

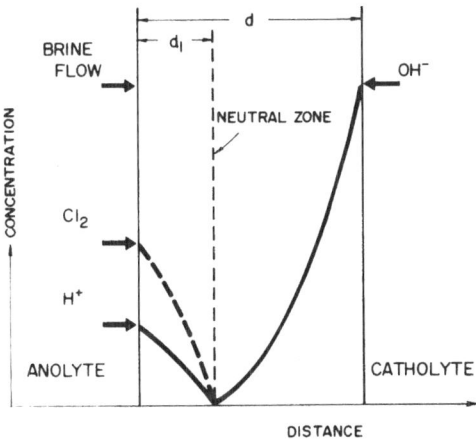

FIGURE 12.14. Mass transfer through the diaphragm and the concentration distribution of OH^-, Cl_2, and H^+ in the diaphragm-type chlor-alkali cell.

For Cl_2:

$$r_{Cl} = -K_{Cl} \frac{dc_{Cl}(x)}{dx} + vc_{Cl} \tag{12.18}$$

For OH^-:

$$r_{OH} = -K_{OH} \frac{dc_{OH}(x)}{dx} + vc_{OH} - t_{OH}c_{OH}i \tag{12.19}$$

and

$$r_H + r_{Cl} + r_{OH} = 0 \tag{12.20}$$

It is assumed that the potential gradient through the diaphragm remains constant, and the coefficients, t, which are related to the transference number (t) of the respective ions, are also constant. With the boundary conditions:

$$c_H = c_H^0 \quad \text{and} \quad c_{Cl} = c_{Cl}^0 \qquad \text{at } x = 0$$

$$c_{OH} = C \qquad \text{at } x = d$$

and

$$c_H = c_{Cl} = c_{OH} = 0 \qquad \text{at } x = d_1 \text{ (at the neutral zone)}$$

we have

$$r_H = \frac{c_H^0(v + t_H i)}{1 - \exp[-(v + t_H i)(d_1/K_H)]} \tag{12.21}$$

$$r_{Cl} = \frac{c_{Cl}^0 v}{1 - \exp[-v(d/K_{Cl})]} \tag{12.22}$$

and

$$-r_{OH} = \frac{C(t_{OH} i - v)}{1 - \exp\{-(t_{OH} i - v)[(d - d_1)/K_{OH}]\}} \tag{12.23}$$

In practical cells, d_1 is very small (less than 10^{-3} cm) compared to the thickness of the asbestos mat d so that d_1 can be neglected in Eq. (12.23).

The amount of NaOH produced, vC, is represented by

$$vC = i - (-r_{OH}) \tag{12.24}$$

Substituting Eq. (12.23) into Eq. (12.24),

$$C = i \Big/ \left\{ v + \frac{t_{OH} i - v}{1 - \exp[-(t_{OH} i - v)(d/K_{OH})]} \right\} \tag{12.25}$$

Therefore, the current efficiency ξ is given by the equation:

$$\xi = (Cv/i) \times 100$$

$$= \cfrac{1}{1 + \cfrac{(t_{OH}i/v) - 1}{1 - \exp[-(t_{OH}i - v)(d/K_{OH})]}} \times 100 \, (\%) \qquad (12.26)$$

Mukaibo's model is in good agreement with the experimental results.[43-46]

Stender et al. calculated the mass transport of OH^- with a simple model.[47] They introduced a parameter which may represent the difference in the velocity and migration:

$$U = t_{OH}i - v \qquad (12.27)$$

Substituting Eq. (12.27) with an assumption $Ud \ll K_{OH}$ into Eq. (12.25), a simple equation for the current efficiency is obtained as follows:

$$\xi = \frac{1}{1 + (K_{OH}/vd)} \times 100 \, (\%) \qquad (12.28)$$

The increased brine flow reduces OH^- at $y = 0$ at the catholyte side, and the back migration of OH^- through the diaphragm to the anode compartment is minimized. The current efficiency increases with increase of the brine flow because the largest factor of the current inefficiency is the back migration of OH^- followed by the reactions with dissolved chlorine in the anode compartment to form hypochlorous acid and/or hypochlorite ions.

Equation (12.26), or more clearly Eq. (12.28), shows that the current efficiency ξ increases when the flow rate v increases.

With Eqs. (12.19) and (12.27), the OH^- concentration in the diaphragm can be obtained:

$$C_{OH} = C \frac{1 - \exp[(U/K_{OH})(d - x)]}{1 - \exp[(U/K_{OH})d]} \qquad (12.29)$$

when $U < 0$, the concentration distribution of OH^- is concave to the flow direction, as is shown in Fig. 12.14. When $U > 0$, on the other hand, the profile becomes convex, and the OH^- migration to the anolyte is stimulated extensively, resulting in current inefficiency, which is not a case in practical cells. Consequently, U is a useful parameter for the performance of diaphragm cells.

We may briefly consider the mass transfer and the ohmic resistance through the micropores of a diaphragm. The flux of viscous fluid through

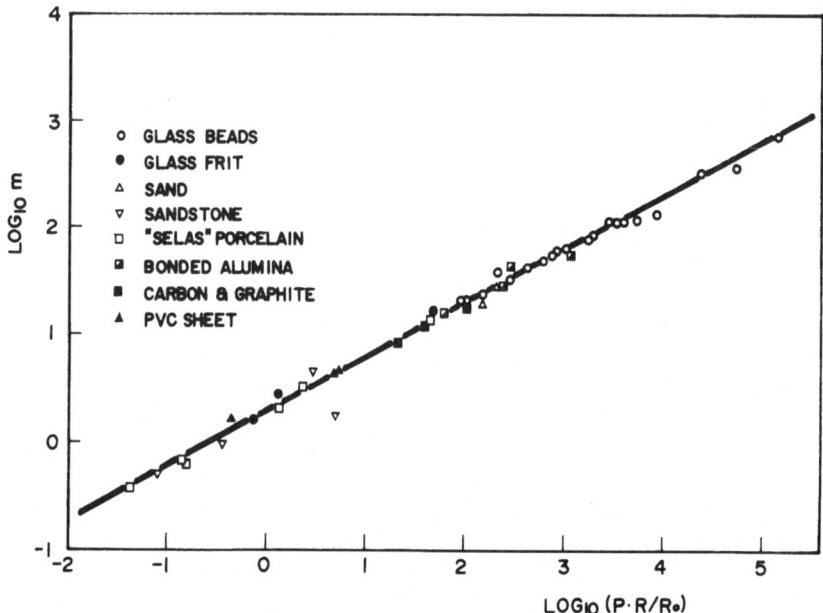

FIGURE 12.15. Correlation plot for MacMullin equation, $m^2 = kPR/R_0$. (By permission of R. B. MacMullin.)

uniformly distributed pores is governed by Poiseuille's law, and is shown by the Kozeny–Cárman equation as follows:

$$P = \frac{\varepsilon^3}{kS^2(1 - \varepsilon)} \tag{12.30}$$

where P is the permeability (m^2), ε is the void fraction, k is a constant, and S is the relative surface area (m^2/m^3).

MacMullin and Muccini obtained a relationship between the void fraction and the resistivity for packed beds with various kinds of particles as shown in Fig. 12.15 and Eq. (12.28)[49]

$$m^2 = kP(R/R_0)$$
$$= 3.666 \pm 0.098 \tag{12.31}$$

where $m = \varepsilon/S$ and is the hydrodynamic radius or the radius of an equivalent pore, R is the resistance of the packed bed containing the solution under discussion, and R_0 is the resistance of the solution having the same volume as the packed bed. It is obvious that the ratio of resistance is reciprocally proportional to the permeability and will be a factor of limitation for design and operation of a diaphragm-type electrolytic cell.

The symbols used in Eqs. (12.17)–(12.29)

$c_H(x)$, $c_{Cl}(x)$, and $c_{OH}(x)$	concentration of H^+, Cl_2 dissolved, and OH^- at $x = x$, respectively	g-eq/m^3
c_H^0 and c_{Cl}^0	concentration of H^+ and Cl_2 dissolved at $x = 0$, respectively	g-eq/m^3
C	OH^- concentration at $x = d$	g-eq/m^3
D	diffusivity	m^2/sec
d	thickness of diaphragm	m
d_1	thickness of acidic layer	m
i	flux of current (=current density/Faraday)	g-eq/m^2 sec
K	constant (=D × effective area/actual area)	m^2/sec
t	coefficient related to transference number (=t/c)	m^3/g-eq
t	transference number	—
r	rate of mass transfer	g-eq/m^2 sec
v	flow rate	m/sec
x	distance from anolyte–diaphragm interface	m
ζ	current efficiency	%

Subscripts

Cl	dissolved chlorine
H	H^+
OH	OH^-

12.3. ANODE MATERIALS

The anode of the alkali water electrolysis cell consists of an electrodeposited coating of nickel on steel because nickel is corrosion resistant to caustic alkali at ambient temperatures, whereas iron is gradually attacked by anodic polarization. The low oxygen overvoltage characteristics of nickel also make it suitable for the anode of the water electrolyzer. Contamination by Cl^- in the electrolytic solution should be avoided since otherwise the anode is seriously pitted.

The graphite anode in the HCl cell is completely durable when the HCl concentration is kept high, i.e., above $3 M$. In dilute HCl solution, the oxygen electrode reaction takes place together with the chlorine formation reaction, and the graphite anode is degraded. The graphite anode has long been employed in chlor-alkali cells, both amalgam and diaphragm cell types, but some consumption of the anode during electrolysis and associated increase of the cell voltage have been serious problems.

As an example of a soluble anode, crude copper dissolves to supply Cu^{2+} into the electrolytic solution in the copper refining cell. The current efficiency for the anodic dissolution is close to 100% in this case. An amount of chloride such as $NiCl_2$ or NH_4Cl is added in the nickel electroplating cell to assist smooth dissolution of the nickel anode because nickel is frequently passivated by anodic polarization in sulfate solution free of Cl^-.

It is evident that sometime we need an insoluble anode, and at other times, the anode must be dissolved efficiently. The following discussion concerns insoluble anodes for electrochemical processes.

Selection of the insoluble anodes for fused salt electrolysis is difficult compared to that for aqueous solution since the availability of corrosion-resistant materials is limited. Iron, nickel, and alloys of those metals can be used as the anode material in caustic alkali melts. The Castner sodium cell using molten NaOH is an example. Carbon and graphite are frequently employed in molten electrolyte composed of metal halides such as in the Downs sodium cell. Most metals and alloys are not resistant to chloride melts while, on the other hand, carbonaceous anodes are seriously attacked by molten electrolytes consisting of hydroxides and oxyacids, similar to behavior in aqueous solutions. The consumption of the carbon anode in the Hall–Heroult-type aluminum cell is large, say 0.4–0.5 ton/ton Al (see Section 9.2), but the carbon contributes efficiently to the reduction of alumina to aluminum as is shown in reaction (9.2). In fact, the carbon is a convenient conductor and participant on the anode reaction which contributes to making the electrolyte invariant on the average. That is, as aluminum-containing ions are reduced, the oxygen-containing ions do not accumulate in the melt. It is helpful to keep the current efficiency high on the cathode.

A significant problem in molten salt electrolysis is the anode effect, in which the high voltages and low currents produced disturb or interrupt electrolysis. The anode effect is thought to be caused by the formation of a nonwetting layer on the carbonaceous anode composed of C-Cl and/or C-F bonds during electrolysis.

Further oxidation of SO_4^{2-} and/or ClO_3^- by means of nonelectrochemical reaction is difficult to achieve, but can be carried out by anodic oxidation on a platinum anode under well-controlled conditions since the oxygen overvoltage of Pt is sufficiently high to retard the oxygen formation reaction. Persulfates and perchlorates are thus produced only by electrolysis of appropriate salt solutions with the Pt anode.

It is obvious that a variety of metals and alloys are being used as the anode material in practical cells because they have adequate properties of mechanical strength, machinability, electrical conductivity, corrosion resistance, and electrocatalytic activity. The materials should not be attacked and should not passivate under the working conditions imposed, particularly at noble oxidizing potentials. It is also important that they be active for the specific anodic reaction under consideration.

Carbon and graphite have good electrical conductivity and sufficient mechanical strength for use. The use of graphite as a practical anode material in chlor-alkali cells has diminished since the invention of the metal anode in the 1960s. Carbon and/or graphite is still the best material for use as an anode in fused salt electrolysis cells so that a process which may not be able to use carbon is considered impractical.

Although metals and alloys are oxidized anodically, they are able to be used as insoluble anodes if the surface oxide is resistant to chemical and physical attack and is also electrically conductive. The lead anode in sulfuric acid is an example. A brown- or chocolate-colored layer of PbO_2 forms on the anode surface and protects the substrate from corrosion. It is well known that Al_2O_3 forms when aluminum is polarized anodically in certain electrolytic solutions such as dilute sulfuric acid free of Cl^-. However, the oxide film has a large electric resistance so that aluminum cannot be used as anode even though it is corrosion resistant.

Since some metal oxides are both electrically conductive and corrosion resistant, they find use as anode material. The positive electrode of lead–acid batteries consists of lead oxide powder supported on a lead alloy grid. The positive of alkaline storage batteries is composed of nickel oxide powder packed into a pocket structure. Nickel oxide is also sintered onto nickel-electroplated iron sheet.

The lead dioxide anode is another type of practical oxide anode. The PbO_2 material is prepared by electrodeposition at the anode in an acidic solution of lead nitrate. The lead dioxide anode material is chemically similar to the material of the lead–acid storage battery positive, but is massive and nonporous. The tetragonal β-PbO_2 is the common modification, whereas the active mass of the battery positive varies its crystal structure with the electrode potential (see Chapter 11 and Fig. 11.5). The as-received lead dioxide electrodeposited layer is used as anode in some cases, but the oxide is also crushed, melted, and cast to prepare anodes having special forms such as cylinders and bars. The lead dioxide anode has been applied in electrolytic cells for the preparation of oxyhalides and to electroorganic processes.

The magnetite anode is prepared by means of casting a molten iron oxide composed of $FeO \cdot Fe_2O_3$. It is resistant to corrosion in acidic solutions of halides and oxyhalides of alkali metals, and is used as anode for chlorate cells. It is interesting that obtaining an equal mole ratio of FeO to Fe_2O_3 is very important according to Itai et al.,[50] since otherwise the anode is seriously attacked.

Beer has invented a unique and ingenious anode material consisting of RuO_2 and TiO_2 thermally deposited on a Ti substrate, named the "dimensionally stable anode" or DSA.[51] Preparation of this electrode is simple. The Ti substrate is painted with a butanol solution consisting of $RuCl_3$ and tetrabutyl titanate acidified with HCl, dried, and heated at about 500°C for few minutes to obtain the mixed oxide layer composed of RuO_2 and TiO_2. The DSA is resistant under the operating conditions of chlor-alkali cells, both amalgam and diaphragm types. The chlorine overvoltage on the DSA is very low. Consequently, the DSA has been rapidly adopted and more than 50% of chlorine is being produced with the DSA throughout the world, although this new electrode material has but a brief history of only some ten years. The DSA is probably the invention of greatest importance to the chlor-alkali industry in this century. Many reports and reviews in technical journals and patents on modifications and applications of the DSA have been published in the last ten years.[52,53]

Although various materials are now available, the selection of an adequate anode for a particular electrochemical process of interest is still difficult. Trace impurities in the electrolytic solution might be a source of difficulty in some cases. Material selection should therefore be based on field tests conducted over an extended period of time rather than based solely on laboratory experiments. Table 12.6 offers a selection guide for screening anode materials for use in various electrolytes.

12.3.1. Example 1: Graphite Anode for Chlorine Evolution

Electrode materials, especially anodes, are selected from several points of view such as:

● geometry and size available on the market,
● mechanical strength and machinability,
● electrical conductivity,
● corrosion resistance to the environment, and
● electrocatalytic behavior.

The cell design is affected greatly by the geometry, the size, the mechanical strength, and the machinability of the materials to be used as electrodes. Electrical conductivity is an important electrode parameter.

A consumable anode such as the copper slab in the copper electrorefining cell must be dissolved at 100% current efficiency in order to supply copper ions to the solution. On the other hand, nonconsumable anodes are employed in chlor-alkali cells and water electrolysis cells. These anodes must be resistant to chemical attack or corrosion by the surrounding electrolyte even at positive potentials.

Because the electrode process is a heterogeneous catalytic reaction accompanied by charge transfer, the reaction and the rate depend strongly on the electrode material, or its catalytic activity with respect to the reaction of interest, as well as on the physicochemical behavior of the electrolyte.

TABLE 12.6

Insoluble Anode Materials for Various Electrolytes

Electrolyte	Anode materials recommended	Comments	Undesirable
Aqueous solutions			
Caustic alkali	Ni, Ni-alloys, Fe, Fe-alloys, carbon steel, stainless steels	Seriously corrodes in halide containing solution	Carbon, graphite, Cu, Cu-alloys
Concentrated chloride, hydrochloric acid	Carbon and graphite	Seriously corrodes in diluted solutions and in solutions containing oxyanions, e.g., SO_4^{2-}	Almost all metals and alloys
	Magnetite, lead dioxide, noble metal oxide coated anodes		
Dilute chloride	Pt, Pt-alloys, Pt-coated, Pt group metals and alloys	Increase in corrosion rate due to active chlorine	Almost all metals and alloys except some Pt group metals
	Magnetite, lead dioxide, noble metal oxide coated anodes such as DSA		
Sulfate, neutral and acidified; chromate	Lead, lead alloys, lead dioxide	Seriously corrodes in halide containing solutions	Carbon, graphite, Ni, Ni-alloys, Fe, Fe-alloys, Cu, Cu-alloys
	Pt, Pt-alloys	Very high overvoltage	
	High-Si cast iron	Relatively large corrosion rate	
Molten salts			
Caustic alkali	Ni, Ni-alloys, Fe, Fe-alloys, carbon steel, stainless steels	Seriously corrodes in halide containing electrolyte	Carbon, graphite
Chloride, fluoride, and other halides	Carbon, graphite	Seriously corrodes when oxyanions and moisture contaminate	Almost all metals and alloys
Sulfate			No suitable materials available

TABLE 12.7

Physical Properties of Carbon and Graphite

	Carbon	Artificial graphite
Density (g/cm^3)		
True	~2	2.2
Superficial		1.6–1.7
Porosity (%)		20–30
Bending strength (kg/cm^2)		180–220
Coefficient of linear expansion (deg^{-1})	0.54×10^{-5}	
Specific heat (cal/°C g)	0.20	0.17
Thermal conductivity (cal/cm sec °C)	8.5×10^{-3}	
Electrical resistivity (ohm cm)	$(3.2 \sim 4.0) \times 10^{-3}$	$(8.0 \sim 8.5) \times 10^{-4}$

Unwanted reactions may take place in parallel with the main reaction on the same electrode, resulting in current inefficiency. This also causes corrosion of the anode in some cases. Consumption of the graphite anode in chlor-alkali cells is an example.

Carbon and graphite are unique materials with "semimetallic" behavior and have good thermal and electrical conductivities, as shown in Table 12.7. These materials are corrosion resistant to various chemicals and environments although they are attacked gradually by oxidizing agents in the electrolyte and by oxygen at high temperatures. Carbon, which is amorphous, is more reactive with oxidizing agents than graphite. The anode of the aluminum smelting cell is composed of carbon, either self-baked (the Söderberg anode) or prebaked, where carbon is consumed by the overall reaction:

$$2Al_2O_3 + 3C = 4Al + 3CO_2$$

to produce metallic aluminum (see Chapter 9.2). The calcium carbide furnace also uses the consumable carbon electrode, although the process is not electrochemical but is electrothermic (see Chapter 9.5).

Graphite is corrosion resistant to concentrated HCl solutions even at the potential range for chlorine evolution. Therefore, commercial HCl cells are equipped with graphite anodes and cathodes, and the anode consumption is almost zero if the HCl concentration is kept high, say at 20% by weight. The graphite anode is degraded gradually when the HCl concentration decreases due to the same reason for consumption of the graphite anode in chlor-alkali cells described below.

Since the oxide-coated titanium anode DSA was invented by Beer, it has gradually replaced the graphite anode in chlor-alkali cells over the last ten years, with benefical reduction of energy consumption and operating labor. However, graphite is still a valuable electrode material in a variety of electrochemical processes.

The graphite anode is consumed by chemical oxidation with physical degradation when it is electrolyzed in alkali chloride solution containing hypochlorite ions due to the reaction[54]

$$C + HClO = CO + HCl$$

and

$$C + 2HClO = CO_2 + 2HCl$$

The source of hypochlorite and hypochlorous acid is the dissolved chlorine in the anolyte

$$Cl_2 + H_2O = HClO + H^+ + Cl^- \tag{12.32}$$

$$K_1 = 4.66 \times 10^{-4} \text{ at } 25°C$$

and

$$HClO = H^+ + ClO^- \tag{12.33}$$

$$K_2 = 3.2 \times 10^{-8} \text{ at } 25°C$$

where K_1 and K_2 are the dissociation constants for the respective reactions. The chemical equilibrium between these reactions depends on the solution pH, as shown in Fig. 12.16.[55] It is well known that part of the hypochlorite

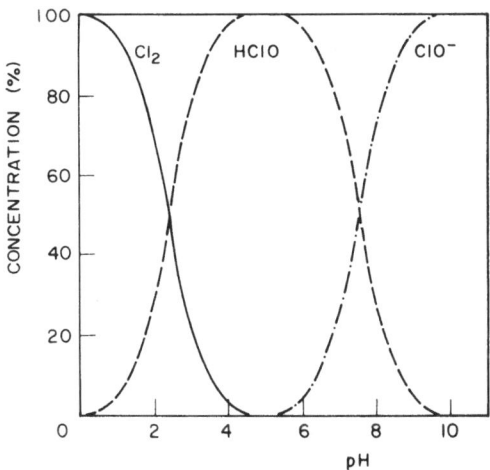

FIGURE 12.16. Concentration ratio of three components of active chlorine in the brine as a function of the pH at 25°C.

is converted further into chlorate by anodic oxidation and/or chemical reaction, depending on the electrode potential, the solution pH, and the temperature. Chlorate also attacks the graphite anode. In practice, the consumption of the graphite anode in the amalgam-type chlorine cell is about 1.5–2.5 kg/ton Cl_2 in comparison with about 3.0–4.0 kg/ton Cl_2 in the diaphragm cell due to differences in NaCl concentration and solution pH.

Generally, acidified brine (pH = 2–3) is fed to the amalgam cell, and is decomposed by only 10–15%. On the other hand, the pH of the feed brine for the diaphragm cell is kept high, say pH = 6–9, since otherwise the asbestos diaphragm is attacked by acid. Also, part of the OH^- migrates from the cathode compartment to the anode compartment through the porous diaphragm, resulting in formation of hypochlorite by reactions (12.32) and (12.33). The low concentration of NaCl in the anolyte is also a large factor in the corrosion of the graphite anode in the diaphragm cell.

Hine et al. investigated the electrochemical consumption of the graphite anode in concentrated NaCl solutions.[2] The concentration of the active chlorine ($Cl_2 + HClO + ClO^-$) and the solution pH were varied individually to study the influence of these factors. Regression analysis of the experimental results showed that the HClO concentration strongly influenced the corrosion rate of graphite anodes, while the solution pH also affected the corrosion rate, as shown in Fig. 12.17. In practical cells, the concentration of dissolved chlorine depends on the NaCl concentration and the solution temperature by Henry's law. The HClO concentration is closely related to the solution pH by reactions (12.32) and (12.33), so that the effects of these factors on the graphite consumption are connected.

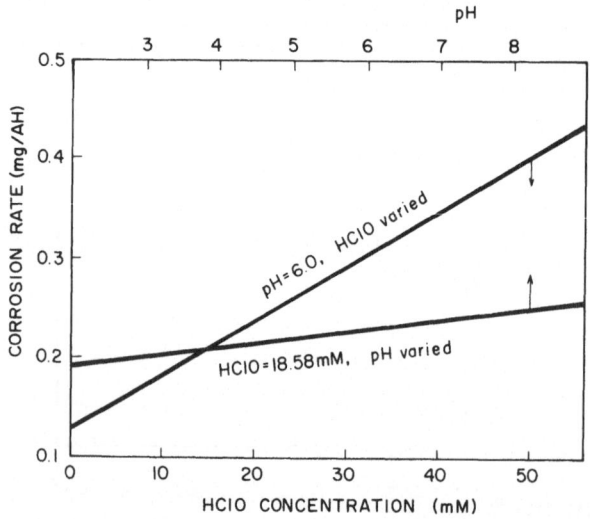

FIGURE 12.17. Corrosion rate of the graphite anode vs. pH and HClO concentration.

In Fig. 12.17, the intersection of the corrosion rate vs. HClO concentration curve shows some weight loss of the graphite anode in a solution free of HClO. For simplicity, we may refer to the standard reversible potential for the competitive reactions at 25°C

$$2H_2O = 4H^+ + O_2 + 4e \tag{12.34}$$

$$E^0 = 1.229 \text{ V vs. SHE}$$

and

$$2Cl^- = Cl_2 + 2e \tag{12.35}$$

$$E^0 = 1.359 \text{ V vs. SHE}$$

These potentials show that the oxygen evolution reaction may take place more easily than the chlorine evolution reaction from the viewpoint of thermodynamics, whereas, in practice, chlorine evolution takes place at high current efficiency, say 96% or more, due to the high overvoltage of the oxygen anode process (Fig.12.18). Once the oxygen formation reaction takes place at the graphite anode, graphite is oxidized by the reaction

$$C + O = CO \tag{12.36}$$

and

$$C + 2O = CO_2 \tag{12.37}$$

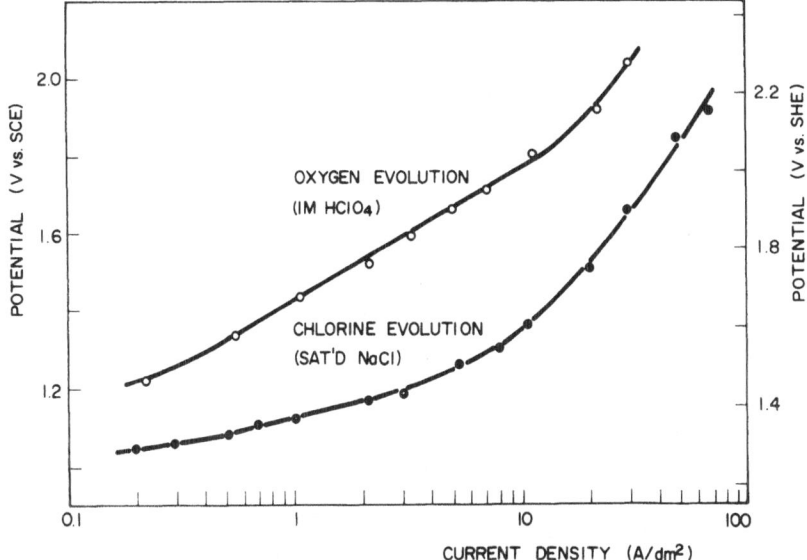

FIGURE 12.18. Steady-state polarization curves of a rotating graphite anode.

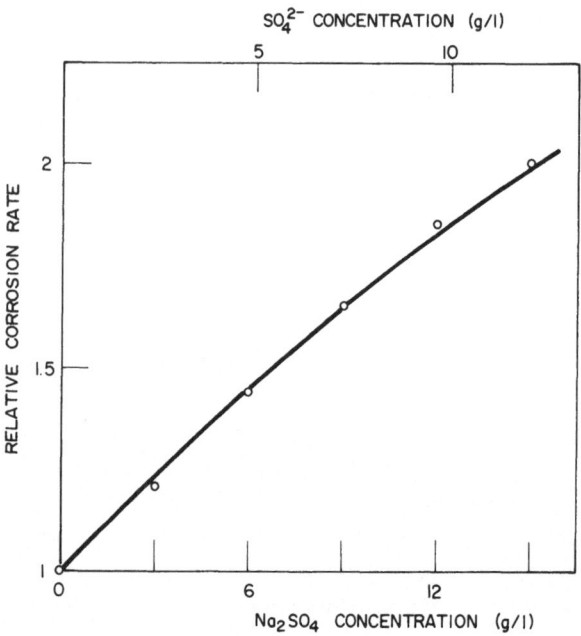

FIGURE 12.19. Corrosion rate of the graphite anode as a function of the Na_2SO_4 concentration in the brine.

The oxygen evolution reaction (12.34) depends greatly on the solution pH, i.e., the higher the pH the larger the reaction rate. In practice, it is well known that oxyanions such as SO_4^{2-} in the brine accelerate the corrosion of the graphite anode, as shown in Fig. 12.19, possibly due to the reaction sequence[56]:

$$SO_4^{2-} = SO_4 \text{ (radical)} + 2e \qquad (12.38)$$

$$C + H_2O + SO_4 \text{ (radical)} = CO + SO_4^{2-} + 2H^+ \qquad (12.39)$$

The NaCl concentration is a major factor of graphite corrosion in chlor-alkali cells. Figure 12.20 shows the CO_2 content in the cell gas of an amalgam-type chlorine cell as a function of the NaCl concentration in the depleted brine leaving the cell.[5] It is clear that the graphite anode is corroded seriously as evidenced by the increase in CO_2 concentration when the NaCl concentration decreases.

Since graphite has many micropores, the chlorine formation reaction in the pore might be limited because of low concentration of Cl^- due to limited mass transfer, resulting in a large increase of the overvoltage. As a result, oxygen formation will be stimulated with an increase in the pH, causing hydrolysis of Cl_2 to form HClO. Therefore, the micropores of graphite must

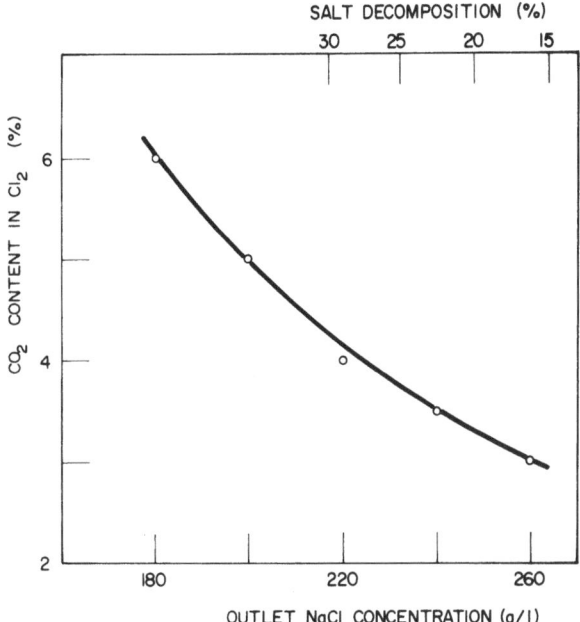

FIGURE 12.20. The CO_2 content in the cell gas as a function of the NaCl concentration in the depleted brine of the amalgam cell.

be minimized to prevent electrochemical consumption. Accordingly, the graphite anodes of the diaphragm-type chlorine cell were impregnated by oil for many years. The oil-treated graphite was inadequate for the amalgam cell for various reasons such as resulting difficulties with the amalgam cathode process and irregular degradation of the anode surface, probably due to very high-current densities, say 60–120 A/dm², compared to 20 A/dm² of the diaphragm cell.

Reactions (12.36) and/or (12.37) suggest the possibility of the formation of an oxide layer as the reaction intermediate on the graphite anode during electrolysis of aqueous solutions. Hine *et al.* conducted experiments with the linear potential sweep method to investigate the oxide layer on carbon and graphite anodes.[55] Figures 12.21 and 12.22 are examples of the potential sweep curves with a glasslike carbon (GC) electrode in 1 M HClO₄, used instead of a conventional graphite electrode because of experimental convenience. The GC material has a surface smooth as glass, thus providing a defined area. Figure 12.21 indicates current maxima at about 0.5 V and 1.3 V vs. CE (calomel electrode) during anodic polarization, associated with the formation of two types of oxide layers,

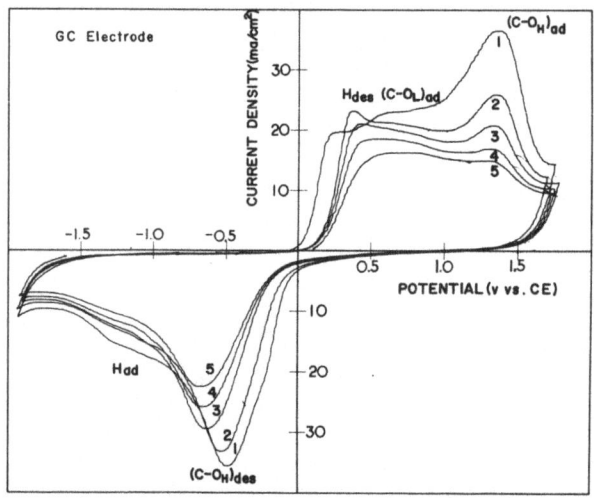

FIGURE 12.21. Potential sweep diagrams with a glasslike carbon electrode in 1 M HClO₄ at 50°C. Sweep rate = 200 mV/sec. Curves are numbered in order of runs.

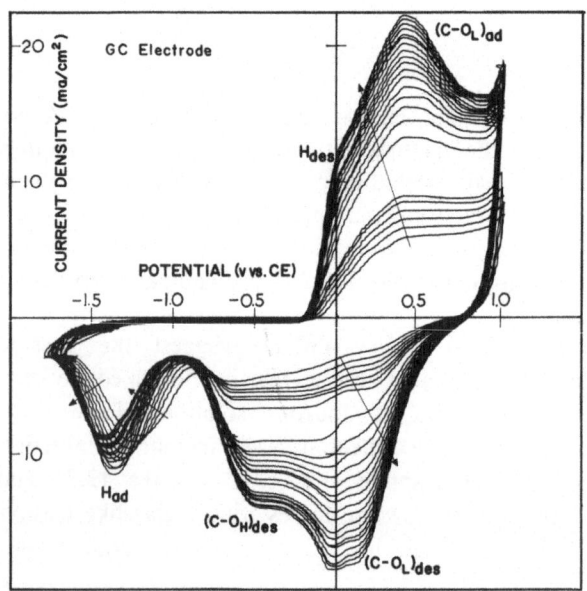

FIGURE 12.22. Potential sweep diagrams with a glasslike carbon electrode in 1 M HClO₄. Sweep rate = 200 mV/sec.

termed the lower oxide $(C-O_L)$ and the higher oxide $(C-O_H)$, respectively. As seen in Fig. 12.22, the desorption current of $(C-O_H)$ is relatively small since the formation of $(C-O_H)$ is limited at the potential range more negative than 1.0 V. The peak current at -0.5 V during cathodic scan, which corresponds to the reduction of $(C-O_H)$, decreases, but not to zero, and the new peak at ca. $+0.1$ V appears. It corresponds to the $+0.5$ V peak of the anodic scan. Since these two potentials of the anodic and cathodic scan are relatively close to each other, the adsorption and desorption of $(C-O_L)$ have reversible reaction behavior. The $+0.1$ V peak of the cathodic scan diminished when the electrode was polarized more positive than 1.3 V vs. CE, which was attributed to an irreversible conversion of $(C-O_L)$ into $(C-O_H)$. The cathodic polarization curve in Fig. 12.21 reveals that $(C-O_L)$ can be reduced easily, but that $(C-O_H)$ is removed at potentials more negative than -0.5 V vs. CE. In other words, $(C-O_L)$ is an unstable intermediate and the process of its formation and reduction is reversible. On the contrary, the formation and reduction of $(C-O_H)$ is quite an irreversible process, suggesting a strong bonding between carbon and oxygen. The authors attribute the corrosion of the graphite anode to the formation of the irreversible carbon oxide $(C-O_H)$ in the positive range of the electrode potential.

12.3.2. Example 2: Precious-Metal-Coated Anodes

Inexpensive solid metals and alloys such as lead alloys are widely employed as the insoluble anode in a variety of electrochemical processes, but the selection of desirable materials is limited. Nickel, for example, resists corrosion in hot concentrated caustic alkali, and its oxygen overvoltage is low. Nevertheless, solid nickel sheet is not used as an anode material in the alkaline water electrolysis cell because of its unacceptably high cost. Instead, nickel-electroplated steel sheet is normally used and is sufficient as an anode because the electrochemical process takes place only at the electrode–electrolyte interface, but not in the bulk of the metal electrode.

These considerations are of utmost importance when the precious metals such as platinum are used in a commercial cell. Although there are some exceptions such as the use of solid Pt foil in the perchlorate cell and the persulfate cell, the use of precious metals must generally be minimized in order to reduce investment cost.

Generally, the corrosion-resistant metals such as Ti, Nb, and Ta are employed as the substrate of the composite anode material. These so-called "valve metals" become passive quickly when they are exposed to the electrolyte, so that polarization current does not flow through at all. The surface of such a metal substrate is modified or covered with a thin layer of the precious metals and/or alloys to ensure electrochemical catalytic activity under operating conditions for the process of interest. A number of methods for loading the active layer have been developed, which include physical procedures such as vapor deposition and sputtering, and chemical procedures such as electroplating and thermal decomposition. Typical precious metal coated anodes are titanium anodes coated with platinum and platinum–iridium alloys. Other metals and alloys of the platinum group metals such as Pd, Rh, and Ru are also used as coatings.

Titanium is a commonly used substrate. Although Nb and Ta are also candidates as substrates, these metals are more expensive than Ti. However, the breakdown potential of the passive film on Nb and Ta is very high, i.e., positive by more than 120 V vs. SCE in 5% NaCl at 30°C, in comparison with 12 V for Ti. The high breakdown potential of the passive film serves to prevent failure of the anode if the applied cell voltage is high.[56]

Precious metal coated anodes, mostly Pt-coated, are used as the counterelectrode of the cathodic protection system for steel structures. Although graphite, high-silicon steel, and lead–silver alloys are also used, it is said that the Pt-coated Ti anode is the most economical because of its durability, i.e., the corrosion rate is only 6–10 mg/yr in sea water, although the initial investment is high.[57,58] The Pt-coated Ti anode also finds application for the cathodic protection of pipe lines buried in soil.[59]

Although the electrowinning cells for various metals such as Cu, Ni, and Zn are usually equipped with lead-alloy anodes, precious metal coated anodes can be used as well.[60,61] However, the investment cost and the service life of these new materials must be evaluated carefully before adoption.

One of the striking applications of precious metal coated anodes is the electrochemical production of sodium chlorate. Since the chlorate cell equipped with platinized titanium anodes was announced by the Olin Corporation,[62] many producers have investigated modern cells with metal anodes as replacements for conventional graphite anodes.[63–68]

Newberry et al. reported that the average consumption of the precious metal coating (4.25 μm thick) was 6.95 g/ton $NaClO_3$ (equivalent to 30 mg/A yr) and that the power consumption was 6300 kWh/ton $NaClO_3$.[62]

Warren et al. reported the performance of the Pt(70%)–Ir(30%) alloy coated anodes in two plants.[63] The precious metal alloy was loaded on the Ti substrate by thermal decomposition (5–10 g/m^2). The corrosion rate was 1.60 mg/A yr in one plant and 0.32 mg/A yr in another. These values differ widely from the above-mentioned Olin data and might be attributed to several factors such as differences in materials and preparation, solution composition, and operating conditions.

The chlorine evolution reaction takes place at the anode by electrolysis, followed by hydrolysis to form hypochlorite. Hypochlorite is further converted into chlorate via two routes: anodic oxidation and chemical oxidation, as follows

$$6ClO^- + 3H_2O = 2ClO_3^- + 4Cl^- + 6H^+ + \tfrac{3}{2}O_2 + 6e \qquad (12.40)$$

and

$$ClO^- + 2HClO = ClO_3^- + 2H^+ + 2Cl^- \qquad (12.41)$$

For 1 mole of ClO_3^-, 9 Faradays are required if ClO_3^- is produced by the electrochemical oxidation (12.40), in comparison with $6F/\text{mol } ClO_3^-$ for the chemical route (12.41).* The chemical oxidation via the disproportionation reaction of hypochlorite is favored in cells equipped with precious metal coated anodes because these anodes retard the anodic oxidation of ClO^-.[69] Operation at higher temperature is also a factor which favors reaction (12.41). (also, see Chapter 15.1.1)

The oxide layer on precious metal anodes has been studied extensively. The structure and the properties of the oxide layer depend on the electrode potential. Two types of the oxide layer have been found on the Pt anode: Oxide I which is a monolayer formed at relatively low potentials, say 0.8 V vs. SHE, and Oxide II which is a multilayer composed of Pt and O at more positive potentials.[70–74] These oxide films may affect the electrochemical kinetics. In chloride solutions, for example, the adsorption of Cl-atom on the Pt anode surface is in competition with the adsorption of the O-atom, depending on the electrode potential. The Pt surface is covered almost completely by Cl-adatom in a potential range more negative than 1.0 V vs. SHE. At more positive potentials, on the other hand, a number of the active sites on the Pt anode are occupied by O-adatoms.[75] Consequently, the chlorine evolution reaction takes place at the oxide-covered metal anode but not at a bare surface.

Electrochemical route:

At the anode:	$6Cl^- = 3Cl_2 + 6e$
In the solution:	$3Cl_2 + 3H_2O = 3HClO + 3H^+ + 3Cl^-$
	$3HClO = 3H^+ + 3ClO^-$
At the anode:	$3ClO^- + \tfrac{3}{2}H_2O = ClO_3^- + 2Cl^- + 3H^+ + \tfrac{3}{4}O_2 + 3e$

Overall:	$6Cl^- + \tfrac{9}{2}H_2O = ClO_3^- + 9H^+ + \tfrac{3}{4}O_2 + 9e$

Chemical route:

At the anode:	$6Cl^- = 3Cl_2 + 6e$
In the solution:	$3Cl_2 + 3H_2O = 3HClO + 3H^+ + 3Cl^-$
	$HClO = H^+ + ClO^-$
	$ClO^- + 2HClO = ClO_3^- + 2H^+ + 2Cl^-$

Overall:	$6Cl^- + 3H_2O = ClO_3^- + 6H^+ + 5Cl^- + 6e$

The corrosion rate of the precious metal coating may be related to the oxide film although these relationships are still unclear. Warren and co-workers classify the degradation of Pt-Ir alloy anodes used for chlorate cells into three types[61,63]

(i) coating dissolution,

(ii) substrate attack, and

(iii) substrate oxidation leading to electrical insulation.

These mechanisms depend to some extent on preparation of materials and the operating conditions. Khodkevich and co-workers reported experiments with platinized Ti anodes in HCl solutions, in which they found smaller corrosion rates with the thicker platinum layers.[76,77] These results suggest that the physical degradation of the coated anodes is significant compared to the chemical attack.

12.3.3. Example 3: The Oxide-Coated Metal Anodes

In a procedure developed by Beer, a Ti sheet is painted with a solution containing Ru salt and Ti salt, dried, and then is heated in air for a few minutes. With this method, Beer obtained an oxide-coated Ti sheet and found that this new material was chemically resistant to a chloride solution saturated with chlorine and also that its chlorine overvoltage was very low.[51,78] (See Chapters 7 and 8, and Fig. 7.6.)

Since the dimensionally stable anode (DSA) based on Beer's invention has been developed commercially by DeNora, the graphite anodes of chlor-alkali cells have been steadily displaced by the DSA. Chlorine cells equipped with the DSA are capable of significant energy saving of more than 20%. Beer's contribution covers not only the economy of the electrochemical industry but also the science of electrochemical processes. His finding has focused attention on the need for strict scientific investigation of metal oxides as electrode materials. There is no doubt that Beer's anode material is the top discovery and invention in the electrochemical field in this century.

Beer's electrode material consists of a mixture of RuO_2 and TiO_2 (or preferably RuO_x and TiO_y because x and y are somewhat less than 2 in general) loaded on the Ti substrate, and its preparation is quite simple. The Ti substrate is treated by hot oxalic acid solutions, for example, to remove the air-grown oxide on the surface, painted with the coating agent containing $RuCl_3$ and tetra-butyl titanate, $Ti(OBu)_4$, and is fired at 400–600°C in air for a few minutes. The coating and firing are repeated when a thick oxide layer is required.

A number of scientific reports and technical reviews of noble metal oxide coated Ti anodes (OCTA), including the DSA, have been published.[79-88] Some features of the RuO$_2$ coated Ti anodes have also been described.[89,90]

The thermal decomposition of the coating agent to form the mixture of RuO$_2$ and TiO$_2$ takes place at 200–400°C in air, depending on the paint composition, and is an exothermic reaction. The oxide layer prepared at relatively low temperatures, 400–600°C, consists of RuO$_2$ and TiO$_2$ of the rutile form, with some anatase-type TiO$_2$. The anatase TiO$_2$ converts to rutile completely when the temperature is high, at about 700°C, thus affecting the physicochemical properties of the oxide electrode. The electric resistance of the oxide coating depends on the firing temperature, as shown in Fig. 12.23. The electric resistance of the oxide layer prepared at low temperatures (300°C) is high, probably due to insufficient decomposition of the paint to oxide. The chlorine overvoltage of such material is also high. At temperatures higher than 600°C, on the other hand, the oxide crystallizes significantly, and the electric resistance increases. The Ti substrate is oxidized gradually by oxygen passing through the oxide layer, also contributing to increased resistance. The optimum range of the temperature is thus 400 to 500°C, depending on the paint composition.

The electric resistance vs. temperature curve of the oxide layer has a negative slope, suggesting a property of semiconducting materials. The electric resistance of the mixed oxide is also affected by the mole ratio of Ru

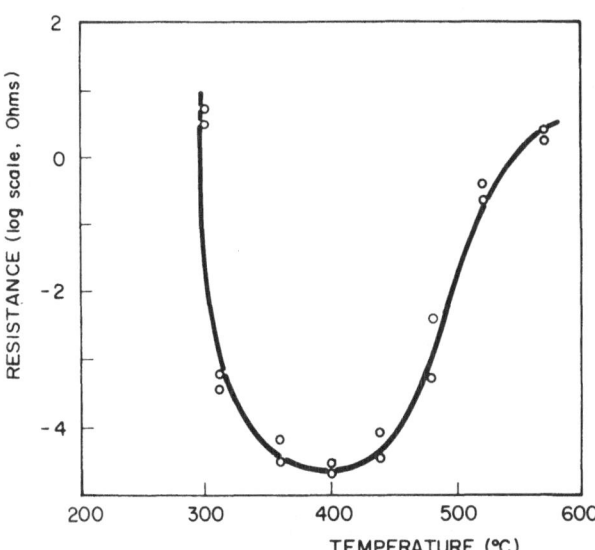

FIGURE 12.23. The electric resistance of the RuO$_2$–TiO$_2$ coatings on Ti substrate at room temperature as a function of firing temperature. RuO$_2$/TiO$_2$ mole ratio = 1/5. Four coats, each fired for 5 min.

to Ti, i.e., the resistance increases with increase of the Ti content. However, the chlorine overvoltage in saturated NaCl is almost independent of the mole ratio of Ru to Ti in the range 1/1 to 1/5 of Ru/Ti.

The OCTA is stable and its chlorine overvoltage is desirably low in concentrated NaCl solutions even at high-current densities, whereas in dilute NaCl solutions, it is attacked gradually and the anode potential becomes high. Figure 12.24 is a scanning electron microscope photograph of the

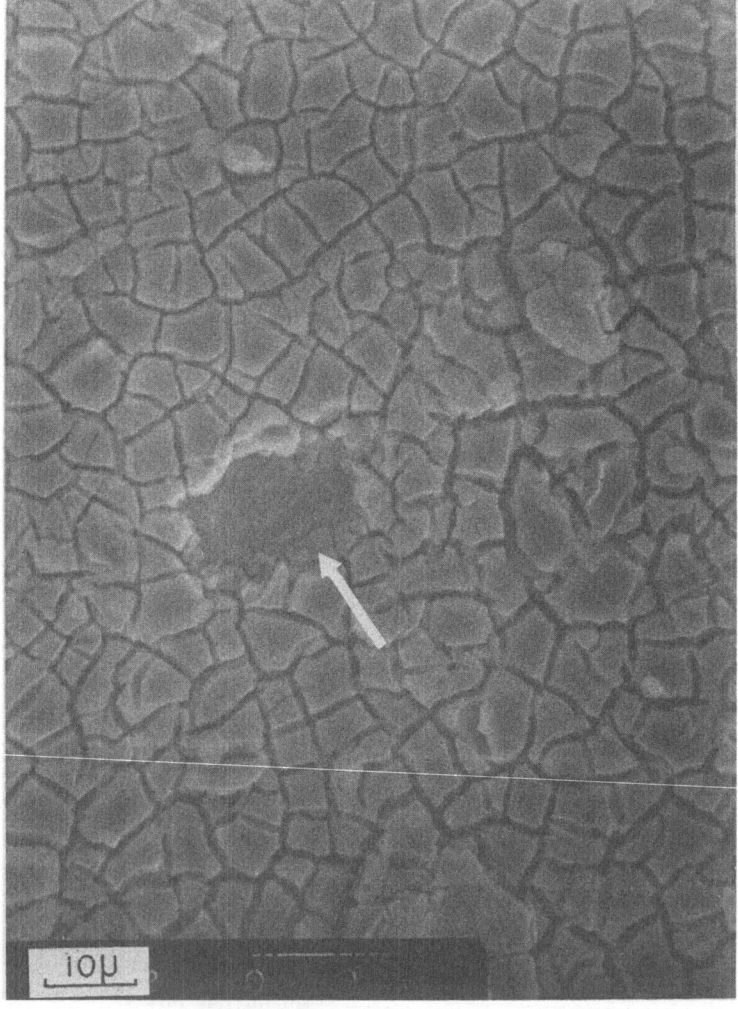

FIGURE 12.24. SEM photograph of an oxide-coated Ti anode. Arrow points to degraded region of normal "mudcrack" surface.

surface of an OCTA. Since there are many cracks, the actual surface area as electrode is very large, which contributes to the low overvoltage characteristics of the OCTA. However, the anode eventually becomes inactive and part of the oxide drops out, as shown by the arrow in Fig. 12.24.

Polarization curves of the OCTA in various solutions are shown in Fig. 12.25. In concentrated NaCl solutions, the anode potential is essentially unchanged at relatively low-current densities, but rises quickly when the current density becomes high, i.e., greater than $300 \, A/dm^2$. In general, the current density required for the potential to rise to a given value, such as 1.25 V vs. SCE, is proportional to the NaCl concentration. The top line shows the polarization curve in plain $HClO_4$ solution where only oxygen forms. The polarization curve for the oxygen electrode process differs from that in solutions containing chloride of some extent because chloride ions may occupy sites on the OCTA active for the oxygen electrode process, thus retarding charge transfer of water molecules and increasing the overvoltage for oxygen formation. In dilute NaCl solutions, 0.1 M NaCl for example, the OCTA polarizes significantly even at low current densities, e.g., $10 \, A/dm^2$, and the oxygen formation reaction rate is comparable to the chlorine formation reaction.

When the anode potential tends to the positive direction, RuO_2 is further oxidized, to RuO_4 for example, followed by the chemical decom-

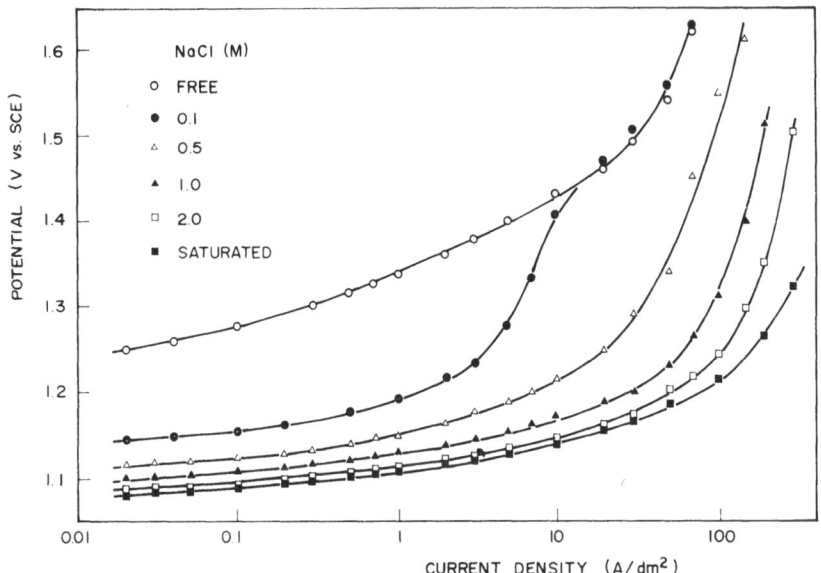

FIGURE 12.25. Polarization curves of the RuO_2–TiO_2-coated Ti anode in mixed solutions of 2 M $HClO_4$ and NaCl in various concentrations at 40°C. RuO_2/TiO_2 mole ratio = 1/2.

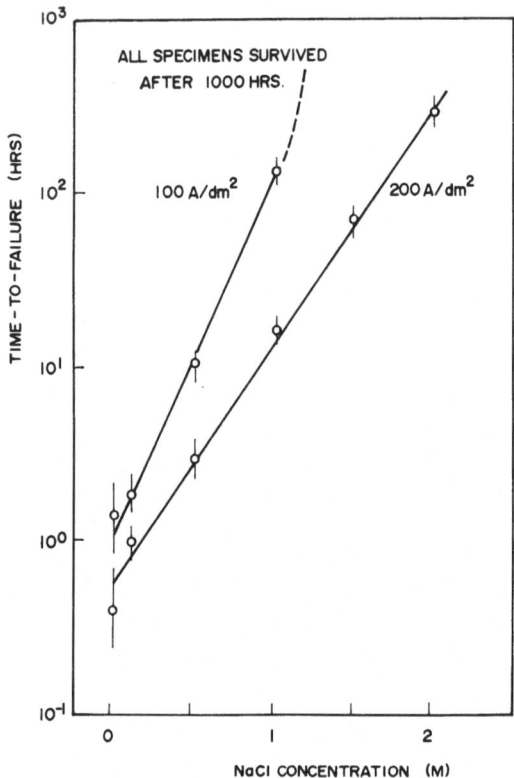

FIGURE 12.26. The time-to-failure vs. NaCl concentration for the oxide-coated Ti anode
(RuO₂/TiO₂ mole ratio = 1/2) in solutions of 2 M HClO₄ and NaCl in various concentrations.

position of RuO_4 to $RuO_2 \cdot 2H_2O$. The Ti substrate is also oxidized,
resulting in mechanical failure of the oxide layer and in electrical insulation.
Consequently, the degradation of the OCTA is related to the competition
between the chlorine formation reaction and the oxygen formation reaction
and is therefore strongly affected by the Cl^- concentration and the current
density. Figure 12.26 shows the experimental results of the life test of the
OCTA under various conditions of solution composition and current density.
The ordinate shows the time to failure (TTF) of the OCTA in hours. It is
clear that the TTF is prolonged at the higher NaCl concentrations and at
lower current densities. On the basis of these experimental results, Hine and
associates proposed a mechanism and sequence for degradation of the
OCTA, as shown in Fig. 12.27. In alkaline solutions, ruthenium dioxide is
oxidized anodically to RuO_4^{2-} at high potential ranges. However, oxygen
formation from OH^- occurs at less positive potentials than from H_2O in
acidic solution. The TTF for the life test of the OCTA with alkaline solution

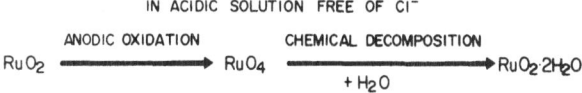

IN ACIDIC SOLUTION CONTAINING Cl⁻

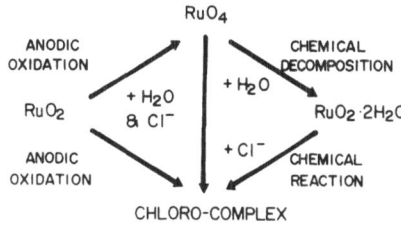

IN ALKALINE SOLUTION

ANODIC OXIDATION

$RuO_2 \longrightarrow RuO_4^{2-}$

FIGURE 12.27. Mechanism of degradation of RuO₂-coated anodes.

free of chloride was about 5 to 6 times longer than that in acidic solutions at equal current density, since the oxygen formation reaction occurs at low potentials, and hence oxidation of the active mass is limited.

In acidic solution free of Cl^-, RuO_2 is oxidized to RuO_4 because the anodic potential is sufficiently high for the reaction, and the higher oxide decomposes immediately to $RuO_2 \cdot 2H_2O$. Although the electrochemical oxidation of RuO_2 and the chemical decomposition of RuO_4 occur even in solutions containing chloride ions, ionic species of Ru combine with Cl^- to form a soluble chloro-complex such as $RuCl_n^{-(n-m)}$,[91] where n is the coordination number and m is the valency of Ru-ions. Furthermore, electrolytic solution penetrates the oxide layer through cracks and the Ti substrate becomes oxidized gradually during electrolysis. The localized corrosion or the subsurface corrosion penetrates into the interface between the Ti and the oxide layer, causing the active mass to drop out (Fig. 12.24) and the anode potential to increase. The active material is then dissolved further and the Ti substrate is more readily oxidized. Electrochemical and chemical dissolution of ruthenium oxide has been investigated,[92-98] as well as oxidation of the substrate.[85,99] Degradation of the OCTA is caused by a combination of these factors, depending on operating conditions and the nature of the materials. Thus, the three types of degradation proposed by Warren[61,63] are applicable to the OCTA under the operating conditions of chlorine cells. (See Section 12.3.2.)

There is continuing interest in nonprecious metal oxide coated anodes as substitutes for expensive noble metal coated anodes. Magnetite, Fe_3O_4, has long been used as the anode of the chlorate cell.[100] Electrodeposited lead dioxide is also a candidate for the chlorine anode.[101]

Recently, Dow Chemical has developed a cobalt oxide spinel coated Ti anode for the chlorine cell.[102,103] The composition of the oxide layer is $Zn_xCo_{3-x}O_4$, which can be modified by substitution of a third metal element such as Zr. The true surface area of the Co-Zn-Zr oxide coating is reported to be high, about $1880 \, cm^2$ per geometrical cm^2, resulting in low overvoltage. It has the advantage of high-current efficiency with respect to chlorine even at high pH such as 5.

12.4. CATHODE MATERIALS

The basis for selecting cathode materials is essentially the same as for the anode, but is somewhat easier because the cathode is safe from corrosion, with some exceptions such as

- at the shutdown time of electrolyzer,
- a part of the metal surface which may not be polarized sufficiently due to uneven current distribution, such as the inside of the box-type electrode, and
- metals and alloys which may form brittle hydride.

Although the cathode potential under electrolysis is more negative than the corrosion potential in general, the material could be attacked by the electrolyte when the current is off at shutdown time. The metallic wall of the electrolyzer is subject to similar difficulties. Combination of dissimilar metals must be carefully avoided to prevent galvanic corrosion. However, it is interesting to note that the Hooker *H*-type cell provides a heavy copper bus in a steel cathode tube so as to reduce the *IR* drop within the metal hardware.[104] No corrosion occurs in this case since the cathode is dried at shutdown times.

The steel cathode of the chlorate cell is protected by a small cathodic current even at shutdown time since otherwise chlorate would seriously attack the cathode. A particular problem of the chlorate cell is scale deposition on the surface of steel cathodes, resulting in high cell voltage. For recently used titanium cathodes in the Diamond Shamrock chlorate cell, the surface is attacked by electrolysis to form a titanium hydride layer which flakes off with unwanted scale being deposited on the cathode surface.[105]

However, this is an unusual application of titanium. Generally, the hydride-forming metals such as Ti, Zr, and Ta are inadequate for use as cathode even though these metals resist general corrosion. Formation of hydride is easy at the negative potentials even if the metal surface is covered by an oxide film.[106] Hydrogen embrittlement of metals and alloys depends on the electrochemical potential, the current density, and the amount of coulombs passed, as well as the electrolyte composition. Therefore, titanium and other related metals can be used as cathodes if the cathode potential is not too negative. In other words, the cathode potential must be selected to avoid the very negative hydrogen discharge region. Typically, in modern cells for making deposited copper foil for printed circuit manufacture, titanium cathodes facilitate deposit stripping and are used with no hydrogen evolution.

Corrosion due to uneven current distribution is troublesome. Such problems can be solved only by cell design to promote current uniformity and by appropriate materials selection.

Polarization behavior is an important factor in selecting a cathode. For example, the dc power consumption of the diaphragm-type chlorine cell could be reduced extensively by use of a low hydrogen overvoltage cathode. (See Section 12.4.2.)

In an amalgam-type chlorine cell, the cathode is composed of pure mercury and/or sodium amalgam. Since the hydrogen overvoltage of mercury is very high, no hydrogen formation takes place even at the negative potentials at which Na^+ discharges on the mercury cathode to form sodium amalgam. Some impurities in the feed brine disturb formation of sodium amalgam on the cathode by several means. Brine treatment is thus of importance in the amalgam cell plant. (See Section 12.4.1.)

The cathodic reaction of organic materials is greatly affected by the nature of the cathode material and the potential.[107] For example, hydroxylamine forms as an intermediate when nitrobenzene is electrolyzed with low hydrogen overvoltage cathodes such as Ni and Pt. On the other hand, hydrazobenzene or aniline is produced by electrolysis with high overvoltage cathodes such as Pb, Sn, Zn, Hg, and monel, as follows:

$$(12.42)$$

Nitrobenzene   Hydroxylamine   p-Aminophenol

$$(12.43)$$

Nitrobenzene   Hydrazobenzene

Monel is a Ni-Cu alloy, composed of 66% Ni, 33% Cu, and some Fe and other elements. The hydrogen overvoltage on monel is high, especially in alkaline solutions, although it is well known that pure nickel has a low hydrogen overvoltage. The cathodic reduction of organic compound is influenced extensively by the electrolyte pH as well as the cathode material. Reaction (12.43) takes place only in a basic solution but not in an acid medium.

The structure of the oxygen depolarized cathode of the air cell is unique. A porous carbon electrode permits oxygen in air to pass through to the electrode–electrolyte interface, where oxygen is reduced to hydrogen peroxide in a region of activity, known as the "three-phase zone". Further reduction of peroxide to OH^- depends on the catalytic activity of the electrode. Both the hydrogen and oxygen electrodes of the fuel cell have the same structure (see Chapter 11.5). The voltage of the chlor-alkali cell could be reduced significantly if the oxygen cathode is applied (see Section 12.4.3).

In the HCl cell, chlorine evolves at the anode and hydrogen at the cathode. Since the demand for hydrogen is limited, the substitution of a process to replace hydrogen formation is of great use in saving energy if that cathodic process proceeds at potentials more positive than hydrogen evolution. The oxygen reduction cathode process is a candidate, although the slow rate is a problem. The catholyte of the Kyoto cell is a mixture of HCl and $CuCl_2$ (see Chapter 6.2). Cupric ions are reduced to cuprous at the cathode, and hydrogen evolution no longer takes place because the reversible potential for the $Cu(I)/Cu(II)$ redox system is positive enough, and the reaction is fast. Cuprous ions are oxidized back to cupric by oxygen in the bulk of solution and return to the cathode surface to form a reaction loop. Although several materials were examined, graphite was selected as having the best cathode properties. Platinum is active for the redox reaction and does not corrode, but is too expensive. Other metals and alloys were attacked by the acid electrolyte at the positive potential of the redox reaction. Although titanium resisted corrosion, the redox reaction on the Ti cathode was slow, probably due to the influence of the surface oxide.

In summary, factors requiring consideration in selecting electrode materials are:

● geometry and size available,
● mechanical strength,
● electrical conductivity,
● corrosion resistance, and
● electrochemical catalysis.

12.4.1. Effects of Brine Impurities on the Amalgam Cathode

Mercury is a unique liquid metal, because

- its purification is easy by means of acid cleaning and distillation,
- it dissolves various metals such as Na, K, Zn, Cu, Au, etc., as the amalgam of these metals, but
- it is not combined with iron and/or steel, and
- the hydrogen overvoltage of mercury is very high.

These properties are desirable for the cathode of the amalgam-type chlor-alkali cell. Mercury is fed to the inlet end-box of the cell, an inclined trough made of steel plate, and flows down to the outlet end-box located at the far end of the trough (see Chapter 7). The Na concentration in the amalgam increases from 0 to about 0.2% by weight along the flow direction of mercury on the steel plate cathode. Sodium ions are reduced with almost 100% current efficiency by electrolysis of purified brine saturated with NaCl, but it is well known that the cathode process of Na-amalgam formation is affected seriously by the brine impurities. These impurities stimulate the hydrogen formation reaction on the amalgam cathode, resulting in the explosion hazard caused by the reaction of H_2 with Cl_2.

The effects of the brine impurities on the amalgam cathode have been well investigated by several authors.[108-112] MacMullin estimated the effectiveness of the brine impurities in the amalgam cell operated at relatively low current densities,[113] and classified them (see Table 7.7) as:

- Very harmful: V, Mo, Cr, Ti, Ta, Mg + Fe
- Moderately harmful: Ni, Co, Fe, W
- Slightly harmful: Ca, Ba, Cu, Al, Mg, graphite
- No effect: Ag, Pb, Zn, Mn

Note that coexistence of certain metals is very harmful, Mg + Fe for example. A trace of fine particles of graphite coming from the anode is a major factor in the formation of so-called "amalgam butter" or "thick amalgam" when the brine is somewhat contaminated with Mg and/or Ca, although graphite particles in highly pure brine are also troublesome.

Hine *et al.* studied the effects of impurities on the amalgam cathode using oscilloscope traces of the potential-time transient curve following current interruption.[111] A summary of the effects of impurities on the amalgam cathode is shown in Table 12.8. The brine impurities may be classified into two groups. The first group includes chromate in low-pH solution at low current densities and vanadate at low current densities which

TABLE 12.8
Summary of Effects of Impurities on the Amalgam Cathode

Impurities	Conditions	Hydrogen evolution	Decay transient on oscilloscope
Pure brine		No	No
Mg^{2+}	Immediately	No	No
	Polarization increase after a few minutes	Yes	Yes
Fe^{3+}		No	No
Ca^{2+}	Low pH, low current density	Slight	No
	high current density	No	No
	High pH, immediately	No	No
	polarization increase with time	Yes	Yes
Cr^{3+}	Low current density	Slight	No
	High current density	No	No
V^{5+}	Low current density	Yes	No
	High current density, polarization increase	Yes	Yes

functioned as activators of the amalgam decomposition process by local cell action. A second group, comprised of calcium ions in alkaline solution and vanadate at high current densities, stimulated the hydrogen formation reaction at the amalgam cathode.

Although the formation of sodium amalgam is a fast reaction, the process is controlled by diffusion of Na in the amalgam, as well as by diffusion of Na^+ through the diffusion layer established in the vicinity of the cathode surface in the solution. Hydrogen evolution may take place at the amalgam cathode even in highly pure solution of NaCl when the Na concentration at the amalgam–electrolyte interface becomes high after electrolysis for a prolonged period of time. The amalgam cathode in a practical cell is well mixed, and the Na concentration is kept low enough during electrolysis so that no hydrogen evolves at all.

If the brine is contaminated by Mg^{2+}, the amalgam cathode is covered by a gray film and mixing of the amalgam slows down, resulting in build-up of the Na concentration at the cathode surface and easy evolution of hydrogen.[112]

Decomposition of the sodium amalgam is stimulated greatly by vanadate ions, as described above. The process associated with vanadate ions is not electrochemical but a chemical or catalytic reaction. The rate constant is almost independent of the solution pH, but is a function of the vanadate ions, with a reaction order of about 0.5 with respect to vanadate.[112] The mechanism of this process is still not fully established.

12.4.2. Low Hydrogen Overvoltage Cathodes in Alkali Solutions

Low-carbon steel plate or screen is used as cathode in the water electrolysis cell and in chlor-alkali cells, both diaphragm-type and membrane type, because steel is corrosion resistant to caustic alkali and has a low hydrogen overvoltage, e.g., 0.4 V at 20 A/dm^2 is 2.5 N NaOH at 90°C.

Recently, water electrolysis has been considered for energy management systems in which a large-scale water electrolysis plant would be operated to utilize the off-peak electricity for "load-leveling". For this purpose, the cell voltage must be minimized to be as low as possible in order to keep the energy efficiency high. At 0.4 V, the hydrogen overvoltage of the steel cathode is still too high for this load-leveling application.

Although at one time the anode overvoltage contributed the largest portion of the terminal voltage of the diaphragm-type chlorine cell, this problem has been solved almost completely by the DSA. Consequently, the cathode overvoltage has become the top target for saving electric energy.

It is well known that the metals which belong to the eighth group of the Periodic Table are active for the hydrogen electrode process.[114] Iron is presumably the best choice of industrial use, but for further reduction of hydrogen overvoltage, the surface of the steel cathode requires modification. The hydrogen overvoltage of the platinized cathode is very low, but the cost is unacceptably great.[115] Consequently, the loading of Ni, Co, and their compounds on a steel substrate is being studied, as described below.

In recent years, advances have been made in the improvement of high-temperature, high-pressure water electrolysis cells with a particular emphasis on increased energy efficiency. The development of the low hydrogen overvoltage cathode in hot concentrated KOH solution (30–40% by weight) at high current densities ($10 \, kA/m^2$) is a key item.[116] Some candidates considered for coating nickel cathode substrates are:

1. *Nickel Sulfide.* The hydrogen overvoltage was 0.33 V at 160°C and 0.36 V at 200°C. However, a part of the coating failed and the overvoltage increased gradually during electrolysis.

2. *Nickel–Zinc Codeposited Layer.* After the Ni-Zn alloy was electrodeposited on the nickel substrate, zinc was dissolved with hot concentrated KOH solution to obtain a porous layer of remaining nickel. It showed very low hydrogen overvoltage.

3. *Nickel–Aluminum Codeposited Layer.* The alloy coating was obtained by thermal diffusion, and was treated with hot concentrated alkali to dissolve Al. The hydrogen overvoltage of this material was somewhat less than that of the Ni-Zn cathode.

4. *Raney Nickel.* Raney nickel powder was bonded with Teflon onto a Ni gauze cathode. Its hydrogen overvoltage was high, however.

Carnell and Needes heated a steel plate coated with Ni and Al at 600°C to obtain a Ni_2Al_3 layer, and leached out aluminum with hot concentrated NaOH. The hydrogen overvoltage at 20 A/dm^2 in a mixed solution of 12% NaOH and 16% NaCl at 96°C was about 25 mV less than that of the steel cathode alone.[117]

Carlin and Darlington prepared (12–15%)W–(3–5%)P-Co alloy coatings that showed low hydrogen overvoltage under the operating conditions of the diaphragm cell.[118]

Malkin and Brannan investigated several materials having low hydrogen overvoltage in caustic soda solution.[119] They said that the cathodes coated with Ni-Zn and Ni-Al had very large surface area, 6,000–12,000 m^2 per geometrical m^2, and the best performance. Diamond Shamrock Corporation is testing these cathode materials in commercial cells.

Man and Tseung treated $NiCo_2O_4$ spinel powder with H_2S at 350°C for 8hr to obtain $NiCo_2S_4$. The cathode, consisting of $NiCo_2S_4$ bonded with Teflon on a Ni screen, showed very low hydrogen overvoltage: only 40 mV at 25 A/dm^2 in 15% NaOH at 85°C, for example.[120,121]

Matsuura *et al.* have developed a low hydrogen overvoltage cathode, named "LHOC", for the diaphragm cell.[122] It consists of Ni and S, and is heated to ensure good bonding with the steel substrate. Tokuyama Soda Company is testing the LHOC in a commercial scale membrane cell.

Electrodeposited nickel containing sulfur, termed "Ni-S" was employed as a low hydrogen overvoltage cathode in industrial water electrolysis cells many years ago in Japan. Patent claims indicate use of $Ni(CNS)_2$ as an electroplating bath.[123,124]

Hine *et al.* studied the physicochemical properties of Ni-S and its polarization characteristics in NaOH solutions. An accelerated life test at high current densities was also conducted for several months. The material obtained by electroplating only is somewhat amorphous, but it crystallizes with heat treatment at temperatures higher than 250°C. At very high temperatures such as 800°C, it decomposes to Ni and Ni_3S_2. Those materials showed very high hydrogen overvoltage in alkali solutions because of deactivation.[125]

12.4.3. Oxygen Cathodes for Chlor-Alkali Cells

The oxygen electrode process has long been studied, mostly from the viewpoint of science rather than practice. Since the fuel cell has received attention as a new energy source, the oxygen cathode process has been investigated extensively in this field. The reversible potential of the oxygen electrode is fairly positive ($E^0 = 1.23$ V vs. SHE at 25°C), and hence the oxygen electrode would be a preferable cathode (positive electrode) for the hydrogen–oxygen fuel cell. The overall reaction of the oxygen cathode is a four-electron charge transfer process for 1 mole of oxygen molecule:

$$4H^+ + O_2 + 4e = 2H_2O \tag{12.44}$$

This reaction is composed of a sequence of various steps, depending on the electrode material and the operating conditions. In aqueous solutions at ambient temperature, oxygen is first reduced to hydrogen peroxide and/or its ion, then further reduced to H_2O and/or OH^-. Although the first step is fast, the electrochemical reduction and the catalytic decomposition of hydrogen peroxide are slow, resulting in a limited rate of the overall reaction.

Development of an active cathode for the process is thus required to increase the reaction rate and the maintain the electrode potential positive. Low solubility of oxygen in electrolytic solution, especially at high temperatures, is also a problem because oxygen becomes depleted at the cathode surface, and the reaction rate is then limited by mass transfer. At present, research and development on electrocatalysis, electrode structure, and other important aspects of the oxygen cathode are being conducted worldwide. Hoare has published an excellent review book on the oxygen electrode.[126]

Recently, the chlor-alkali industry has shown interest in the oxygen cathode. In diaphragm cell and membrane cell plants, hydrogen liberated at the steel cathode is presently burned as fuel to recover the heat value. Of course, there are valuable uses for the hydrogen other than as fuel, but these are unusual.

It is presumed that the power consumption for brine electrolysis can be reduced extensively when the chlor-alkali cell is equipped successfully with the oxygen cathode instead of the hydrogen formation cathode. The thermodynamic decomposition voltage for electrolysis of NaCl solution with the oxygen reduction cathode would be about 1.0–1.1 V in comparison with the 2.2 V of a conventional cell. In practice, however, there are many difficulties with the oxygen cathode for chlor-alkali cells.

Formation of hydrogen peroxide intermediates is one of the biggest problems, as described above. The specially designed cathode, mostly composed of activated carbon with catalyst, is deactivated by small amounts of hydrogen peroxide in hot, concentrated caustic soda, so that the intermediate must be decomposed perfectly as rapidly as possible. Contamination by CO_2 is also troublesome.

The chlor-alkali cell is normally operated at current densities of 20–40 A/dm^2, which is too high for a traditional fuel cell cathode. Hydrogen formation could result if the cathodic reduction of oxygen is limited at a current density lower than the operating current density, with explosion as a possible result.

Case Western Reserve University and Diamond Shamrock Corporation have jointly investigated the air cathode for chlor-alkali cells, supported by the U.S. Department of Energy. The target for the air cathodes for the membrane cell is as follows[127]:

Current density:	30 A/dm^2
O_2 source:	air, CO_2 free
Electrolyte:	30% NaOH (membrane cell grade)
Operating potential:	>0.8 V vs. RHE
Temperature:	85°C
Life:	>10,000 hr

The research at Case Western is on basic concepts of electrocatalysis of various materials such as transition metal complexes as well as on traditional platinum catalysts, while Diamond Shamrock has constructed and operated a pilot scale cell equipped with the air cathode.[128]

The cathode of the Diamond Shamrock cell operates at a potential of 600 mV vs. RHE (reversible hydrogen electrode in the sample solution) at a current density of 31 A/dm^2. This corresponds to a saving of approximately 0.85 V over a steel cathode with hydrogen evolution at the same current density. Plans have been indicated for commercialization of the oxygen cathodes in membrane cells by 1984–1985.

REFERENCES

1. F. Hine, M. Yasuda, R. Nakamura, T. Noda, *J. Electrochem. Soc.* **122**, 1185 (1975).
2. F. Hine, M. Yasuda, I. Sugiura, and T. Noda, *J. Electrochem. Soc.* **121**, 220 (1974).
3. C. L. Mantell, *Electrochemical Engineering*, 4th ed. p. 268, McGraw-Hill, New York (1960).
4. A. Schmidt, *Angewandte Elektrochemie*, p. 156, Verlag Chemie, Weinheim (1976).
5. F. Hine, unpublished work.
6. S. Okada, S. Yoshizawa, and S. Fujioka, *Soda and Chlorine* **6**, 93 (1955).
7. M. Murozumi, *Electrochem. Technol.* **5**, 236 (1967).
8. J. C. Davis, *Chem. Eng.* p. 44 (July 7, 1975).
9. F. Beck and H. Guthke, *Chem. Ing. Tech.* **41**, 943 (1969).
10. F. Beck, *J. Appl. Electrochem.* **2**, 59 (1972).
11. M. M. Baizer, *J. Electrochem. Soc.* **111**, 215 (1964).
12. M. M. Baizer and J. D. Anderson, *J. Electrochem. Soc.* **111**, 223 (1964).
13. D. N. Bennion and J. Newman, *J. Appl. Electrochem.* **2**, 113 (1972).
14. G. B. Adams, R. P. Hollandsworth, and D. N. Bennion, *J. Electrochem. Soc.* **122**, 1043 (1975).
15. R. S. Wenger, and D. N. Bennion, *J. Appl. Electrochem.* **6**, 385 (1976).
16. A. K. P. Chu, M. Fleischmann, and G. J. Hills, *J. Appl. Electrochem.* **4**, 323 (1974).
17. A. K. P. Chu and G. J. Hills, *J. Appl. Electrochem.* **4**, 331 (1974).
18. A. T. Kuhn and R. W. Houghton, *J. Appl. Electrochem.* **4**, 69 (1974).

19. A. T. Kuhn and R. W. Houghton, *Electrochim. Acta* **19**, 733 (1974).
20. R. Alkire and P. K. Ng, *J. Electrochem. Soc.* **121**, 95 (1974).
21. F. Coeuret, *Electrochim. Acta* **21**, 185, 203 (1976).
22. B. G. Ateya, E. A. S. Arafat, and S. A. Kafafi, *J. Appl. Electrochem.* **7**, 107 (1977).
23. F. Coeuret, D. Hutin, and A. Gaunand, *J. Appl. Electrochem.* **6**, 417 (1976).
24. J. R. Backhurst, J. M. Coulson, F. Goodridge, and R. E. Plimley, *J. Electrochem. Soc.* **116**, 1600 (1969).
25. M. Fleischmann and D. Pletcher, *Chem. Br.* **11**, 50 (1975).
26. M. Fleischmann, J. W. Oldfield, and L. Tennakoon *J. Appl. Electrochem.* **1**, 103 (1971).
27. J. N. Hiddleston and A. F. Douglas, *Electrochim. Acta* **15**, 431 (1970).
28. L. J. J. Janssen, *Electrochim. Acta* **16**, 151 (1971).
29. D. C. Carbin and D. R. Gabe, *Electrochim. Acta* **19**, 645 (1974).
30. D. C. Carbin and D. R. Gabe, *Electrochim. Acta* **19**, 653 (1974).
31. C. J. H. King, K. Lister, and R. E. Plimley, *Trans. Inst. Chem. Engs.* **53**, 20 (1975).
32. R. W. Houghton and A. T. Kuhn, *J. Appl. Electrochem.* **4**, 173 (1974).
33. H. G. Grube, *Z. Elektrochem.* **44**, 640 (1938).
34. Fiat Final Report, No. 831 (1946).
35. C. Jackson, B. A. Cooke, and B. J. Woodhall, Diaphragms and electrolytes, in *Industrial Electrochemical Processes*, edited by A. Kuhn, p. 575, Elsevier, New York (1971).
36. See D. A. MacInnes, *The Principles of Electrochmistry*, p. 232, Reinhold, New York (1939).
37. F. Hine, *Topics in Pure and Applied Electrochemistry*, p. 167, Society for the Advancement of Electrochemical Science and Technology (SAEST), India (1975).
38. F. Hine, *Denki Kagaku* (J. Electrochem. Soc., Jpn.) **43**, 687 (1975).
39. F. Hine, *MOL* **13**(3), 25 (1975).
40. M. Yasuda, *Soda and Chlorine* **25**, 225 (1974).
41. R. B. MacMullin, *Electrochem. Technol.* **1**, 5 (1963).
42. T. Mukaibo, *Denki Kagaku* **20**, 482 (1952).
43. F. Hine, M. Yasuda, and K. Fujita, *J. Electrochem. Soc.* **128**, 2314 (1981).
44. F. Hine, M. Yasuda, and T. Tanaka, *Electrochim. Acta* **22**, 429 (1977).
45. A. Korezynski and R. Dylewski, *Angew. Chemie* **12**(2A), 183 (1968).
46. V. N. Lazarev, *Khimiya Tekhnol.* **14**, 12 (1968).
47. V. V. Stender, O. S. Ksenzhek, and V. N. Lavarev, *Z. Prikl. Khim., Leningrad* **40**, 1293 (1967).
48. O. S. Ksenzhek and V. M. Serehitskii, *Elektrokhimiya* **4**(12), 1439 (1968).
49. R. B. MacMullin and G. A. Muccini, *AIChE J.* **2**, 393 (1956).
50. R. Itai, S. Shibuya, T. Matsumura, and G. Ishi, *J. Electrochem. Soc.* **118**, 1709 (1971).
51. H. B. Beer, British Patent, 6490/67 (1967).
52. O. deNora, *Chem. Ing. Tech.* **42**, 222 (1970); **43**, 182 (1971).
53. J. Horacek and S. Puechaver, *Chem. Eng. Prog.* **67**(3), 71 (1971).
54. L. E. Vaaler, *Electrochem. Technol.* **5**, 170 (1967).
55. F. Hine, M. Yasuda, and M. Iwata, *J. Electrochem. Soc.* **121**, 749 (1974).
56. J. B. Cotton, *Chem. Ind.* p. 492 (April 26, 1958).
57. P. B. Byrne, *Mater. Prot.* **10**(3), 21 (1971).
58. R. M. Skomoroski, R. Baboian, R. Zobbi, and E. K. Camp, Plating **60**, 1115 (1973).
59. T. H. Lewis, Jr., Paper No. 194, Corrosion/79, Atlanta, GA, March 1979.
60. M. A. Warne and P. C. S. Hayfield, *Trans. Inst. Met. Finish.* **45**, 83 (1967).
61. I. H. Warren, D. Wensley, and K. Seto, Paper presented at the 78th meeting, AIChE, Salt Lake City, UT, August 1974.
62. J. R. Newberry, W. C. Gardiner, A. J. Holmes, and R. F. Fogle, *J. Electrochem. Soc.* **116**, 114 (1969).

63. I. H. Warren, G. E. Olsen, and K. Seto, Paper presented at the Fourth Joint Chemical Engineering Conference of CSChE/AIChE, Vancouver, Canada, September 1973.
64. M. Baker, J. Cormier, and R. Scalliet, Abstract No. 442, meeting of Electrochemical Society, Sattle, WA, May 1978.
65. T. J. Navin, Abstract No. 444, meeting of Electrochemical Society, Seattle, WA, May 1978.
66. F. Greweldinger, Abstract No. 445, meeting of Electrochemical Society, Seattle, WA, May 1978.
67. T. K. Suonpera and J. P. T. Karikko, Abstract No. 446, meeting of Electrochemical Society, Seattle, WA, May 1978.
68. N. W. Meyers, Abstract No. 461, meeting of Electrochemical Society, Seattle, WA, May 1978.
69. D. Landolt and N. Ibl, *J. Appl. Electrochem.* **2**, 201 (1971).
70. S. Shibata and M. P. Sumino, *Electrochim. Acta* **16**, 1089 (1971).
71. S. Shibata, *Electrochim. Acta* **17**, 395 (1972).
72. W. Visscher and M. Blijlevens, *Electrochim. Acta* **19**, 387 (1974).
73. J. P. Hoare, *J. Electrochem. Soc.* **127**, 1758 (1980).
74. J. P. Hoare, *Electrochim. Acta* **26**, 225 (1981).
75. D. V. Kokoulina, Yu. I. Krasoviskaya, and L. I. Krishtalik, *Sov. Electrochem.* **7**, 1105 (1971).
76. S. D. Khodkevich, N. E. Veselovskaya, L. M. Yakimenko, and O. L. Danilova, *Sov. Electrochem.* **7**, 336 (1971).
77. S. D. Khodkevich, I. E. Veselovskaya, L. M. Yakimenko, and L. A. Gus'kova, *Sov. Electrochem.* **6**, 126 (1970).
78. H. B. Beer, Metal anodes, in *Diaphragm Cells for Chlorine Production*, p. 11, Society of Chemical Industry, London (1977).
79. O. de Nora, *Chem. Ing. Tech.* **42**, 222 (1970); *Chem. Ing. Tech.* **43**, 182 (1971).
80. J. Horacek and S. Puechaver, *Chem. Eng. Prog.* **67**(3), 71 (1971).
81. G. Faita and G. Fiori, *J. Appl. Electrochem.* **2**, 31 (1972).
82. A. T. Kuhn and C. J. Mortimer, *J. Appl. Electrochem.* **2**, 283 (1972).
83. E. A. Kalinovskii, R. U. Bonder, and N. N. Meshkova, *Sov. Electrochem.* **8**, 1430 (1972).
84. R. C. Erenburg, L. I. Krishtalik, and V. I. Bystrov, *Sov. Electrochem.* **8**, 1690 (1972).
85. A. T. Kuhn and C. J. Mortimer, *J. Electrochem. Soc.* **120**, 231 (1973).
86. D. Galizzioli, F. Tantardini, and S. Trasatti, *J. Appl. Electrochem.* **4**, 57 (1974).
87. K. J. O'Leary and T. J. Navin, Abstract No. 257, meeting of Electrochemical Society San Francisco, CA, May 1974.
88. *Electrodes of Conductive Metallic Oxides*, edited by S. Trasatti, Parts A and B, Elsevier, Amsterdam (1980, 1981).
89. F. Hine, M. Yasuda, and T. Yoshida, *J. Electrochem. Soc.* **124**, 500 (1977).
90. F. Hine, M. Yasuda, T. Noda, T. Yoshida, and J. Okuda, *J. Electrochem. Soc.* **126**, 1439 (1979).
91. J. W. Mellor, *A Comprehensive Treatise on Inorganic and Theoretical Chemistry*, Vol. 15, p. 520, Longman, London (1970).
92. C. Iwakura, K. Hirao, and H. H. Tamura, *Electrochim. Acta* **22**, 329 (1977).
93. J. Llopis, I. M. Tordesillas, and J. M. Alfayate, *Electrochim. Acta* **11**, 623 (1965).
94. J. Llopis and M. Vázzuez, *Electrochim. Acta* **11**, 633 (1965).
95. J. Llopis, J. M. Gamboa, and J. M. Alfayate, *Electrochim. Acta* **12**, 57 (1967).
96. V. V. Gorodetskii, M. M. Pecherskii, Ya. B. Skuratnik, M. A. Dembrovskii, and V. V. Losev, *Sov. Electrochem.* **9**, 856 (1973).
97. A. T. Kuhn and P. M. Wright, *J. Electroanal. Chem. Interfacial Electrochem.* **41**, 329 (1973).

98. M. Antler and C. A. Butler, *Electrochem. Technol.* **5**, 126 (1967).
99. T. Loucka, *J. Appl. Electrochem.* **7**, 211 (1977).
100. T. Matsumura, R. Itai, M. Shibuya, and G. Ishi, *Electrochem. Technol.* **6**, 402 (1968).
101. A. T. Kuhn and P. M. Wright, Electrodes for industrial processes, in *Industrial Electrochemical Processes*, edited by A. Kuhn, p. 525, Elsevier, Amsterdam (1971).
102. D. L. Caldwell and M. J. Hazelrigg, Cobalt spinel based chlorine anodes, in *Modern Chlor-Alkali Technology*, edited by M. O. Coulter, p. 121, Ellis Horwood, Chichester, England (1980).
103. D. L. Caldwell, Production of chlorine, in *Comprehensive Treatise of Electrochemistry*, edited by J. O'M. Bockris, B. E. Conway, E. Yeager, and R. E. White, Vol. 2, p. 105, Plenum, New York (1981).
104. M. P. Grotheer and C. J. Harke, Chlorine Bicentennial Symposium, p. 209, Electrochemical Society (1974).
105. T. J. Navin, Abstract No. 444, meeting of Electrochemical Society, Seattle, WA, May 1978.
106. F. Hine, M. Yasuda, and H. Sato, *Corrosion NACE* **33**, 181 (1977).
107. M. R. Rifi and F. H. Covitz, *Introduction to Organic Electrochemistry*, pp. 85, 183, Marcel Dekker, New York (1974).
108. S. Yoshizawa, M. Matsuda, R. Hiramatsu, and M. Nishida, *Kogyo Kagaku Zasshi* (J. Chem. Soc. Jpn., Ind. Chem. Section) **51**, 89 (1948); **52**, 18 (1949).
109. G. Angel, T. Lunden, S. Dahlerus, and R. Brännland, *J. Electrochem. Soc.* **99**, 435 (1952); **102**, 124, 246 (1955).
110. K. Hass, *Electrochem. Technol.* **5**, 246 (1967).
111. F. Hine, S. Matsuura, and S. Yoshizawa, *Electrochem. Technol.* **5**, 251 (1967).
112. F. Hine, M. Yasuda, F. Wang, and K. Yamakawa, *Electrochim. Acta* **16**, 1519 (1971).
113. R. B. MacMullin, *Chlorine*, ACS Monograph No. 154, p. 151 (1962).
114. H. Kita, *Electrode Processes*, p. 79, edited by E. Yeager and E. Eisenmann, Electrochemical Society (1970).
115. G. Kissel, F. Kulesa, C. R. Davidson, and S. Srinivasan, Abstract No. 490, meeting of Electrochemical Society, Seattle, WA, May 1978.
116. M. Prigent, L. J. Mas, and F. Verillon, Abstract No. 491, meeting of Electrochemical Society, Seattle, WA (May 1978).
117. D. W. Carnell and C. R. S. Needes, Abstract No. 260, meeting of Electrochemical Society, Boston, MA, May 1979.
118. W. W. Carlin and W. B. Darlington, Abstract No. 261, meeting of Electrochemical Society, Boston, MA (May 1979).
119. I. Malkin and J. R. Brannan, Abstract No. 262, meeting of Electrochemical Society, Boston, MA (May 1979).
120. M. C. M. Man and A. C. C. Tseung, Abstract No. 263, meeting of Electrochemical Society, Boston, MA (May 1979).
121. British Patent 1,556,452.
122. S. Matsuura, Y. Ozaki, K. Motani, Y. Ohashi, and Y. Onoue, Abstract No. 417, meeting of Electrochemical Society, St. Louis, MO, May 1980.
123. Japanese Patent Sho 25-2505.
124. K. Kanzaki and K. Fukatsu, *Denki Kagaku* (J. Electrochem. Soc. Jpn.) **19**, 255 (1951); **23**, 169 (1955).
125. F. Hine, M. Yasuda, and M. Watanabe, *Denki Kagaku* **47**, 401 (1979).
126. J. P. Hoare, *The Electrochemistry of Oxygen*, Wiley-Interscience, New York (1968).
127. E. Yeager, paper presented at the Japan Soda Industry Association in the Kansai Area, Osaka, April 18, 1980. Japanese translation: *Soda and Chlorine* **31**, 147 (1980).
128. V. H. Thomas, paper presented at the 5th Symposium on Soda Industry, Kyoto, November 1981.

Chapter 13

CURRENT DISTRIBUTION
AND POTENTIAL DISTRIBUTION

13.1. PRIMARY CURRENT DISTRIBUTION

Much attention has been given to the design of electrolytic cells in the field of modern chemical engineering. One of the most important problems is the distribution of potential or current in such cells. The yield of products and the energy consumption are closely related factors. In some cases, quality of products depends on the potential distribution. In the case of electrochemical oxidation and/or reduction, for example, unwanted products may be generated as a result of uneven distribution of potential and current on the working electrode. In the case of electroplating and also of electrochemical winning of metals, the distribution of deposit, the current efficiency, and the soundness and structure of deposited metal are functions of the current distribution. Beginning with the pioneering efforts of Kasper, Wagner, and others,[1-7] the mathematical basis for predicting the distribution of potential and current has been firmly established. Because of mathematical difficulties associated with boundary conditions that describe polarization phenomena at the electrode surface, a completely general solution has not yet been obtained.

For mathematical simplicity, the effect of polarization may be neglected in the present analysis of the electric field. The resulting "primary distribution" of potential and/or current is thus an approximation, but it provides results that are useful and which serve as a firm basis for more complex problems.

According to electromagnetic theory, the distributions of potential and current in the electric field are described by the Laplace differential equation. The solution of the differential equation is a function only of the geometry of the field if potential is uniform over the electrode surface, thus giving the primary distribution of potential and/or current, as mentioned above. Solutions yield the "secondary distribution" when the effect of polarization at electrodes is mathematically taken into account, although calculation of the Laplace equation with these conditions is generally complicated.

The primary distribution of current and/or potential is obtained from solution of the Laplace equation

$$\text{div grad } V = 0 \qquad (13.1)$$

The local current density i_x is determined by Eq. (13.2)

$$i_x = -\kappa \text{ grad } V \qquad (13.2)$$

where κ is the conductivity of electrolytic solution. Equation (13.2) is a type of Ohm's law.

We may consider a rectangular cell having a pair of parallel flat-plate electrodes, as shown in Fig. 13.1. The potential of the two electrodes are assumed to be 0 and V_0, respectively. The Laplace equation and the boundary conditions are as follows:

$$\frac{\partial^2 V}{\partial x^2} + \frac{\partial^2 V}{\partial y^2} = 0 \qquad (13.3)$$

$$V = 0 \qquad \text{at } x = 0$$

$$V = V_0 \qquad \text{at } x = a$$

$$\frac{\partial V}{\partial y} = 0 \qquad \text{at } y = 0$$

$$\frac{\partial V}{\partial y} = 0 \qquad \text{at } y = b$$

The solution of this equation is represented by a Fourier series, and is relatively simple because of the simple geometry that bounds the electric field.

In the more general case, mathematical difficulties arise because the domain of the electric field under consideration is limited by the electrolytic cell. In fact, if the field were unrestricted in all directions, analysis would be much simpler. However, in the case of parallel-plate electrodes in a rectangular cell, which a commonly used type of bath in electrochemistry

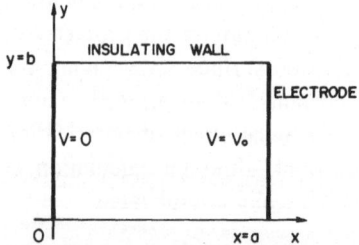

FIGURE 13.1. A rectangular cell with parallel flat-plate electrodes.

and in the electrochemical industry, the mathematical treatment becomes more difficult due to the presence of insulating tank walls surrounding the electrodes.

A two-dimensional field for electrodes of relatively simple geometry can be obtained by the method of conformal transformation. We may assume an equation

$$w = \log z \tag{13.4}$$

where w and z are the complex function:

$$w = u + iv \tag{13.5}$$

and

$$z = x + iy \tag{13.6}$$

where i is the imaginary unit. Equation (13.4) is rewritten as follows:

$$u + iv = \frac{1}{2} \log(x^2 + y^2) = i \tan^{-1}\left(\frac{y}{x}\right) \tag{13.7}$$

Therefore,

$$x^2 + y^2 = e^{2u} \tag{13.8}$$

and

$$\frac{y}{x} = \tan v \tag{13.9}$$

Equation (13.8) appears as concentric circles on the $x - y$ plane with a parameter u, while Eq. (13.9) represents straight lines radiating from the origin, with slope $\tan v$, as shown in Fig. 13.2. A straight line on the w plane, $u = u_1$, for example, corresponds to a circle on the z plane, and $v = v_1$ is related to a radial line on the z plane. Therefore, if two electrodes are positioned at a pair of circles, shown by the heavy lines in Fig. 13.2, the concentric circles between the two electrodes denote equipotential lines having an equal interval, and the radial lines are the flow lines of electric current. As shown in Fig. 13.2, the smaller the diameter, the smaller is the gap between neighboring circles, indicating that the potential gradient becomes large as the diameter decreases.

The conformal transformation is a method for calculating a complex potential field by means of transmapping between two holomorphic functions such as z and w shown above. The potential field on the z plane can be obtained if a function $w = f(z)$ is known. Several examples of these mappings can be found in textbooks of mathematics.

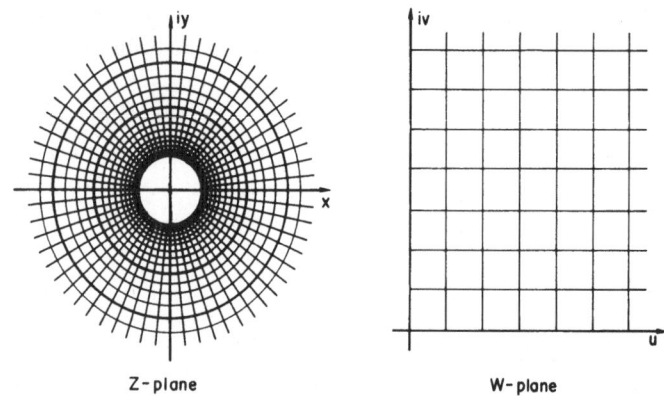

FIGURE 13.2. Conformal transformation with $w = \log z$.

Often, the potential field under consideration is not always represented by a known function. More complicated fields may then be analyzed by the Schwarz–Christoffel transformation. The outline of this treatment is as follows.

The upper domain of the ζ plane in the inside of the convex polygon on the z plane shown in Fig. 13.3 is mapped by the equation:

$$z = C_1 \int \prod_{i=1}^{n} (\zeta - \xi_i)^{-\lambda_i} d\zeta + C_2 \tag{13.10}$$

$$\sum_{i=1}^{n} \lambda_i = 2 \tag{13.11}$$

where C_1 and C_2 are constants, and λ_i is the external angle at point A_i of the polygon. An example of the transformation is described in Fig. 13.4. The z plane illustrates an electric field consisting of a semi-infinite flat-plate electrode parallel to an infinite flat-plate electrode. The upper part of the ζ plane

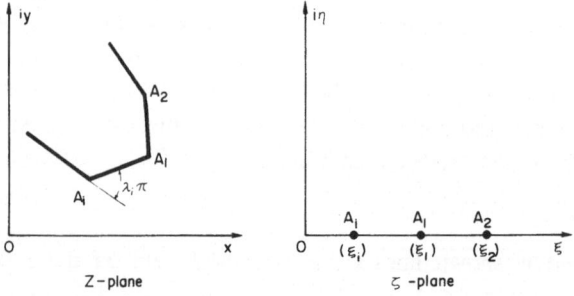

FIGURE 13.3. Schwarz–Christoffel transformation.

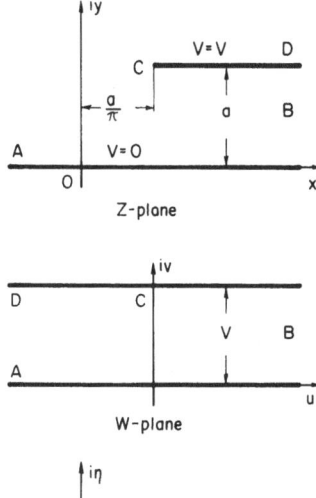

FIGURE 13.4. Schwarz–Christoffel transformation of a cell equipped with a semi-infinite electrode in parallel with an infinite electrode.

is mapped into the polygon $ABCD(A)$ on the z plane by the Schwarz–Christoffel transformation

$$z = C_1 \int \frac{\zeta - 1}{\zeta} \, d\zeta + C_2 = C_1(\zeta - \log \zeta) + C_2 \qquad (13.12)$$

because the following conditions are satisfied:

$$
\left.
\begin{aligned}
\lambda &= 1 \quad \text{at point B} \\
\lambda &= -1 \quad \text{at point C} \\
\lambda &= 1 \quad \text{at point D} \\
\lambda &= 1 \quad \text{at point A}
\end{aligned}
\right\} \quad \sum \lambda = 2
$$

For a small circle $\zeta = re^{i\theta}$ at point B on the ζ plane, y on the z plane changes from 0 to a when θ changes from π to 0. It is clear that $z = a/\pi + ia$ when $\zeta = 1$. Consequently,

$$z - ia = \frac{a}{\pi}(\zeta - \log \zeta) \qquad (13.13)$$

is obtained. The relationship between w and ζ can be obtained in a similar manner, with the result

$$w - iV = -\frac{V}{\pi} \log \zeta \qquad (13.14)$$

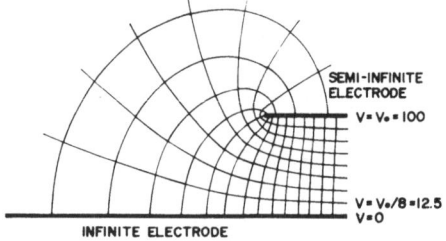

FIGURE 13.5. Distributions of the potential and current near the semi-infinite electrode.

where V is the terminal voltage. Thus the function between z and w can be obtained from Eqs. (13.13) and (13.14). Figure 13.5 illustrates the equipotential lines and the lines of current flow near the semi-infinite flat-plate electrode. This figure shows that current density is quite uniform on the lower (infinite) electrode in the region facing the upper electrode, but decreases steadily in the region beyond. Since, according to Eq. (13.2), the local current density i_x is proportional to the local gradient of potential in the y direction, perpendicular to the electrode, a rough measure may be obtained from the vertical spacing between the lines for $V = 0$ and $V = V_0/8$. For example, at a location on the infinite electrode equal to the distance a (interelectrode spacing) measured from the position of the edge of the upper (semi-infinite) electrode, the local current density is (approximately) only 38% of the uniform current density found in the region between electrodes. The crowding of potential lines at the edge of the semi-infinite (upper) electrode indicates steep gradients and thus very large local current densities, rising theoretically to infinite just at the edge.

Figure 13.6 illustrates two typical configurations of electrolytic cells: one is a rectangular cell having a pair of flat-plate electrodes (A), and the other is a cell with multicouples of electrodes (B). If the effects of four walls

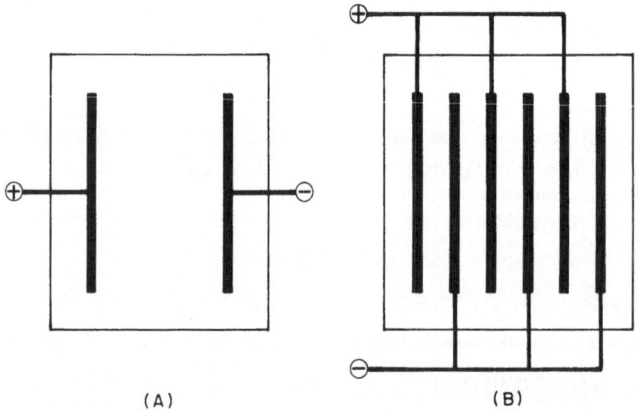

(A)   (B)

FIGURE 13.6. General configurations of an electrolytic cell.

in a rectangular cell are considered, a hyperelliptic integral appears in the equation of the Schwarz–Christoffel transformation, and the analysis is beyond possibility. However, reasonable approximations of the geometry can provide simplified models and useful results. For example, arrangement (A) in Fig. 13.6 may be treated as having remote walls front and back so that boundary conditions for the field may be simplified. Similarly, for electrodes which are not wide, the potential field may be considered as uninfluenced by the side walls of the cell. Another useful approximation is obtained by considering the field near a corner only. Similar considerations apply for the arrangement shown in Fig. 13.6(B), where simplification is possible by considering a single pair of electrodes or an end electrode.

13.1.1. Example 1: Effects of the Side Wall of an Electrolytic Cell on the Current Distribution

It is known from experience that the walls facing the back of electrodes have relatively little influence on the distribution of potential and current,

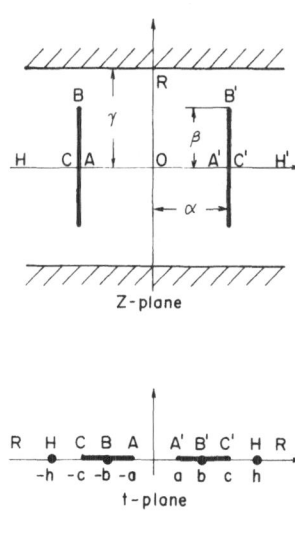

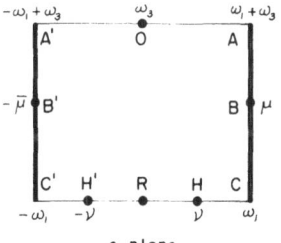

FIGURE 13.7. Conformal mapping of a cell with the parallel flat-plate electrodes.

especially when the electrodes extend close to the side walls. Accordingly, the present treatment considers only the effect of the side walls on the current distribution[6] and assumes that the back walls are far removed from the electrodes, as shown in Fig. 13.7.

Applying the methods of conformal mapping, the interior of the polygon $(OA'B'C'H'RHCBAO)$ formed at the upper half of the z plane in Fig. 13.7 can be transformed on the upper half of the t plane. Here, 2α, 2β, and 2γ are the distances between the electrodes, the width of the electrodes and of the cell, respectively.

The transformation from the z plane to the t plane can be made by the equation

$$\frac{dz}{dt} = M \frac{t^2 - b^2}{(t^2 - a^2)^{1/2}(t^2 - c^2)^{1/2}(t^2 - h^2)} \tag{13.15}$$

where M is a constant and a, b, c, and h are also constants corresponding to the positions of the points A, B, C, and H on the t plane, respectively.

Using the equation[8]

$$t^2 = \mathscr{P}(s) - e_3 \tag{13.16}$$

the t plane is transformed within the rectangle of the s plane with $2\omega_1$ and ω_3/i as the lengths of its adjacent sides. If, of the two periods $2\omega_1$ and $2\omega_3$ of the Weierstrass $\mathscr{P}$-function, $\mathscr{P}(s)$, the former is a real number, then the latter is purely imaginary, and $\omega_1 > 0$, $\omega_3/i > 0$. Therefore, from Eqs. (13.15) and (13.16) and the use of the relation

$$\mathscr{P}'(s) = -2(\mathscr{P}(s) - e_1)^{1/2}(\mathscr{P}(s) - e_2)^{1/2}(\mathscr{P}(s) - e_3)^{1/2}$$

$$e_1 = \mathscr{P}(\omega_1), \qquad e_2 = \mathscr{P}(\omega_2), \qquad e_3 = \mathscr{P}(\omega_3)$$

and

$$e_1 + e_2 + e_3 = 0$$

we obtain

$$\frac{dz}{ds} = M \frac{\mathscr{P}(s) - \mathscr{P}(\mu)}{\mathscr{P}(s) - \mathscr{P}(v)} \tag{13.17}$$

where μ and v are the coordinates of the points B and H on the s plane, respectively.

Employing the addition formula of the Weierstrass ζ-function

$$\frac{\mathscr{P}'(u)}{\mathscr{P}(u) - \mathscr{P}(v)} = \zeta(u - v) + \zeta(u + v) - 2\zeta(u)$$

we find from Eq. (13.17)

$$\frac{dz}{ds} = M \frac{\mathscr{P}(v) - \mathscr{P}(\mu)}{\mathscr{P}'(v)} \left[\zeta(\mu + v) - \zeta(\mu - v) - \zeta(s + v) + \zeta(s - v) \right] \quad (13.18)$$

Considering the relation

$$\zeta(u) = \frac{d}{du} \log \sigma(u) \quad (13.19)$$

we obtain

$$z = M \frac{\mathscr{P}(v) - \mathscr{P}(\mu)}{\mathscr{P}'(v)} \left[\{\zeta(\mu + v) - \zeta(\mu - v)\} s - \log \frac{\sigma(s + v)}{\sigma(s - v)} \right] + C \quad (13.20)$$

To obtain the constants M and C, adequate boundary conditions are substituted in Eq. (13.20), and finally we obtain

$$z = \frac{\gamma}{\pi} \left[\frac{2 v \eta_3}{\omega_3} (s - \omega_3) + \log \frac{\sigma(s - v)}{\sigma(s + v)} \frac{\sigma(\omega_3 + v)}{\sigma(\omega_3 - v)} \right] \quad (13.21)$$

$$v = \frac{\alpha}{\gamma} \frac{\omega_3}{i} \quad (13.22)$$

$$-\alpha + i\beta = \frac{\gamma}{\pi} \left[\frac{2 v \eta_3}{\omega_3} (\mu - \omega_3) + \log \frac{\sigma(\mu - v)}{\sigma(\mu + v)} \frac{\sigma(\omega_3 + v)}{\sigma(\omega_3 - v)} \right] \quad (13.23)$$

Because of the uniform distribution of potential, as is clear from the nature of the s plane in Fig. 13.7, Eq. (13.21) represents directly the relation between the z plane and the w plane. It gives the potential function, $s = f(z)$, of the electric field, which is depicted in the z plane and may be obtained by solving Eq. (13.21).

If the conjugate complex is denoted by $\mathfrak{R}$, the intensity of the electric field $\mathfrak{F}_s$ can be represented as

$$\mathfrak{F}_s = \mathfrak{R} \left[i \frac{ds}{dz} \right]_s \quad (13.24)$$

Substituting Eqs. (13.17) and (13.21) in Eq. (13.24)

$$\mathfrak{F}_s = -i \frac{\pi}{\gamma} \frac{\mathscr{P}(v) - \mathscr{P}(\mu)}{\mathscr{P}'(v)} \frac{\mathscr{P}(s) - \mathscr{P}(v)}{\mathscr{P}(s) - \mathscr{P}(\mu)} \quad (13.25)$$

Figure 13.8 shows an example of an approximate numerical calculation of the electrical field intensity on the electrode surfaces in an electrolytic cell

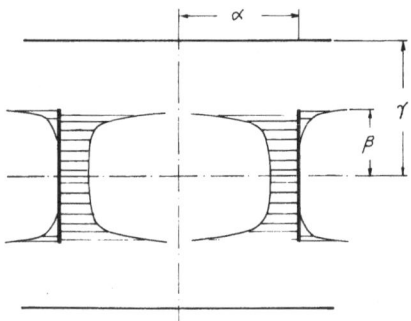

FIGURE 13.8. An example of the current distribution in a cell with flat-plate electrodes. $\beta/\alpha = 0.4981$, $\alpha = \gamma$. The total current is 90% of a cell where the electrodes touch the side walls, i.e., $\beta = \gamma$. The share of current at the inner surface is 74.94% of the total current.

whose width (2γ) is equal to the distance between the electrodes (2α). Although it is qualitatively appreciated that the larger the electrodes and the nearer they are to the walls, the more uniform will be the electrical field intensity and the current distribution, this is shown in concrete and quantitative form in Table 13.1. The percent of current allotted to the inner and the outer surface of the electrode is given; total current in percent is referred to overall current when the electrodes touch the side walls, i.e., $\beta = \gamma$.

The calculations shown in Fig. 13.8 were made for the special case where $\gamma/\alpha = 1$. Because each new cell design requires an independent set of laborious and time-consuming calculations, Fig. 13.9 has been prepared as a graphical approximation for practical use.

On the abscissa of Fig. 13.9 is scaled, from left to right, the percentage of the width of the electrode to that of the cell ($\beta/\gamma \times 100\%$), so that the top scale from right to left shows the percentage of the gap between the edge of

TABLE 13.1
Current Distribution in Cells with $\alpha = \gamma$ (see Fig. 13.7)

Electrode width β/γ(%)	Total current (%)	Percent of current allotted (%)	
		Inner surface	Outer surface
100	100	100	0
94.95	99.90	97.48	2.52
84.05	99.00	92.02	7.98
72.35	97.00	86.50	13.51
57.90	93.00	78.93	21.07
49.81	90.00	74.94	25.06

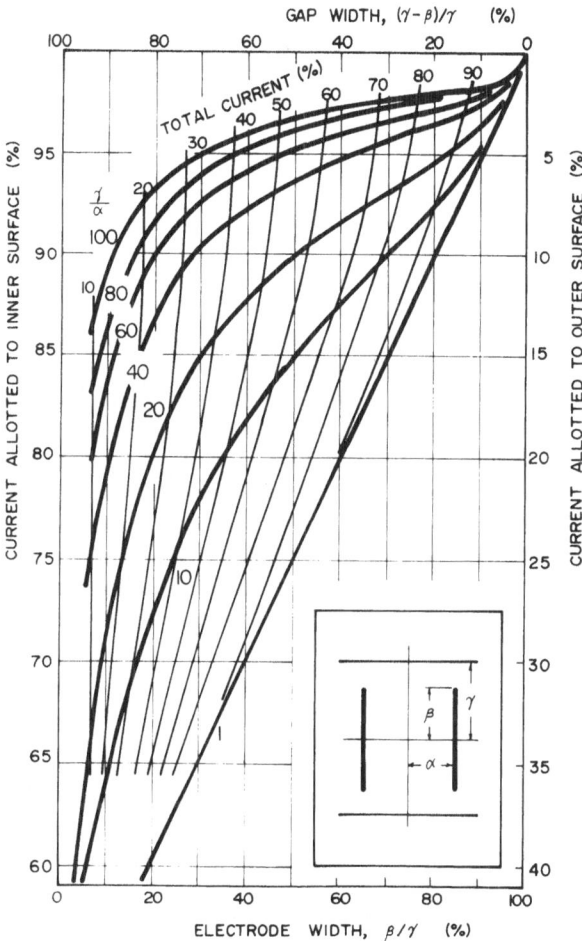

FIGURE 13. 9. Current distribution at the electrode surface.

the electrode and the insulating wall, $(\alpha - \beta)/\gamma \times 100\%$. The ordinate indicates the percentage of the electric current at the inner surface to the total current on the upward scale and that at the outer surface on the downward scale. The inner surface must carry 100% of the current when the electrode is extended to the cell wall; the share of both surfaces are equal in the extreme case of a point electrode. The current distribution is shown by the S-shaped lines, each of which represents a constant value of γ/α as labeled, from 1 to 100.

The finer lines numbered from 10 to 90 are the lines of equipercentage of the total current referred to the current when the electrodes touch the side walls. This percentage shows a characteristic decrement as the gap between electrode and wall grows larger.

The use of Fig. 13.9 can be clarified by consideration of a practical example. Assume electrodes 40 cm in width, 2 cm apart, immersed in a cell 50 cm wide. Then

$$\beta/\gamma \times 100 = (40/50) \times 100 = 80\%$$

$$\gamma/\alpha = 50/2 = 25$$

Although the diagram lacks a line at $\gamma/\alpha = 25$, one estimates the value: share of the inner surface, 95.5%; share of the outer surface, 4.5%. The current is about 85% of its value for the widest electrodes, since the point located above lies between the finer lines for 80% and 90%.

As shown in Fig. 13.8, one may conclude that when cells of the same size and form are in question, larger electrodes cause better uniformity in the current distribution at the electrode surface, as far as the inner surface is concerned. This is consistent with practical experience and the mathematical analysis serves as an exact and concrete basis.

Electrodes which fit closely at the cell wall would seem at first glance to be ideal, since the current distribution is best and the total current is highest. However, in actual practice, technical difficulties arise such as insufficient agitation of solution around the electrodes. When gases are evolved and must be expelled rapidly, as in the horizontal amalgam-type chlorine cell, it is advisable to have wider gaps between electrodes and cell walls than otherwise necessary. Such dilemmas are often encountered, and other view-points are necessary in order to arrive at the most economical cell design.

With regard to this question, the S-curves in Fig. 13.9 afford insights. The ratio γ/α becomes nearly 100 in an electrolyzer 40 to 50 cm wide, with the electrodes as close to each other as 0.5 cm, as is exemplified in the horizontal amalgam cell. This type of horizontal cell requires special consideration, since the mercury cathode covers the entire cell profile, and β and γ may be nearly equal. In this case, neither the total current nor the distribution at inner and outer anode surfaces change much for the ratios of γ/α between 50 and 100, as seen in Fig. 13.9. If β/α becomes smaller than 95%, however, the ratio γ/α has more importance. Ordinary electrolytic cells have γ/α equal to 10 to 30, and there are scarcely any industrial cells with γ/α as low as 1. Examples might be found in cells for electroplating, but cathodes are of various shapes and auxiliary anodes are often used, making the geometric form of the cell quite different from the present model. It is not rare in laboratories to encounter a case where γ/α is 0.1, as in the Haring cell used to test throwing power in electroplating, or when electrodes are simply inserted in a beaker. Although such a case is not shown in Fig. 13.9, it is almost identical with the line for $\gamma/\alpha = 1$.

13.1.2. Example 2: Effects of the Backside Wall of an Electrolytic Cell on the Current Distribution

For simplicity, we may assume that the backside wall of the cell is located close to the electrode, but that the side wall is remote, as shown by the z plane in Fig. 13.10.[7]

The Schwarz–Christoffel transformation between the t plane and the z plane involves Jacobian elliptic functions and is as follows:

$$\frac{dz}{dt} = A \frac{1 - k^2 \operatorname{sn}^2 a\, t^2}{(1 - k^2 \operatorname{sn}^2 \mu\, t^2)(1 - t^2)^{1/2}(1 - k^2 t^2)^{1/2}} \tag{13.26}$$

Substituting

$$t = \operatorname{sn}(u, k) \tag{13.27}$$

and adequate boundary conditions into Eq. (13.26), we have

$$\frac{z}{\gamma} = \frac{2}{\pi}\left(\frac{\operatorname{cn} v \operatorname{dn} v}{\operatorname{sn} v(1 - k^2 \operatorname{sn}^2 a \operatorname{sn}^2 v)} u - \Pi(u, \mu)\right) \tag{13.28}$$

where a, μ, and v are the coordinates shown in Fig. 13.10, and $\Pi(u, \mu)$ is the Π-function:

$$\Pi(u, \mu) = k^2 \operatorname{sn} \mu \operatorname{cn} \mu \operatorname{dn} \mu \int_0^u \frac{\operatorname{sn}^2 u \, du}{1 - k^2 \operatorname{sn}^2 \mu \operatorname{sn}^2 u} \tag{13.29}$$

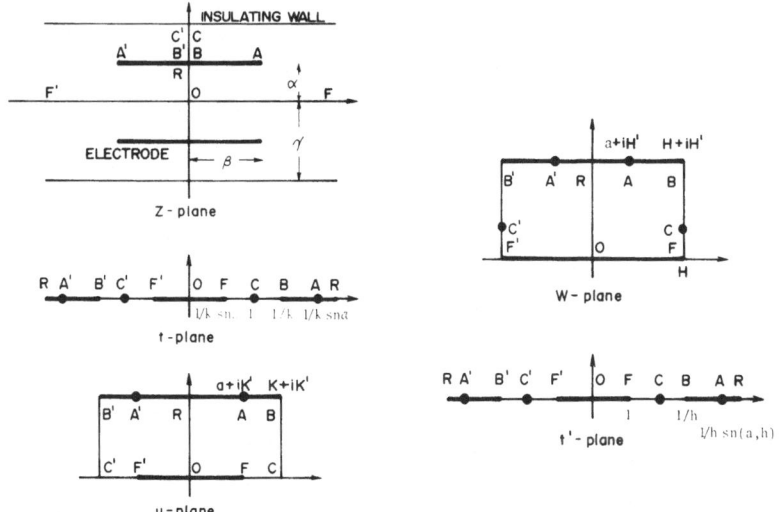

FIGURE 13.10. Conformal mapping of a cell equipped with parallel flat-plate electrodes.

In the u plane, a small electrode (FOF') is located at a cell wall (COC') and is parallel to a large electrode, which extends to the cell width (BRB'). Although the u plane has a relatively simple form, calculation of the current distribution in this plane is still complicated. Therefore, we may consider an additional mapping into the w plane, in which both electrodes extend all the way across the cell so that the current distribution is uniform. The upper part of a parameter plane t' can be transformed to the inside of the w plane by a simple equation:

$$t' = \text{sn}(x, h) \tag{13.30}$$

It is clear that the t' plane resembles the t plane, but these functions have different moduli, k and h, as shown by Eqs. (13.27) and (13.30).

Although it is difficult to derive an explicit form between z and w, the point-to-point correspondence on both planes can be obtained numerically, and thus the electric field in the z plane is obtained.

Assume that an electrolytic cell is equipped with a pair of parallel flat plate electrodes in which the interelectrode gap is 2α, and that both electrodes extend all the way across the cell width, which is 2β. In this case, it is clear that the current distribution is uniform, and the total current at a given terminal voltage is proportional to β/α. We may call it the standard cell.

On the other hand, the current passing through the cell shown by the z plane in Fig. 13.10 is not a linear function of β/α but is proportional to H/H', where H and H' are the complete elliptic integrals of the first kind (see the w plane). The ratio $(H/H')/(\beta/\alpha)$ represents the total current for the cell under discussion (the z plane) referred to the standard cell at a given voltage, and a/H is the current allotted to the inner surface of the working electrode. The ratio $(H/H')/(\beta/\alpha)$ is always larger than unity, and increases with the decrease of β/α. Figure 13.11 shows the total current $(H/H')/(\beta/\alpha)$ as a function of the electrode width β/α and the cell size γ/α. The current allotted to the inner surface is represented by a/H and is shown by a vertical scale in Fig. 13.11. The current allotted to the outer surface of the working electrode may be determined as $(H - a)/H$.

In the case of $\gamma/\alpha = \infty$, a pair of parallel electrodes is positioned within a very large cell, with no effect of the cell wall on the current distribution. Calculation was made only for γ/α values of 1.1 and 2.5 because of mathematical inconvenience when γ/α is large. However, the effect of the side wall is more important than that of the backside wall when γ/α is large.

For a small laboratory cell with dimensions $\alpha = 5$ cm, $\beta = 5$ cm, $\gamma = 6$ cm, the total current passing through the working electrode shown in the z plane is about 1.6 times larger than that of the standard cell, with 83 % being allotted to the inner surface of electrode. Thus, the average current density on the inner surface is as large as 1.33 times ($= 1.6 \times 0.83$) the uniform current density in the standard cell. For an electrode of greater

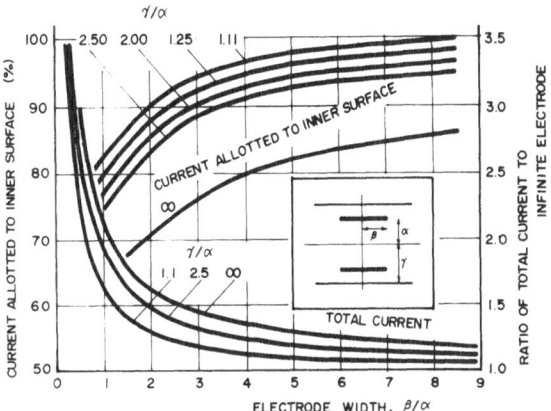

FIGURE 13.11. Current distribution at the electrode surface and the ratio of the total current to that of the infinite electrode.

width with $\beta = 25$ cm, the total current is about 1.1 times that of the standard cell, or the current increases to about 3.5 times ($=5 \times 1.1/1.6$) that of a small electrode, whereas the electrode is five times as wide.

The scale-up of an electrolytic cell can be made in two ways: increase of the current density, and increase of the surface area of the working electrode. The interelectrode gap remains generally unchanged, or else the gap is shortened in order to minimize the cell voltage. Consequently, the scale-up of the cell geometry is two-dimensional and therefore the current distribution profile is affected. It is evident that the current distribution becomes uniform (with the exception of the profile at the electrode edge) since the normalized width β/α increases with scaling up of the cell geometry.

Figure 13.6B illustrates a monopolar-type cell equipped with a number of electrode pairs. The electric field provided by a couple of these electrodes (with the exception of the end electrodes) is illustrated for this case by the z plane in Fig. 13.12. The case of $\alpha = \gamma$ of the cell shown by the z plane in Fig. 13.10 is similar, but mathematical treatment of Fig. 13.12 is simpler.

The Schwarz–Christoffel transformations between z and t, and between t and u are as follows:

$$z = \frac{\alpha}{\pi} \log \frac{1+t}{1-t} \tag{13.31}$$

$$t = \text{sn } u \tag{13.32}$$

Therefore, the local current density i_x on the working electrode is represented by the equation

$$i_x = \left| \frac{\kappa\pi \text{ dn } u'}{2\alpha k \text{ cn } u'} \right| \tag{13.33}$$

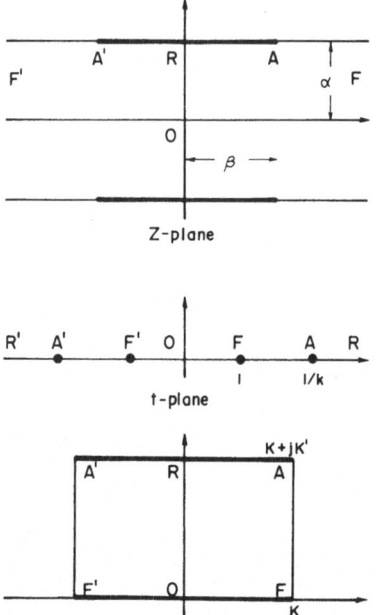

FIGURE 13.12. Schwarz–Christoffel transformation of a cell equipped with parallel electrodes.

where κ is the conductivity of the electrolytic solution, k is the modulus of the elliptic function, and u' is the real part of the complex number u.

Figure 13.13 shows an example of calculation of the current distribution on the working electrode. The dotted line is the current density on the infinite electrode working at the same voltage as that of the electrode under discussion. Since, in this case, the electrode is small in comparison with the electrode gap ($\beta/\alpha = 0.25$), the current density is unevenly distributed and is significantly high.

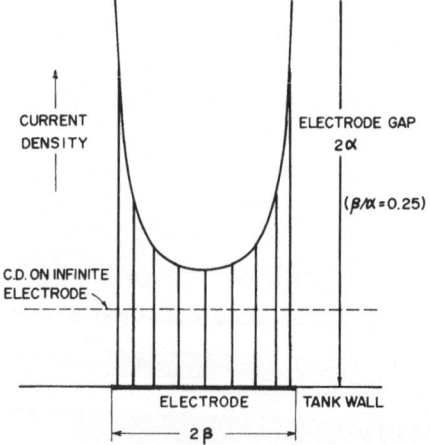

FIGURE 13.13. Current distribution on a flat-plate electrode.

13.2. EFFECT OF THE ELECTRODE RESISTANCE

In the preceding section, the electric resistance of the electrode was assumed to be negligible. In practice, however, there are some ohmic voltage drops within the electrode, and the current distribution in the cell is affected. These subjects have been investigated by Komagata,[9] Tobias and Wijsman,[10] and others.

Consider that a flat-plate resistant electrode (A) is positioned in an electrolytic cell in parallel with a conductive counter electrode (B), also flat plate, as shown in Fig. 13.14. Also assume that the potential of electrode B is 0, and is a basis. The symbols and the units used here are listed:

a	width of electrode A	cm
I	current	A
$I_0 = i(0)\,a\lambda$	current	A
I_m	current in electrode A	A
i	electrolytic current density	A/cm^2
$i(0)$	i at $x = 0$	A/cm^2
$i_m = \dfrac{I_m}{a}$	current in the working electrode per unit length	A/cm
$i_m(0)$	i_m at $x = 0$	A/cm
L	interelectrode gap	cm
$R = \rho \dfrac{L}{ax}$	electric resistance of electrolytic solution	ohm
$R_m = \rho_m \dfrac{x}{at}$	electric resistance of electrode A	ohm
$R_s = Rax = \rho L$		ohm cm^2
t	electrode thickness	cm
V	voltage	V
x	distance from the top end	cm
ρ	electrical resistivity of solution	ohm cm
ρ_m	electrical resistivity of electrode A	ohm cm
$\lambda^2 = \dfrac{R_m}{R_s} = \dfrac{R_m}{\rho L}$	resistance ratio	1/cm^2

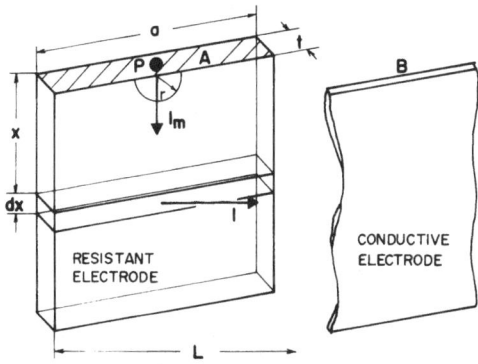

FIGURE 13.14. A resistant electrode in parallel with a perfectly conductive electrode.

First, we may discuss a case where the electric current is fed from the surface of the resistant electrode A, hatched in Fig. 13.14. The voltage drop in an element dx is represented as follows:

$$-dV = R_m \, dI_m$$

Assuming that i_m is unchanged in a small distance dx, we have

$$-\frac{dV}{dx} = R_m i_m \tag{13.34}$$

The decrement of the electric current $-dI_m$ in dx is equal to the electrolytic current:

$$-dI_m = ia \, dx = \frac{I}{x} \, dx$$

Substituting

$$R_s = Rax = \rho L \tag{13.35}$$

into the above equation, we have

$$-dI_m = -a \, di_m = \frac{V}{R} \frac{dx}{x} = \frac{Va}{R_s} \, dx$$

and hence

$$-\frac{di_m}{dx} = i = \frac{V}{R_s} \tag{13.36}$$

With Eqs. (13.34) and (13.36), we have

$$\frac{d^2 i_m}{dx^2} = \frac{1}{R_s}\left(-\frac{dV}{dx}\right) = \frac{R_m}{R_s} i_m \tag{13.37}$$

or

$$i_m = Ae^{-\lambda x} + Be^{\lambda x} \tag{13.38}$$

where

$$\lambda^2 = \frac{R_m}{R_s} = \frac{R_m}{\rho L} \tag{13.39}$$

Substituting the boundary conditions:

$$i_m \equiv 0 \qquad \text{at } x = \infty$$

$$i_m = i_m(0) \qquad \text{at } x = 0$$

into Eq. (13.38), we obtain

$$i_m(x) = i_m(0)\, e^{-\lambda x} \tag{13.40}$$

Substituting Eq. (13.40) into Eq. (13.36), the electrolytic current density i is obtained:

$$i = i_m(0)\, \frac{1}{\lambda}\, e^{-\lambda x} \tag{13.41}$$

If we put

$$i = i(0) \qquad \text{at } x = 0$$

the current density is represented by a simple form

$$\frac{i}{i(0)} = e^{-\lambda x} \tag{13.42}$$

and is illustrated in Fig. 13.15 as a function of a parameter λ, or the resistance ratio R_m/R_s.

It is clear with Eq. (13.39) that λ increases with decrease of the interelectrode gap L, resulting in uneven current distribution along the working electrode. In other words, it is desirable to use a more conductive electrode when two electrodes are brought to closer each other so as to minimize the IR drop in the electrolyte.

The total current I can be obtained by integration of Eq. (13.42) with respect to x, i.e.,

$$I = I_0(\cdots - e^{-\lambda x}) \tag{13.43}$$

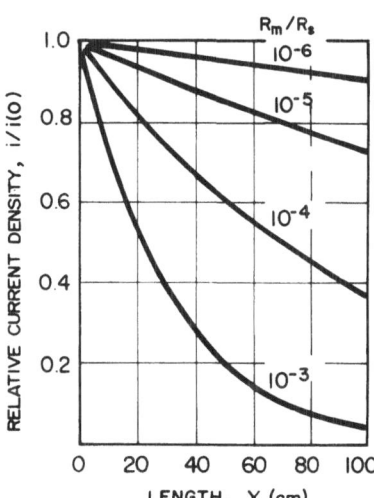

FIGURE 13.15. Current distribution on a resistant electrode. Electric power is supplied from the top end of the electrode.

It shows that the current per unit length along the electrode decreases with increase of λ.

Now, we may consider another case where the electric current is supplied from point P shown in Fig. 13.14. Thus, the current flows radially in electrode A. The differential equations corresponding to Eqs. (13.34) and (13.36) are as follows:

$$-\frac{dV}{dr} = \frac{R_m I_m}{2\pi r} \tag{13.44}$$

$$-\frac{dI_m}{dr} = \frac{2\pi r}{R_s} V \tag{13.45}$$

From these equations, we have

$$\frac{d^2 V}{dx^2} + \frac{1}{r}\frac{dV}{dr} = \lambda^2 V \tag{13.46}$$

A general solution for Eq. (13.46) is represented by an equation

$$V = AJ_0(j\lambda r) + BH_0(j\lambda r) \tag{13.47}$$

where J_0 and H_0 are the Bessel function and the Hankel function, respectively and j represent the unit of imaginary number in this section.

It is clear that $A = 0$ since V is finite at $r = \infty$. Assuming that

$$I_m = I_m(0) \qquad \text{at } r = r_0$$

a constant B is obtained:

$$B = \frac{I_m(0)}{(2\pi r_0/\lambda)jH_1(j\lambda r_0)}$$

Therefore, we have

$$i_m = i_m(0)\frac{I_m/2\pi a}{I_m(0)/2\pi r} = i_m(0)\frac{H_1(j\lambda r)}{H_1(j\lambda r_0)} \tag{13.48}$$

and

$$i = \frac{I}{2\pi r}\left(-\frac{dI_m}{dr}\right) = \frac{I_m(0)jH_0(j\lambda r)}{2\pi r_0[-H_1(j\lambda r_0)]} \tag{13.49}$$

Substituting $i = i_0$ at $r = r_0$ into Eq. (13.49)

$$\frac{i}{i_0} = \frac{iH_0(j\lambda r)}{iH_0(j\lambda r_0)} \tag{13.50}$$

With an approximation:

$$\lim_{r \to 0} jH_0(j\lambda r) = \frac{2}{\pi} \log \frac{2}{\gamma \lambda r}$$

in a small range of r, we have

$$i \simeq \log \frac{2}{\gamma \lambda r} \tag{13.51}$$

where γ is Euler's constant, and is

$$\gamma = \lim_{m \to 0} \left(\frac{1}{1} + \frac{1}{2} + \cdots + \frac{1}{m} - \log m \right) \tag{13.52}$$

For large r, on the other hand,

$$jH_0(j\lambda r) \simeq \sqrt{\frac{2}{\pi \lambda r}} \, e^{-\lambda r} \tag{13.53}$$

and hence, we have

$$i \propto \frac{e^{-\lambda r}}{\sqrt{r}} \tag{13.54}$$

Figure 13.16 illustrates an example of the electrolytic current density as a function of the radius and the resistance ratio, R_m/R_s, with $r_0 = 5$ cm.

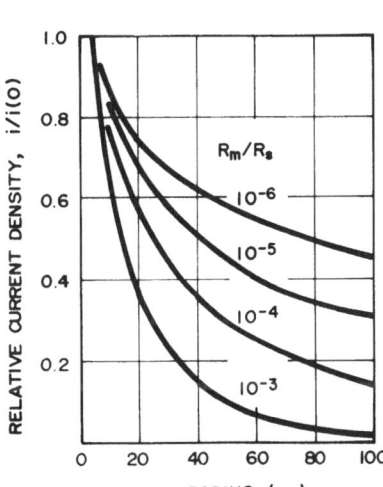

FIGURE 13.16. Current distribution on a resistant electrode. Electric power is supplied from a point on the top of the electrode.

13.3. SECONDARY CURRENT DISTRIBUTION

The current distribution is influenced only by the cell geometry if the potential is uniform over the electrode surface, as described in Section 13.1. The profile of the current distribution in the electrolytic cell may deviate from this primary current distribution when the working electrode is polarized and the overvoltage η is a function of the current density i:

$$\eta = f(i) \tag{13.55}$$

13.3.1. Effect of the Overvoltage on the Current Distribution on a Resistant Electrode

We may consider the effect of the overvoltage on the current distribution on a resistant, flat-plate electrode parallel to a counterelectrode, also flat-plate, as shown in Fig. 13.14. For simplicity, a counterelectrode does not polarize at all.

The overvoltage η is a function of the current density i and is represented by the Tafel equation:

$$\eta = \frac{RT}{\beta F} [\log i - \log i_0] \tag{13.56}$$

where i_0 is the exchange current density, and β is the transfer coefficient. For simplicity, however, a linear relation between η and i is assumed:

$$\eta = \eta_0 + ki \tag{13.57}$$

where η_0 is an intersection of the η vs. i curve at $i = 0$, and k is the slope of the curve in ohm cm^2.

The differential equations governing the subject under discussion are essentially the same as for Eqs. (13.34) and (13.36), except $V - \eta$ instead of V in Eq. (13.36), i.e.,

$$i = -\frac{di_m}{dx} = \frac{V - \eta}{R_s} \tag{13.58}$$

Substituting Eq. (13.57) into Eq. (13.58),

$$-\frac{di_m}{dx} = \frac{V - \eta_0}{R_s[1 + (k/R_s)]} \tag{13.59}$$

With Eqs. (13.34) and (13.59), we have

$$\frac{d^2 i_m}{dx^2} = \frac{R_m}{R_s[1 + (k/R_s)]} i_m \tag{13.60}$$

which is essentially the same as Eq. (13.37). Therefore, we obtain

$$i_m(x) = i_m(0)\, e^{-\lambda_1 x} \tag{13.61}$$

$$\lambda_1^2 = \frac{R_m}{R_s[1 + (k/R_s)]} = \frac{R_m}{\rho L(1 + k/\rho L)} \tag{13.62}$$

$$\frac{i}{i(0)} = e^{-\lambda_1 x} \tag{13.63}$$

where $\lambda_1 < \lambda$ since $k > 0$. The current distribution becomes uniform when the slope k of the polarization curve increases.

It is clear that

$$\mathrm{Wa} = \frac{k}{\rho L} \tag{13.64}$$

is an important nondimensional factor for determining the effect of polarization on the current distribution, and is called the Wagner number, although Wagner himself attached importance to k/ρ, but not $k/\rho L$.[4]

13.3.2. Effect of the Overvoltage on the Current Distribution on a Finite-Plate Electrode

The current distribution on a finite-plate electrode such as shown by the z plane in Fig. 13.12 has been described by Wagner.[4] He uses a linear relationship [Eq. (13.57)] of η vs. i for simplicity. That is, the terminal voltage V is represented as follows:

$$V = E + \eta_0 + \frac{k}{\rho}\left(\frac{dV}{dn}\right) \tag{13.65}$$

where E is the thermodynamic potential, and n is the direction of current.

Wagner showed a general solution of the Laplace equation, (13.2), with a Fredholm integral equation of second kind:

$$V = \int_{-1}^{+1} g_1(Z) \log[(X - Z)^2 + Y^2]\, dZ + C_2 \tag{13.66}$$

$$X = x/a, \qquad Y = y/a$$

Since the boundary conditions are given by

$$V = E + \eta_0 + \frac{k}{\rho}\left(\frac{\partial V}{\partial y}\right) \qquad \text{at } |x| < a,\ y = 0$$

$$\frac{\partial V}{\partial y} = 0 \qquad \text{at } |x| > a,\ y = 0$$

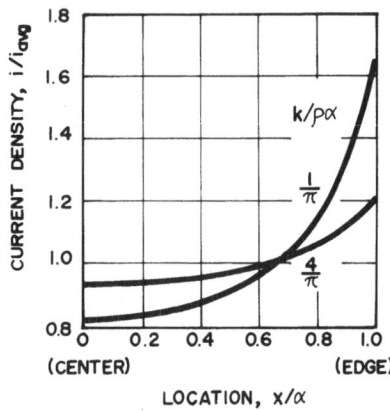

FIGURE 13.17. Effect of polarization on the current distribution on the electrode, by Wagner.

a singular solution is obtained:

$$g_1(X) = \frac{\alpha\rho}{2\pi k}(E + \eta_0 + C_2)\, g_1^*(X) \tag{13.67}$$

$$g_1^*(X) = 1 - \frac{\alpha\rho}{\pi k}\int_{-1}^{+1} g_1^*(Z)[-\log|X - Z|]\, dZ$$

$$\simeq \left[1 + \frac{\alpha\rho}{\pi k}\{2 - (1 + X)\log(1 + X) - (1 - X)\log(1 - X)\}\right]^{-1} \tag{13.68}$$

The local current density on the electrode is thus represented by an equation

$$\frac{i}{i_{avg}} = \frac{g_1^*(X)}{\int_0^1 g_1^*(X)\, dX} \tag{13.69}$$

and is illustrated as a function of $k/\rho\alpha$, as shown in Fig. 13.17. It shows that uniform current distribution can be performed when the Wagner number is large enough.

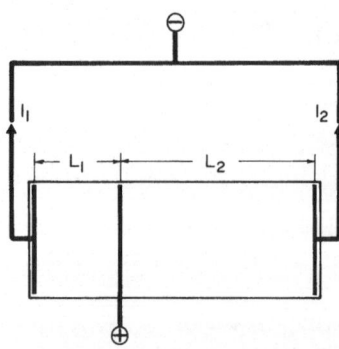

FIGURE 13.18. Haring cell for testing the throwing power of electroplating solution.

The Haring cell shown in Fig. 13.18 is used for estimating the throwing power of electroplating solutions. There are two cathodes, one at each end of the cell. The anode is located in a noncentral position between them so that the anode–cathode distances differ from each other: L_1 on one side and L_2 on the other side. The current passing through the two cathodes would be reciprocally proportional to the distance if the electrode polarization were negligible. Since the overvoltage increases with increase of the current density, the current allotted to cathode 1 decreases when $L_1 < L_2$. Wagner described the throwing power as follows:

$$\frac{I_1}{I_2} = \frac{L_2 + k}{L_1 + k} \qquad (13.70)$$

I_1/I_2 tends to unity when k increases.

The current distribution and the throwing power are practically important in the electroplating industry. Accordingly, several test methods have been developed, including the Haring cell. The Hull cell has a trapezoidal shape whose parallel sides are the insulated walls. The anode is positioned at the wall perpendicular to the insulating walls, while the cathode is located at the angled wall. Because the anode–cathode spacing varies, the current density on the cathode is uneven, and influences the quality of the deposited metal on the cathode. For example, the brightness of nickel electroplate changes along the cathode strip, and hence the electroplater can evaluate an adequate range of operating current density.

Skwirzynski and Huttly developed a mathematical treatment to obtain the current distribution in the Hull cell, but did not consider the effect of the polarization.[11]

Methods for the calculation of current distribution in electrolytic cells using computer techniques have been described.[12,13]

In conclusion, the current distribution is of importance for electrochemical system design and operation, although strict analysis is usually complicated and time-consuming. Complete analytic or computer-aided solutions for the potential field are desirable for the design of electrolytic cells. However, direct experimental methods are also useful for rapidly providing an estimate of current distribution of sufficient accuracy to aid in cell design.

REFERENCES

1. C. Kasper, *Trans. Electrochem. Soc.* **77**, 353, 365 (1939); **78**, 131, 147 (1940).
2. S. Ishizaka, *Denki Kagaku* (J. Electrochem. Soc. Jpn.) **17**, 1 (1949).
3. S. Ishizaka and H. Matsuda, Report of the Government Chemical Industrial Research Institute, Tokyo, **46**(9) (1952).

4. C. Wagner, *J. Electrochem. Soc.* **98**, 116 (1951).
5. J. T. Waber, *J. Electrochem. Soc.* **101**, 271 (1954); **102**, 344, 420 (1955); **103**, 64, 138 (1956).
6. F. Hine, S. Yoshizawa, and S. Okada, *J. Electrochem. Soc.* **103**, 186 (1956).
7. F. Hine, *Kogyo Kagaku Zasshi* (J. Chem. Soc. Jpn., Ind. Chem. Section) **56**, 875 (1956).
8. E. T. Whittaker and G. N. Watson, *A Course of Modern Analysis*, 4th ed. p. 448, Cambridge University Press, Cambridge (1927).
9. S. Komagata, Researches of the Electrotechnical Laboratory, Ministry of Communications, Tokyo, No. 294 (1930).
10. C. W. Tobias and R. Wijsman, *J. Electrochem. Soc.* **100**, 459 (1953).
11. J. K. Skwirzynski and M. Huttly, *J. Electrochem. Soc.* **104**, 650 (1957).
12. J. A. Klingert, S. Lynn, and C. W. Tobias, *Electrochim. Acta* **9**, 297 (1961).
13. J. S. Newman, *Electrochemical Systems*, Chap. 13, p. 340, Prentice-Hall, Englewood Cliffs, NJ (1973).

Chapter 14

OPTIMUM DESIGN
OF AN ELECTROLYTIC CELL

In the early 1950s, the capacity of a typical Japanese chlor-alkali plant was in the range of 200 to 300 tons per month. After thirty years of continuous improvement, many giant plants with capacities of 1000–1500 tons/day are now being operated.

It is well known that the manufacturing cost in the process industry is reduced as a result of increase in plant capacity. The advantage of scale is a basic concept of process industry economics. The investment cost of a giant plant is, however, very large. Today, some 10 billion yen or 50 million dollars or more are required even for a plant of medium size. Therefore, mistakes in design and fabrication are to be avoided, and optimum design is the most important target. The optimum design is not pursuit of a cheap plant, but is the investigation of a suitable design of a plant where the overall manufacturing cost of the products, including the fixed charge and the operating cost, must either be minimized or kept acceptably low.

In practice, there are several uncertain factors in the evaluation of plant economics such as:

1. The investment depends greatly on the price of individual equipment, the construction materials, the fabrication cost, and the plant location.

2. A detailed feasibility study is possible only with the data of practical operation. In other words, accurate estimation of economy before start-up is very limited.

3. The investment is affected by the plant layout and by local regulations and law.

4. There are several methods for estimating the plant investment cost and the operating cost, occasionally yielding substantially different results. Thus, it is difficult to find a suitable technique for cost estimation. Furthermore, other types of problems should be anticipated, such as:

5. Considerable delay of start and schedule of the plant construction due to action of nearby residents and approval and/or rejection of the local government. During the delay, the costs of construction materials and

equipment may rise considerably, money interest rates may increase, and marketability of the intended products may be affected seriously.

Consequently, the reader is cautioned that the plant design economics described in the following pages can only be considered as an example. For simplicity, we may assume that

1. The manufacturing cost consists of the investment or the fixed charge and the operating cost.
2. Depreciation, tax, insurance, maintenance, and some other factors such as supplies are proportional to the investment.
3. The fixed charge Fc is represented as follows:

$$Fc = \frac{J}{365\,P}\frac{\beta}{100} \qquad (\$/\text{ton}) \qquad (14.1)$$

where J is the investment in dollars, P is the production capacity in tons/day, and β is the interest rate of the investment in percent.
4. The method for evaluating the investment is not discussed here, but a typical range is covered.
5. Since β depends greatly on the financial situation of each company, details are not discussed.

For electrochemical process plants, several factors must be considered:

(i) the size and the number of cells to be installed;
(ii) the operating conditions such as the current density;
(iii) the degree of instrumentation and automation, and the number of employees;
(iv) the percentage of operation at less than full-design capacity, and the balance between production and demand; and
(v) the size and degree of provision for the safety of employees and for the prevention of hazards.

The labor cost per unit production is greatly affected by several factors such as the size, layout, location, and situation of the plant, the type and number of cells, the degree of automation and labor rate, and the labor quality. Estimation of the labor cost Lc is therefore complicated.

A de Nora cell plant is operated by only three men per shift, and the operating man-hours for various sizes of plant up to 36 cells is almost constant.[1] Gardiner points out that the man-hours per unit production in an Olin cell plant decrease considerably with increasing plant capacity, but he also indicates that the labor cost varies with the location and situation even though the plant capacity and the cell size are much the same.[2]

Anode adjustment of the amalgam-type chlorine cell is of importance for reducing the cell voltage, but is a labor-intensive procedure.[3] With the use of a recently developed computer-aided adjustment system, it is possible to save on both the power consumption and the operating labor costs. Introduction of this type of control system in an amalgam-type chlorine plant is effective in cost reduction, but there is an optimum point because the equipment costs are high, especially for small plants.

Figure 14.1 shows the trend of man-hours vs. capacity in Japanese chlor-alkali plants for the years 1955–1966. The logarithm of man-hours vs. the logarithm of plant capacity is a straight line with a slope of about −1.5 instead of −1.0, which is a significant result.[4] A reasonable explanation of this result is that large plants equipped with desirable automation systems are new, whereas small plants have operated for many years and modernization of these old plants is limited. Otherwise, the slope of the logarithm of man-hours vs. logarithm of plant size might be about unity if plants of the same design, but of different sizes, were compared.

In conclusion, the labor cost Lc is roughly shown by the equation:

$$Lc = \overline{Lc}/P \tag{14.2}$$

where $\overline{Lc}$ is Lc for a plant of unit production.

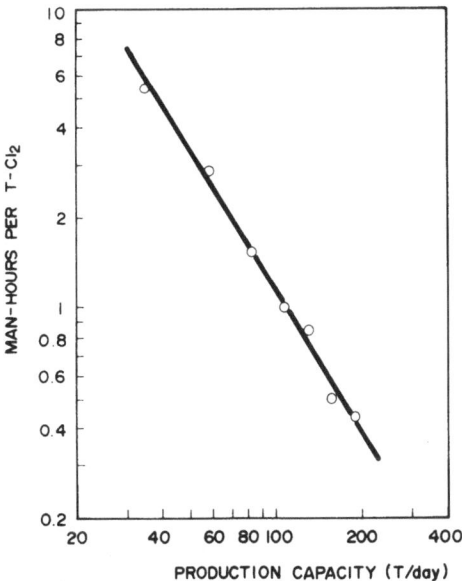

FIGURE 14.1. Man hours vs. plant capacity in the amalgam-type chlorine plant.

The labor cost is also affected by the plant location, policy of labor unions, training and morale of employees, the system of management, and the ability of managers. Of these, the factors related to human relations are difficult to change over a short period of time, but they must be improved as a long-range goal.

Another important factor of plant economy is the percent operation at less than full capacity because the scale merit in a large process plant is based on a reasonable balance between production and demand. Therefore, marketing research must be continued on an ongoing basis.

14.1. SIZE OF ELECTROLYZER AND OPTIMUM CURRENT DENSITY—A CASE STUDY

Deciding on the plant capacity, the number of cells to be installed, and the optimum conditions of operation in a plant depends on the market situation, the power cost, the labor rate, the cell type, and other factors. The plant size should meet the market, but too small a plant is not economical. In general, the manufacturing cost decreases with increasing plant size if the plant is operated at nearly full capacity. According to the author's experience, the manufacturing cost Mc of chemical products, including electrochemical products, can be represented by the equation:

$$Mc = a - b \log P \quad (\$/ton) \tag{14.3}$$

where a and b are coefficients, and P is the production capacity.

The number of cells, N, or conversely the cell current load, I, in kA, is a major factor in optimizing a plant having the capacity of P tons/day by Faraday's law, as follows:

$$P = k\xi IN \tag{14.4}$$

where k is the electrochemical equivalent,* and ξ is the current efficiency of the cell in percent.

---

* For chlorine production,

$$k = \frac{M}{nF} = \frac{35.5 \ (\text{kg/kg mol})}{26.8 \ (\text{KAh/kg mol})} \times 24 \ (\text{hr/day}) = 31.8 \ (\text{kg/kA day})$$

or

$$= 31.8 \ (\text{kg/kA day}) \times \frac{1}{1000} \ (\text{ton/kg}) = 0.0318 \ (\text{ton/kA day})$$

where M is the molecular weight, n is the number of charge transfer, and F is Faraday's constant.

It is well known that the investment cost is a function of the plant size, as follows:

$$\text{Plant cost} = \text{const (size)}^\alpha \qquad (14.5)$$

where α is a coefficient and is about 0.6 in general.[5]

While the plant size is a major factor in the cost of an electrochemical plant, the operating current density must also be considered because the production capacity is a function of current density, that is,[6]

$$\text{Plant cost} = \text{const (size)}^\alpha \text{ (current density)}^\gamma \qquad (14.6)$$

Assume that the sum J' of the investment for brine purification, handling of products, and some other parts is a function of the plant size, but is independent of the current density of the cell.

On the other hand, the cost J consists of the costs for the rectifier, the cells, and the cell room including connecting bus and piping. Thus, J varies with the current density i. The amalgam-type chlorine cell plant will be described as a case study.[4,6]

Cost was considered for three cells of different size, 7.5, 15, and 30 m^2 cathode area, operated under various current densities, 55, 70, and 85 A/dm^2, with vertical decomposers in plants having capacities of 50–400 metric tons chlorine a day.[6]

The cost of the cell room and rectifier area was calculated by Chilton's method.[7] An example showing the percentage distribution of the investment for a 100 ton/day plant having thirty 15 m^2 cells operated at 70 A/dm^2 is shown in Table 14.1.

Rectifier cost, representing 10–20% of plant cost, depends on the type of equipment, the manufacturer, and dc voltage. Costs per kW for equipment of the same type were found to vary with voltage as follows:

$$\text{Rectifier cost per kW} = \text{const (voltage)}^{-\delta} \qquad (14.7)$$

where $\delta = 0.15$–0.16.

TABLE 14.1

*Example of Investment for a Chlorine Plant Having
100 Metric Tons Chlorine per Day Capacity with
30 Cells of 15 m^2 Operated at 70 A/dm^2 (U.S. Cost)*

Cell cost	56.3 %
Rectifier cost	17.0
Bus	8.2
Piping	5.3
Buildings and others	13.2
TOTAL	100.0%

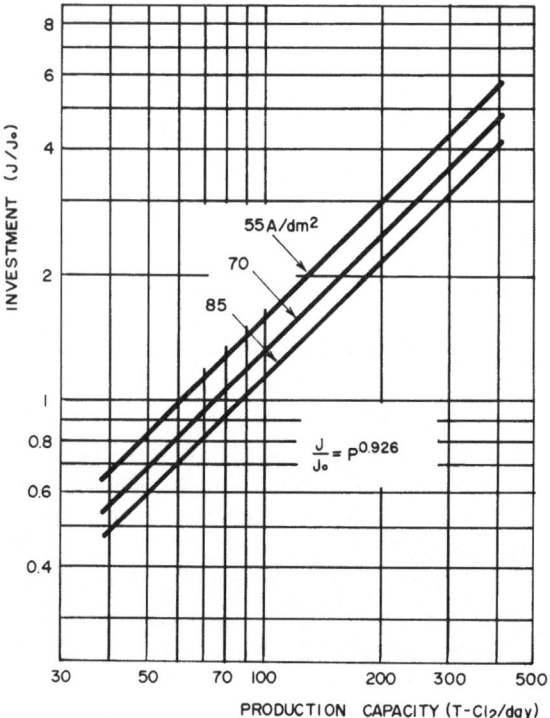

FIGURE 14.2. Investment vs. production capacity of the amalgam-type chlorine plant. Basis: 100 tons/day plant operated at 100 A/dm².

The investment J was plotted as a function of plant size P as shown in Fig. 14.2. The ordinate shows the logarithm of the index of investment, J/J_0, and the abscissa is the logarithm of the production capacity in metric tons of chlorine per day, where J_0 is the investment for a 100 ton/day plant operated at $100\ A/dm^2$ as a basis for calculation.

The three lines for various current densities are parallel and have a slope of 0.926 per decade, instead of 0.6. Thus, the cell size is a small factor in the investment in this case, and no significant difference is found in comparing the three cells.

The investment is also plotted as a function of current density, as shown in Fig. 14.3. The ordinate is the logarithm of the index of the investment and the abscissa is the logarithm of current density in A/dm^2. The parallel lines have a slope of -0.740 per decade, so that the following equation can be derived from the figures

$$J = J_0 \left(\frac{P}{100} \right)^{0.926} \left(\frac{i}{100} \right)^{-0.740} \qquad (\$) \qquad (14.8)$$

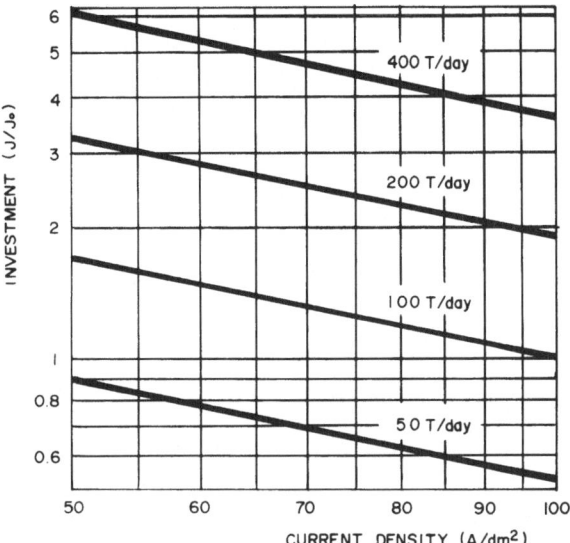

FIGURE 14.3. Investment vs. current density of the amalgam-type chlorine plant as a function of production capacity. Basis: 100 tons/day plant operated at 100 A/dm².

where J is the plant cost in dollars, J_0 is the cost, also in dollars, of a 100 ton/day plant operated at 100 A/dm², P is the plant size in metric tons in chlorine per day, and i is the current density in A/dm².

The total fixed charge, $Fc(\text{total})$, per unit production in \$/metric ton Cl_2, which consists of depreciation, tax, maintenance, and supplies, is as follows

$$Fc(\text{total}) = Fc + Fc' = \frac{J(\text{total})}{365P} \frac{\beta}{100} \tag{14.9}$$

$$Fc = \frac{P}{365P} \frac{\beta}{100} \tag{14.10}$$

$$Fc' = \frac{J'}{365P} \frac{\beta}{100} \tag{14.11}$$

where Fc and Fc' are the fixed charges based on the investments J and J', respectively, for a plant of P metric tons per day, J and J' are the investments depending on and independent of the current density, respectively, and β is a coefficient depending on the location and situation, and is about 15–40%.

Assume that the manufacturing cost Mc is the sum of the total fixed charge, $Fc(\text{total})$, and the total operating cost, $Oc(\text{total})$, as follows:

$$Mc = Fc(\text{total}) + Oc(\text{total}) \tag{14.12}$$

The total operating cost Oc(total) consists of various factors, which are classified into two types: The power cost for electrolysis and the cost for raw materials are proportional to the rate of production. On the other hand, utility cost, for example, is almost independent of the production and the rate of operation.

Since Fc and Oc depend on various factors, Mc is also a function of them. Therefore, the optimum value of a factor x, which will minimize Mc, is given by the equation:

$$\frac{\partial Mc}{\partial x} = 0 \tag{14.13}$$

14.1.1. Optimum Size and Number of Cells

The total investment of a chlorine plant includes the costs of the cell room, the rectifier, the brine treatment system, the handling of crude salt, chlorine and caustic, and other auxiliaries. Of these, the investment for the cell room and the rectifier room varies with the number of cells, but other costs are almost independent and are only a function of the plant capacity.

The cell room cost depends on the plant size, the number of cells, or the amperage of the circuit, and the operating current density of the cells, as described above. It increases slightly with the number of cells, because the cost of a small cell is rather high.

Since the rectifier cost per kilowatt is shown by Eq. (14.7), we have

$$
\begin{aligned}
\text{Rectifier cost} &= \text{const}(\text{kA of cell, } I) \\
&\quad \times (\text{number of cells, } N) \\
&\quad \times (\text{voltage per cell, } v) \\
&\quad \times (Nv)^{-\delta} \\
&= \text{const}(P/\xi)(v)^{1-\delta}(N)^{-\delta} \quad \text{(dollars)} \tag{14.14}
\end{aligned}
$$

where v is the voltage of a cell, including intercell bus bars, in volts.

Assume that Eq. (14.6) or (14.8) is valid for the cost estimation per unit cell, J_1, in dollars, that is,

$$J_1 = J_0 \left(\frac{P/N}{100}\right)^\alpha \left(\frac{i}{100}\right)^\gamma \quad \text{(\$/cell)} \tag{14.15}$$

Since the chlorine plant of P tons/day in capacity has N cells, the plant cost J also in dollars, is N times J_1 as follows:

$$J = J_0 N \left(\frac{P/N}{100}\right)^\alpha \left(\frac{i}{100}\right)^\gamma = J_0 (N)^{1-\alpha} \left(\frac{P}{100}\right)^\alpha \left(\frac{i}{100}\right)^\gamma \quad \text{(dollars)}$$

$$\tag{14.16}$$

The cost of electric power per unit of production Ec in \$/ton Cl_2, is shown by the equation

$$Ec = 0.756 \times 10^5 \left(\frac{c'}{\xi}\right) v$$

$$= 0.756 \times 10^5 \left(\frac{cv}{\xi}\right)\left(\frac{100}{m'}\right) \qquad (\$/\text{ton } Cl_2) \qquad (14.17)$$

where c' and c are the prices of dc and ac power in \$/kWh, respectively, and m' is the conversion efficiency of the rectifier in percent, which is a function of the terminal voltage, V, in volts. It also depends on the type of rectifier as follows

$$\frac{100}{m'} = \frac{1}{1 - (m/V)} \simeq 1 + \frac{m}{V} \qquad (14.18)$$

where m is a characteristic value of the rectifier. Substituting this into Eq. (14.17),

$$Ec = 0.756 \times 10^5 \left(\frac{cv}{\xi}\right)\left(1 + \frac{m}{V}\right)$$

$$= 0.756 \times 10^5 \left(\frac{c}{\xi}\right)\left(v + \frac{m}{N}\right) \qquad (\$/\text{ton } Cl_2) \qquad (14.19)$$

The operating cost, Oc(total), consists of the power cost, the labor cost, and other expenses, Oc(other), which is almost independent of the number of cells installed, that is,

$$Oc(\text{total}) = Ec + Lc + Oc(\text{other}) \qquad (14.20)$$

Substituting Eqs. (14.16) and (14.20) into Eq. (14.17), we have the optimum number of cells, N(opt), by the differentiation of Mc with respect to N, that is,

$$\frac{dMc}{dN} = \frac{dFc(\text{total})}{dN} + \frac{dOc(\text{total})}{dN}$$

$$= \frac{1}{365P}\frac{\beta}{100}\frac{dJ}{dN} + \frac{dEc}{dN} = 0 \qquad (14.21)$$

$$N(\text{opt}) = \left[\frac{1-\alpha}{276 \times 10^5}\frac{J_0}{P}\frac{\beta}{100}\left(\frac{P}{100}\right)^\alpha\left(\frac{i}{100}\right)^\gamma\left(\frac{\xi}{cm}\right)\right]^{1/(\alpha-2)}$$

$$= \left(\frac{276 \times 10^5}{1-\alpha}\frac{cm}{\beta \overline{F}c}\right)^{1/(2-\alpha)} \qquad (14.22)$$

where $\overline{Fc}$ is the annual fixed charge corresponding to J. Thus, the larger the annual fixed charge, the smaller is the optimum number of cells to be installed.

On the other hand, if the cost of dc electric power is high, the number of cells should be increased. The slope of the investment vs. plant capacity curve, α, is also a strong influence on N(opt). When α is small, N(opt) becomes small. For α close to 1, a small variation of α or $(1 - \alpha)$ affects N(opt) greatly.

Generally speaking, variations of ξ and m are small, but c varies from 2 to 10 ¢/kWh. Both $\overline{Fc}$ and α are functions of the cell type, the plant location, the plant layout, and other conditions. From Eq. (14.22), the optimum number of cells is a function of α and the ratio of the ac power cost to the annual fixed charge. Figure 14.4 shows N(opt) as a function of $cm/\xi\overline{Fc}$ or $c/\overline{Fc}$ at constant values of m and ξ (assuming $m = 1.5$ and $\xi = 95\%$ in this case) for a reasonable range of α.

It is clear that the value of α for the plant under consideration is a very important factor although, in practice, there are additional complexities so that experience in plant design and construction must be depended on for an

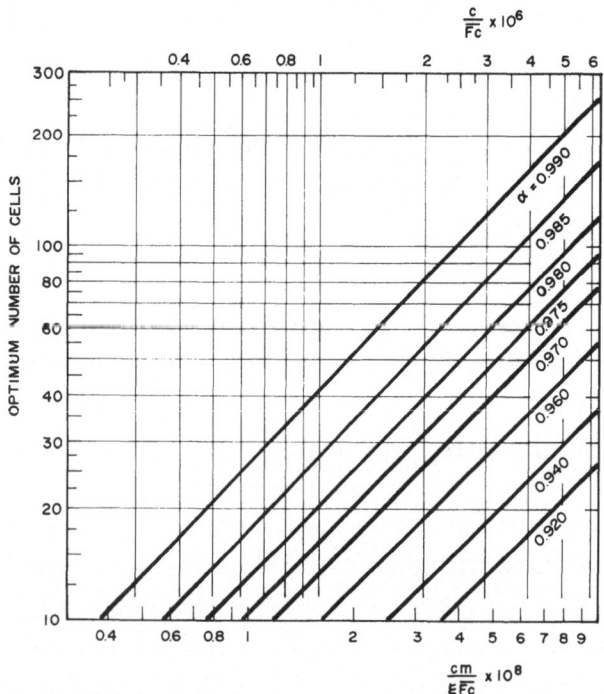

FIGURE 14.4. The optimum number of cells as a function of annual fixed charge and the ac power cost. The top scale is based on $m = 1.5$ and $\xi = 95\%$.

answer. In 1970, with values typical at the time of $\overline{Fc} = \$2000$, $c = 5$ mils/kWh, $\xi = 95\%$, and $m = 1.5$, the optimum number of cells from Eq. (14.20) was found to be 14 and 51 for $\alpha = 0.93$ and 0.98, respectively. Since that time, energy costs have sky-rocketed due to the oil crisis in 1974–1975, and the economic situation in the electrochemical industry, including chlor-alkali, has changed greatly. Construction costs have also increased signifi-cantly. With current prices, we may assume that $\overline{Fc} = \$10,000$ and $c = 8¢/kWh$, with the same values for ξ, m, and α shown above. The optimum number of cells increases to be 38 and 157, respectively. It is reasonable that the conversion efficiency of a rectifier increases with increasing terminal voltage, so that a number of cells must be operated in series when the price of electricity is high, as stated above.

14.1.2. Optimum Current Density

Because 35.5 kg of chlorine requires 26.8 kAh of dc energy by Faraday's law, the cost of electricity for a unit ton of chlorine, Ec, in \$/metric ton chlorine, is represented as follows:

$$Ec = \frac{26.8(kWh)}{35.5 \times 10^{-3}(ton)} \frac{1}{[\xi(\%)]/100} v(volt)\, c \quad (\$/kWh)$$

$$= 0.756 \times 10^5 (c/\xi)v \quad (\$/ton\ Cl_2) \tag{14.23}$$

$$= 0.756 \times 10^5 (c/\xi)(a + bi) \quad (\$/ton\ Cl_2) \tag{14.24}$$

The cell voltage v is a linear function of the current density at high-current density, with slope b. The term a is the voltage projected to zero current density and is about 3.1 V (see Fig. 7.5). This value is not the ther-modynamic decomposition voltage, but is, nevertheless, a useful factor for engineering purposes. The value Ec is the cost of electricity for electrolysis only, and does not include any other electric power consumption, as for example, pumps.

If the plant size is the same, the operating costs Oc', excluding the power for electrolysis, include the costs of the brine purification system, raw materials, and some other auxiliaries, are almost the same. The labor cost is, of course, a function of the number of cells, or is dependent on the operating current density. However, it is a small factor in the total operating cost per unit ton of chlorine produced by a modern cell.

The total operating cost, $Oc(total)$, therefore, consists of Ec and Oc' as follows:

$$Oc(total) = Ec + Oc' \tag{14.25}$$

The manufacturing cost may consist of the total fixed charge and the total operating cost as shown by Eq. (14.12). A part of the manufacturing cost Mc is a function of current density, but the other part is independent. Therefore, the optimum current density, i(opt), can be obtained by the differentiation of Mc with respect to i, that is,

$$i(\text{opt}) = 2.65 \times 10^{-8}(\overline{Fc}) \left(\frac{\xi}{bc} \right) \qquad (\text{A/dm}^2) \qquad (14.26)$$

where

$$\overline{Fc} = 365Fc = \frac{J}{P} \left(\frac{\beta}{100} \right) \qquad (14.27)$$

The larger the annual fixed charge, $\overline{Fc}$, in dollars for unit production, and the

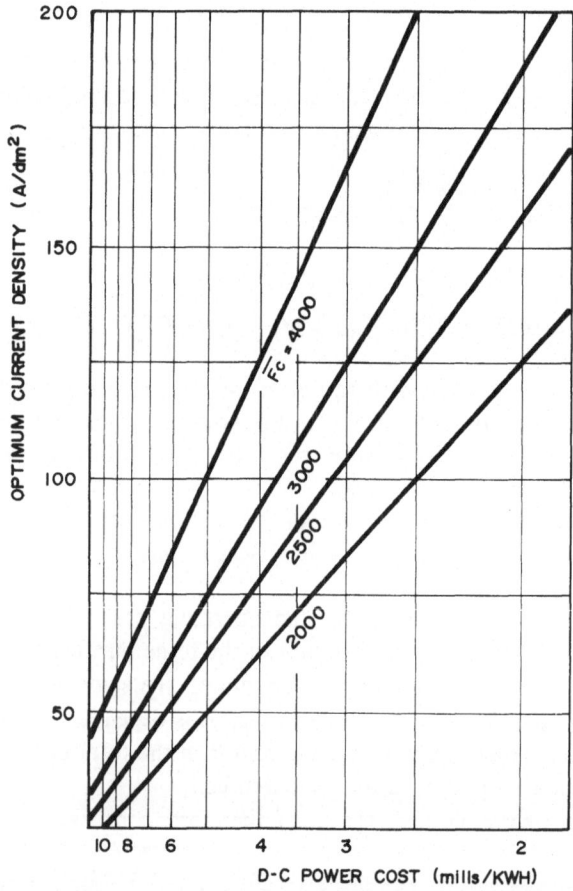

FIGURE 14.5. The optimum current density as a function of dc power cost and the annual fixed charge. Basis: 95% current efficiency for cell gas and $b = 2 \times 10^{-2}$ V/A dm^2.

cheaper the power cost, the higher is the optimum current density, i(opt). Plant size, P, is a small factor for the annual fixed charge because the coefficient α is almost unity. Figure 14.5 shows the optimum current density as a function of the power cost under the assumption of 95% current efficiency for cell gas and $0.02\,\text{V/A/dm}^2$ for the slope b of cell voltage vs. current density. The annual fixed charge in 1970 was about \$3000/ton of chlorine and electricity cost was as low as 6 mil/kWh. Therefore, the optimum existed between 60 and 65 A/dm^2. However, the price of electrical energy has risen by a factor of ten since 1970, while plant cost has also increased almost three times. Consequently, the optimum current density has become small so as to minimize electrical power consumption.

It is clear from Eqs. (14.23) and (14.24) that the cost of electrical energy decreases with decrease of v. Since a in this equation is almost constant, b must be minimized to be as small as possible. As is shown in Fig. 4.7, the chlorine overvoltage and the solution IR drop including the bubble effects are the major factors. Of these, the chlorine overvoltage has been reduced significantly by introducing the DSA. Also, the metal anode helps to minimize the bubble effects because of sufficient open area for escape of gas bubbles from the electrode gap. Recently, the electrode gap of the amalgam-type chlorine cell has been controlled automatically by a computer-aided device. Even so, the electrical power cost contributes a large percentage in the production cost of chlorine and caustic soda.

14.1.3. Optimum Number of Stand-by Electrolyzers

Because an electrolytic cell utilizes low voltage and large amperage dc power, a number of cells must be operated simultaneously in series so as to maintain a sufficiently high conversion efficiency of the rectifier. The productivity per unit volume of the cell, or per unit floor space, is relatively small because the electrochemical reaction takes place only at the electrode surface, and hence a number of cells must be operated to cover the demand for product. Therefore, we need to discuss the optimum number of cells to be installed in a plant, as described in the preceding section.

Generally speaking, the configuration of the electrochemical cell is not simple: it is equipped with a number of pairs of electrodes and separators in a narrow space. Feedstock and products may pass through the channel with complicated flow patterns. These chemicals are sometimes corrosive. It is well known that electrochemical oxidation at the anode and reduction at the cathode are very strong reactions. These circumstances and conditions may affect the life of facilities. Therefore, maintenance of the electrolyzer must be conducted at adequate intervals to prevent unwanted and unexpected troubles during operation. Consequently, we need to have some extra cells, or stand-by cells.

The economics of multiple units in chemical process industries has been treated by Quigley.[8] Assume that the investment J_1 of a unit facility of capacity p is shown by the equation

$$J_1 = Kp^\alpha \tag{14.28}$$

and that the total capacity P is represented by pN, where N is the number of units. Also, we assume that the average capacity of a unit cell is shown by $P/(N-1)$ when one unit stands by. Therefore, the total investment J is

$$J = KN \left(\frac{P}{N-1} \right)^\alpha \tag{14.29}$$

The optimum number of cells, N(opt), is thus obtained by differentiation of J with N, $dJ/dN = 0$, i.e.,

$$N(\text{opt}) = \frac{1}{1-\alpha} \qquad \alpha < 1 \tag{14.30}$$

and

$$\frac{J}{J(\text{opt})} = \frac{N}{(N-1)^\alpha} (1-\alpha) \left(\frac{\alpha}{1-\alpha} \right)^\alpha \tag{14.31}$$

Figure 14.6 illustrates J/J(opt) as a function of the Williams coefficient α and the number of cells. Since α for an electrochemical plant is large, the minimum point of the reduced investment, J/J(opt) exists in the region of a large number of cells.

Alternative approaches to the economics of electrochemical process industries have been carried out by several authors.[9-13]

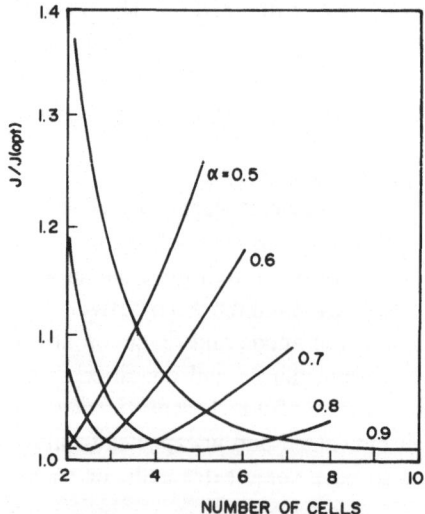

FIGURE 14.6. Investment vs. the number of cells with a standby cell.

14.2. ENERGY SAVING IN THE CHLOR-ALKALI INDUSTRY

Although energy saving for electrolysis is not a new subject for chlorine engineers, it has become increasingly important because of worldwide depletion of primary energy sources and of sky-rocketing prices in the last few years. This situation will continue according to long range predictions. Consequently, extensive efforts for saving energy are being made, and new concepts and/or innovations are being sought and examined for feasibility.

The present situation for saving energy in the chlor-alkali industry is described as a case study for energy saving in the electrochemical industries.[14]

Since the electrochemical process is part of an energy-consuming industry, not only the electricity but also other type of energy, i.e., the overall consumption of energy for processing, must be minimized. There are three methods, namely:

(i) reduction of the power consumption for electrolysis,
(ii) reduction of the overall energy consumption, e.g., electrolysis and processing, and
(iii) reduction of the price of energy, e.g., use of off-peak energy.

The power cost for electrolysis, Ec, is represented as follows

$$Ec = K \left(\frac{cV_T}{\xi} \right) \left(\frac{100}{m} \right) \tag{14.32}$$

and

$$V_T = E_d + \eta_A + (-\eta_C) + \Delta V_S + \Delta V_M \tag{14.33}$$

where V_T is the terminal voltage, E_d is the decomposition voltage, η_A and η_C are the anodic and cathodic overvoltage, respectively, ΔV_S is the solution IR drop, and ΔV_M is the IR drop in hardware.

In practice, the terminal voltage is roughly shown by the linear equation:

$$V_T = a + bi \tag{14.34}$$

With Eqs. (14.32) and (14.34), several ways for reducing the electric power cost may be proposed:

(i) reduction of the price, c,
(ii) improvement of the current efficiency, ξ,
(iii) improvement of the conversion efficiency of rectifier, m,
(iv) reduction of the slope of the volt–ampere curve, b, and
(v) operation at low current densities.

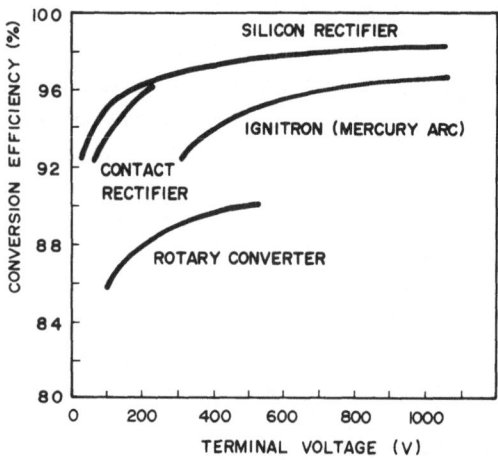

FIGURE 14.7. Conversion efficiency of rectifiers as a function of the terminal voltage.

The conversion efficiency of modern rectifiers composed of semiconducting silicon diodes and/or thyristors is high. Also, both the iron and copper losses in the system are controlled to be as small as possible through adequate design in order to keep the overall conversion efficiency high. Recently available options for improving the conversion efficiency of rectifiers are thought to the worthwhile since extra investment charge can be returned within a relatively short period. Figure 14.7 illustrates the conversion efficiency of various types of rectifiers as a function of the terminal voltage. Generally, the conversion efficiency increases with increasing terminal voltage. It is a reason why we need to operate a number of electrolytic cells in series because the voltage for a unit cell is only in the range 1–10 V. The conversion efficiency of the ignitron, a kind of the mercury arc rectifier, is much higher than those of conventional mercury arc rectifiers and of the mechanical rectifiers such as the rotary converter. Nevertheless, ignitron efficiency is only 95–96% at 600 V or higher. The contact rectifier, a special design of the mechanical rectifier, has a high efficiency at relatively low voltages, but was troublesome in operation. Consequently, this type of equipment is no longer used. Because of high efficiency even at low voltages, the silicon rectifiers are widely employed in electrochemical industries at present, from large plants such as aluminum refineries and chlorine plants to small factories such as electroplaters. Various types of silicon rectifiers have been developed to improve performance and operation. The thyristor is useful equipment for the electrochemical field because of high conversion efficiency at low voltages, for example.

Since the conversion efficiency of the rectifier increases as the terminal voltage is increased, a number of cells must be connected and operated in series. However, there is an optimum size and number of cells to be used, which is related to the bus bars required, the energy loss in the bus, labor for operation, maintenance, and other factors (see Section 14.1).

The basic concept for reducing the energy consumption of an electrochemical process consists of:

    (i) optimization of design, installation, and operation, and

    (ii) detailed analysis of the components of the cell terminal voltage.

14.2.1. Use of Off-peak Electricity

Seasonal and daily variations of demand of electricity are troublesome, and the smoothing of the supply by load-leveling is being given importance. Although storage of electric energy is discussed extensively, straight use of electricity is most desirable. Electrochemical processes could contribute to the national policy for energy conservation via use of the off-peak energy. In return, the company could receive an abundant supply of electricity at relatively low price if the electrochemical cell is capable of variation of load in order to utilize the off-peak electricity. Unfortunately, the aluminum refining cell, which consumes a great amount of electricity, is affected seriously by changing the current load. The electric current must be kept constant since otherwise the shape of the frozen crust of electrolyte in the cell changes due to change of the heat balance. It is well known that the water electrolytic cell can tolerate load variation. Many water electrolysis plants have been installed to utilize the off-peak electricity generated at night. The calcium carbide furnace also has an adaptability for a large variation of power supply, and it is able to utilize electricity of inferior quality.

According to experience, the electrolytic current of the amalgam-type chlorine cell can be changed over a relatively wide range without serious difficulty. It is said that change of the current of the diaphragm-type chlorine cell is difficult compared to the amalgam cell since the vacuum-deposited asbestos diaphragm is affected. Adjustment of the electrolyte flow and the percent decomposition of NaCl are also somewhat complicated. The physical properties of the deposited asbestos diaphragm are affected considerably by the brine flow and the electrolytic current. Such troubles have been reduced extensively by use of the modified asbestos diaphragm. Thus, a chlorine plant using modern diaphragm cells may be able to utilize off-peak electricity, but how much the power price can be discounted remains a problem.

According to Meredith, most diaphragm cells are capable of a 10% load variation from the rated capacity without significant effects, and some cells may be capable of 20% variation for short periods as long as the rectifiers are not overloaded. For further variation, it is necessary to change the number of cells under operation, and the operating conditions for the cells on line must be optimized.[15]

Hooker Chemicals and Plastics Corporation is utilizing off-peak electricity in their Taft, Louisiana, plant, where Hooker H-4 cells of 150 kA at the normal rate are operated. Hooker pays 9 mils for a kWh in the daytime and about half for the 8-hr off-peak energy supply in the night. The company is studying further economical operation including procedures for conservation of electricity and of energy for caustic evaporation.

The membrane cell is said to accomodate to variation of the electric current, but as yet, there is no practical proof.

Utilization of seasonal off-peak energy must be examined more carefully prior to practical operation since the plant economy is related to various factors such as investment, operating cost, and marketing. Meredith has studied the feasibility of seasonal off-peak energy for the production of chlorine under several assumptions. A plant, in his study, model 1, is 20 years old, and is operated at a constant rate for a year. On the other hand, model 2 is also a 20-year-old plant having the same capacity as for model 1, but is operated for only 9 months of the year. To cover the demand for 3 months during shut-down time, a new cell room having a capacity of one-third of the old facility is installed, and is operated in parallel with the old one. The break-even point with two models shown above depends on the price of electricity and the selling price of chlorine and caustic soda. As an example, if the selling price of chlorine is $150/ton and the normal price of electricity is 2¢/kWh, the chlorine plant must receive a 58% reduction in the price of the off-peak electricity to profitably convert from model 1 to model 2, according to the evaluation by Meredith.

14.2.2. Reduction of Consumption of the Overall Energy for Processing

While the electric power for electrolysis is the largest part of the total energy required for electrochemical processing, energy requirements for other parts, such as utilities, are not small. An example is the steam consumption for concentrating caustic liquor in the diaphragm cell plant. Comparison of the energy consumption in the amalgam cell plant and the diaphragm cell plant is shown in Table 14.2. It is clear that the power consumption for electrolysis with the diaphragm cell is smaller than that with the amalgam cell. However, the total consumption of energy in the diaphragm cell plant is about 10% larger than that in the amalgam cell plant because about 4 tons steam/ton NaOH are required for caustic evaporation. One ton of steam is assumed to be equivalent to 250 kWh of the ac electric power.

The steam consumption for caustic evaporation depends greatly on the design and the operation of the evaporator. Operation at the normal rate

TABLE 14.2

Utilities for the Amalgam-Cell Plant and the Diaphragm-Cell Plant[a]

		Amalgam-cell plant	Diaphragm-cell plant	
			Actual	Conversion
Electrolysis	kWh	3,200	2,500	2,500
Steam	ton	0	4	1,000[b]
Auxiliary	kWh	100	150	150
(Sum)	kWh	3,300		3,650
Water	ton	20	137	

[a] Basis: Per ton NaOH and 60,000 ton NaOH/year.
[b] 1.0 metric ton steam = 250 ac kWh.

under steady-state conditions and increased intervals between cleaning are important. The interval for cleaning is affected by the circulation velocity of the solution flow and the purity of the cell liquor.

The steam consumption per unit production of caustic soda decreases when the number of effects of the evaporator increases:

$$S_T = \frac{W}{N} + S_e \qquad (14.35)$$

where S_T is the steam consumption, W is the amount of water to be evaporated, N is the number of effects, and S_e is the steam consumption of the ejector.

The steam consumption for a unit volume of water increases with an increase in the concentration of caustic soda, whereas the total amount of water to be evaporated decreases. Therefore, an optimum concentration of caustic soda for minimizing the steam consumption may exist. The caustic concentration is also controlled by the cell performance so that a large variation of the caustic strength in the cell is not desirable. Generally, the current efficiency of the membrane cell is reduced with an increase of the NaOH concentration in the cathode compartment. Therefore, the cell performance, and, in particular, the energy consumption, depends greatly on the nature of the membrane being used.

Although the steam consumption in the multi-effect evaporator is small, the temperature difference in the successive stages decreases, and operation becomes difficult. Triple-effect evaporators are popular in chlor-alkali plants worldwide. The first evaporator of quadruple effects was installed in the Battleground, Texas, plant of Diamond Shamrock Corporation in 1975. This giant plant, 1200 tons/day, is controlled by an IBM computer. The results of computer simulation have shown an estimated consumption of steam to be 5000–6000 lb/ton NaOH, which is about 20% less than that for the conventional evaporator.[16] Recently, Asahi Chemical has tried successfully to recover the joule heat of the membrane cell. It is reported that the steam consumption for evaporation is only 0.7 ton/ton NaOH.[17,18]

14.2.3. Reduction of the Power Consumption for Electrolysis

Although the fixed charge increases, the power consumption for electrolysis can be reduced directly if the cell is operated at low current densities. The optimum current density for the amalgam cell has been described in Section 14.1.2, and the decrease in the optimum current density for energy conservation has been suggested.

It is clear that i(opt) becomes large when the coefficient b in Eq. (14.25) is small. This coefficient is an important parameter for evaluating cell performance, and hence b must be minimized to be as small as possible. Of course, any extra charge of Fc and the labor for minimizing the b-coefficient must be avoided or at least minimized.

Operation at low current densities is not always economical. From the standpoint of industrial economy, the manufacturing cost must be minimized, but not the power cost alone. To achieve minimum manufacturing cost, the consumption of electric power may increase, or a sufficient amount of electricity must be utilized in some cases. Of course, high current density operation of some electrolyzers is uneconomical if there is a low percentage of operation of the existing capacity and/or the number of cells installed is more than sufficient. In this case, all the cells should be operated at low current density, and the cell voltage kept low to reduce the energy consumption per unit production. Detailed consideration of strategies for operation rate and the use of standby facility is of great importance under conditions of worldwide slump of business activity and economy downturn.

Improvement of the current efficiency is also important for reducing the electric energy consumption. In the amalgam cell plant, mercury recycled from the decomposer is washed thoroughly with water prior to being returned to the cell in order to avoid a rise in brine pH, since otherwise a part of dissolved chlorine is converted into hypochlorite. The consumption of graphite anodes may increase extensively when the hypochlorite content in the brine is high. The DSA is also affected by brine of high pH level. Generally, the pH of the feed brine is kept low at about 3–4 to improve the current efficiency by 1–2%.

The effect of the chlorine bubbles generated at the metal anode (DSA) facing downward is relatively small compared to that of the graphite anode because the openings in the DSA structure are large enough to permit easy exit of gas bubbles. This reduces to opportunity of contact of chlorine with amalgam, and the current efficiency is improved by 1%.

The content of hypochlorite in the anolyte of the diaphragm cell is high compared to that in the amalgam cell, so that the consumption of the graphite anode is significantly greater in comparison. This problem has been solved almost completely by use of the DSA. However, a problem for the oxide-coated anodes, such as the DSA, is a low oxygen overvoltage. The anode gas of the diaphragm cell equipped with the DSA is said to contain some oxygen, and is a factor of current inefficiency. Oxygen in the cell gas is troublesome for chlorine liquefaction, for example. The oxygen content in the cell gas of the amalgam cell is lower than that of the diaphragm cell even with the DSA, probably due to high overvoltage for oxygen evolution at high-current densities in the amalgam cell ($80-120 \ A/dm^2$).

There is backmigration of OH^- from the cathode compartment through the asbestos layer in the diaphragm cell, causing loss of OH^- and Cl_2 since hypochlorite and chlorate form more readily in the anode compartment with a rise of solution pH. Operation at high current density and at high temperature is desirable to minimize these problems although the current efficiency of the diaphragm cell depends not only on the current density but on several other factors as well. There is alkali loss to some extent in the evaporation system as crystallized salt is washed sufficiently to recover alkali. Lengthened intervals between cleanings of the evaporator help to minimize alkali loss.

The performance of the membrane cell is related directly to the characteristics of the membrane being used. Permeation of OH^- through the membrane should be minimized to improve the current efficiency of the process.

There is no doubt that minimization of the slope of the volt–ampere curve is the most important method available for reducing electric power consumption in chlor-alkali cells of all kinds.

14.2.4. Voltage Balance

An example of the voltage balance of an amalgam-type chlorine cell is shown in Fig. 4.17. The cathode overvoltage and the IR drop of metal hardware are small, but the chlorine overvoltage on the graphite anode is very large. On the other hand, the anode overvoltage in the modern chlorine cell equipped with the DSA is only a small portion of the terminal voltage, as shown by the dotted line in the Fig. 4.17. The largest factor of the cell voltage is the IR drop in the brine gap containing a great amount of chlorine bubbles which increase the electrical resistivity of the solution.

The anode overvoltage on the DSA is also very small in the diaphragm cell, as listed in Table 8.2. The cathode overvoltage, the electrolyte IR drop, and the IR drop through the diaphragm must be minimized for further reduction of the cell voltage. The bubble effect is one of the most important factors.

The hydrogen overvoltage on the iron cathode in alkali solution is preferably low, presumably the lowest among commercial metals and alloys. However, the hydrogen overvoltage is still a significant factor since it is as large as 0.3 V, as shown in Table 8.2. Reduction of the overvoltage by 0.1 V would represent a great saving of electric energy, e.g., about 12×10^6 kWh/year for a 500 ton/day plant or 200×10^6 kWh/year in Japan. A number of patents on low hydrogen overvoltage cathodes have already been issued. Most of these activated cathodes consist of surface treatment of the iron substrate. The surface area of these cathodes is very large, so that the actual current density is reduced significantly while the electrocatalytic activity is also increased by such treatment (see Section 12.4.2).

Since the IR drop through the asbestos diaphragm is one of the largest factors of the terminal voltage of the modern chlorine cell, it should be reduced to be as small as possible. On the other hand, the thickness of the asbestos mat is closely related to the anolyte permeation, backmigration of OH^-, separation of hydrogen and chlorine, and other factors of the cell operation. Substitution for asbestos mat by new materials and change of the thickness of the diaphragm are complicated tasks in practice.

Development of the modified asbestos diaphragm, which is a polymer-bonded asbestos mat, has been of great value for decreasing voltage drop. Unwanted swelling and degradation of the vacuum-deposited asbestos layer during electrolysis have been almost stopped. The thickness of the asbestos layer can be reduced in this manner, and hence the solution IR drop between the anode and the diaphragm may be reduced significantly.

14.2.5. Bubble Effects

The electrical resistance and the IR drop of the electrolytic solution increase considerably when nonconductive gas bubbles disperse in the solution. Electrolytic gas bubbles also cover the surface of the working electrode, causing a rise of the current density and of the overvoltage. Consequently, bubble effects involve two factors: one is to increase the solution IR drop, called the dispersion effect, and the other to decrease the electrode area, known as the shield effect.

The bubble effects are, of course, significant at high-current densities in a narrow channel. Hine and co-workers[19-22] and MacMullin[28] studied the solution IR drop in the chlorine cell, and proposed improved shapes for the graphite anodes and their arrangement in the horizontal-type amalgam cell.

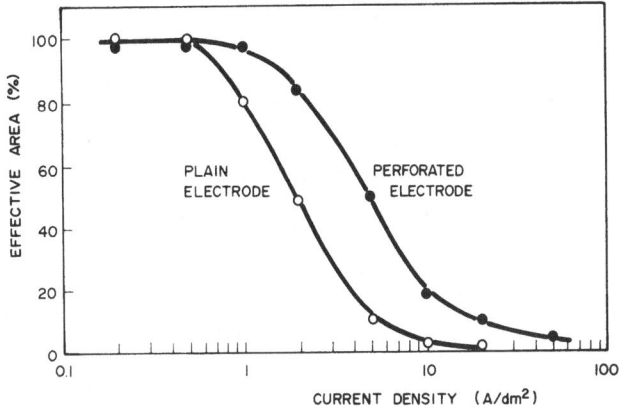

FIGURE 14.8. Effective area of graphite anodes in saturated NaCl (pH = 4) at 55°C.

The IR drop of the electrolytic solution containing gas bubbles depends on the flow rate of electrolyte when the cell is designed and operated under forced convection of the electrolytic solution.[24–26] MacMullin indicates that "the vertical-type diaphragm cell such as the Hooker cell is a good pump, with no moving parts, and produces enormous flows of very low heads."[27,28] The chlorine bubbles under the graphite anode are removed by the brine flow, which is accelerated by rapid flow of the mercury cathode, so that the voltage of the amalgam cell is low even at high-current densities[29,30] (see Section 4.1.4).

The shield effect of the electrode surface by gas bubbles is also significant. More than 90% of the surface of the graphite anode in a horizontal cell is covered by gas bubbles at high current densities, as shown in Fig. 14.8, and causes the increase in the overvoltage.[19–22]

14.2.6. New Technology

Although the chlor-alkali industry has a long history of over 90 years, there are still problems to be solved. Accordingly, scientific research and practical innovation are of continuing importance for developing improvements to existing methods. On the other hand, new concepts of great interest have been proposed. Application of the oxygen cathode in the chlorine cell, both diaphragm- and membrane-type, is an example. The oxygen cathode for the fuel cell has been investigated extensively. Some fuel cells can be operated at surprisingly high-current densities with the oxygen cathode, say 50 A/dm^2 or more. It would be a great advance for energy saving in chlor-alkali cell practice if the oxygen cathode can be utilized successfully. However, its economy is still uncertain, and safety is also of serious concern.

The projected use of the oxygen cathode has a sound fundamental basis. The decomposition voltage, E_d, in Eq. (14.33) is a specific value of the process under discussion, which varies only in a small range depending on operating conditions such as activity, pressure, and temperature. Application of the oxygen cathode to the chlorine cell is an attempt to reduce the thermodynamic decomposition voltage through conversion of the overall reaction from hydrogen evolution to oxygen reduction (see Section 12.4.3).

Other technical innovations involve utilization of beta-alumina, direct production of soda ash in the chlor-alkali cell, various applications of the sodium amalgam, and others, which are presently in various stages of advanced development. Beta alumina is a unique material having a permselectivity for Na^+ alone in fused salt, but not in aqueous solution, at high temperatures. It is feasible to employ as a separator in an electrolysis cell of NaCl melt, but the cell design would be troublesome because of its mechanical properties.

Of these, feasible techniques involve the use of the porous diaphragm and application of the solid polymer electrolyte (SPE). A porous sheet of synthetic polymer could substitute for the conventional diaphragm composed of vacuum-deposited asbestos mat. The mass transfer through the porous sheet differs greatly from that in the asbestos mat. With improved understanding of the mechanism of mass transfer, the porous diaphragm can be of great value in various fields of electrochemical processes.

The SPE has been developed by the G.E. research group as a part of fuel cell technology for the space program in the U.S.A. General Electric and de Nora have jointly attempted to utilize the SPE as a new technology of chlor-alkali cell.

These new concepts and technologies can contribute significantly to energy saving in the chlorine industry, and can find a variety of applications in other electrochemical fields as well.

In summary, present technologies for decreasing energy consumption in the chlor-alkali industry and the feasibility of some new technologies have been discussed. It is believed that the energy crisis will continue and that shortages will scarcely be improved even in the future. Consequently, we need to develop and innovate new technology with our knowledge in order to save reserves of energy and resources on a worldwide basis. Reconsideration of present processes and technologies and the optimization of plants now under operation are of great importance. Trial lengthening of the service life time of asbestos diaphragms and of anodes, both graphite and DSA, is an example. Systematic reinvestigation of the whole process from the viewpoint of both science and engineering must be conducted.

There is no doubt that energy saving is one of the most important tasks for the electrochemical industries because the industry consumes a great amount of electricity, say more than 6% of the total generation in industrialized countries. Electrochemical engineers have made efforts for reducing energy consumption in their plants for many years. Here we need to consider a balance of industrial economy as well as energy conservation. Our target is how to minimize the manufacturing cost rather than saving of energy only. In some cases, the operating conditions for minimizing cost require more energy. Collins presented the case history of the development of the amalgam cell in ICI for a period of over 80 years.[31] The current density has been increased extensively, by more than a factor of 10 times, during the period, whereas the energy consumption has been kept almost constant. This case provides a good example of the essence of electrochemical engineering, that is investment cost, and hence the fixed charge, being reduced significantly while the operating cost remains unchanged. However, under present world conditions, energy saving is one of the most direct methods for reducing the manufacturing cost of electrochemical products because the energy price continues to climb at present and will presumably continue to do so into the future.

14.3. USE OF COMPUTERS

Utilization of computers in the electrochemical process industry involves: (1) design of electrolyzers and electrochemical systems, (2) design and simulation of the process flowsheet from the crude material to the products, and (3) process control and management during operation. Of these, the first item is use of the computer for designing the hardware to be installed. The second and the third items are the optimization of the system under discussion. The material balance and the energy balance are of great importance.

A computer simulation for the diaphragm-type chlor-alkali cell has been carried out by MacMullin, who has evaluated the operation data of Hooker S-3 cells equipped with graphite anodes.[27] A similar discussion of the material balance in the diaphagm cell has also been presented by Nagy.[32]

Adjustment of the anode-to-cathode gap of the amalgam-type chlorine cell (called "anode adjustment" in the plant) is of great importance for minimizing the cell voltage. Generally, the anode blades, either graphite or metallic, are fixed to the bus bar located on the cell top. These anode assemblies are driven up and down by a motor, which is controlled by the process computer. Variations of the cell voltage and/or the amperage are determined and these data are sent to the computer. Several papers on anode adjustment have been published. For example, Ralston points out the major features of the control system for Olin cells as being:[33]

(i) automatic anode adjustment with a back-up system for remote manual adjustment,

(ii) an independent system to automatically detect high current and to raise anodes,

(iii) a control room console with meters that can be switched to display current and voltage readings anywhere in the cell room.

Amalgam cell rooms operated by this system yield a resulting voltage reduction of 0.4–0.6 V per cell.

Another example of the application of computers in the chlor-alkali plant is process control of the caustic evaporation system in diaphragm-cell plants. Diamond Shamrock Corporation has installed the world's first quadruple-effects evaporator in a giant plant having 1200 ton/day capacity at Battleground, Texas.[16] The system is controlled by an IBM computer.

The steam consumption per ton of caustic soda produced decreases with an increase of the number of effects of the evaporator, whereas the operational procedure becomes complicated due to the decrease of temperature difference in the individual tanks. Consequently, conventional diaphragm cell plants generally make use of the triple effects evaporator system. According to the computer simulation on the system at the Battleground plant, the steam consumption is 5000–6000 lb/ton NaOH, which is about 20% less than that of the conventional triple-effects evaporator system.

Sokol and McLarty of Hooker Chemical Corporation emphasize several actual benefits from the use of computers modeling in developing chlor-alkali plant design:[34]

1. Probably the major advantage of computer calculation is that it facilitates the following of trace impurities in the chlorine stream throughout a complex process arrangement replete with many recycle streams.

2. Computer calculations give assurance that correct physical property data has been used in computation.

3. Computer simulation facilitates the evaluation of alternate process conditions or arrangements in sufficient detail to uncover subtleties that might later prove important.

4. The process of preparing the process input data forces a logical review of the process and interacting forces.

5. The computer simulation program permits less experienced personnel to perform sophisticated calculations.

6. The input data is concise and can be easily reviewed by supervisory personnel. The input data contains a complete statement of the calculations. It reveals the flow sheet, control schemes, input data and calculation instructions.

7. The output data is in a readable form which permits quick checks for accuracy. The output data can be used by anyone without interpretation.

At present, there is no doubt that the use of the computer for the designing and operation of the electrochemical systems is increasing rapidly, and that benefits are significant.

REFERENCES

1. M. M. Silver, Abstract No. 103, meeting of Electrochemical Society Cleveland, OH, May 1966.
2. W. C. Gardiner, private communication.
3. F. B. Grosselfinger and J. Schuecker, *Chem. Age India* 12, 540 (1961).
4. F. Hine, *J. Electrochem. Soc.* 117, 139 (1970).
5. R. W. Williams, Jr., *Chem. Eng.* 54(12), 124 (1948).
6. F. Hine, *Electrochem. Technol.* 6, 69 (1968).
7. H. F. Rase and M. H. Barrow, *Project Engineering of Process Plants*, 2nd ed., pp. 51 and 53, Wiley, New York (1961).
8. H. A. Quigely, *Chem. Eng.* p. 97 (August 29, 1966).
9. N. Ibl, *Electrochim. Acta* 22, 465 (1977).
10. N. Ibl and E. Adam, *Chem. Ing. Techn.* 37, 573 (1965).
11. A. Schmidt, *Chem. Ing. Techn.* 39, 692 (1967).
12. A. Schmidt, *J. Electrochem. Soc.* 118, 2046 (1971).
13. R. Keller, *Electrochim. Acta* 25, 303 (1980).
14. F. Hine, *Denki Kagaku* (J. Electrochem. Soc. Jpn.) 47, 704 (1979).
15. R. E. Meredith, *The Use of Electric Energy and Load Management Possibilities in Industrial Electrolytic Processes*, reported to the Electric Power Research Institute, 1975–1976, p. 68.
16. D. J. Harvey and J. R. Fowler, *Chem. Eng. Prog.* 72(4), 47 (1976).
17. M. Seko, paper presented at the 20th Chlorine Plant Managers' Siminar, The Chlorine Institute, New Orleans, LA, February 1977.
18. M. Seko, paper presented at the 22nd Chlorine Plant Managers' Seminar, The Chlorine Institute, Atlanta, GA, February 1979.
19. F. Hine, S. Yoshizawa, and S. Okada, *Denki Kagaku* 24, 370 (1956).
20. S. Okada, S. Yoshizawa, F. Hine, and Z. Takehara, *Denki Kagaku* (J. Electrochem. Soc. Jpn.) 26, 165; 211 (1958).
21. S. Yoshizawa, F. Hine, Z. Takehara, and M. Yamashita, *Denki Kagaku* (J. Electrochem. Soc. Jpn.) 28, 205 (1960).
22. F. Hine, S. Yoshizawa, S. Okada, and T. Uesugi, *Kogyo Kagaku Zasshi* (J. of Ind. Chem., Chem. Soc. Jpn.) 57, 554 (1954).
23. R. B. MacMullin, Electrolysis of brines in mercury cells, in Chlorine, edited by J. S. Sconce, ACS Monograph 154, Reinhold, New York (1962).
24. F. Hine, M. Yasuda, R. Nakamura, and T. Noda, *J. Electrochem. Soc.* 122, 1185 (1975).
25. F. Hine and K. Murakami, *J. Electrochem. Soc.* 127, 292 (1980).
26. F. Hine and K. Murakami, *J. Electrochem. Soc.* 128, 64 (1981).
27. R. B. MacMullin, *Electrochem. Technol.* 1, 5 (1963).
28. R. B. MacMullin, *Denki Kagaku* (J. Electrochem. Soc. Jpn.) 38, 570 (1970).
29. M. Murozumi, *Electrochem. Technol.* 5, 236 (1967).
30. H. Shibata and Y. Yamasaki, *Electrochem. Technol.* 5, 239 (1967).

31. J. H. Collins, Abstract No. 439, meeting of Electrochemical Society, Seattle, WA, May 1978.
32. Z. Nagy, *J. Electrochem. Soc.* **124**, 91 (1977).
33. R. W. Ralston, Chlorine Bicentennial Symposium, p. 145, Electrochemical Society (1974).
34. E. L. Sokol and C. S. McLarty, Chlorine Bicentennial Symposium, p. 285, Electrochemical Society (1974).

FEASIBILITY OF ELECTROCHEMICAL PROCESSES

Electrochemical processes provide several unique features which permit specific reactions to be performed which are otherwise difficult to achieve by conventional chemical methods. Production of chlorates and perchlorates is an example, in which chloride ions are oxidized anodically. An electrochemical route for producing adiponitrile has also stirred interest and is being operated on a large scale today.

15.1. PRODUCTION OF CHLORATE

The use of sodium chlorate is increasing at a healthy rate in North America, and plant capacity is thereby being expanded. A number of basic studies have been reported in the last 15 years and new designs for chlorate cells also have been announced by several companies. This section deals with the situation of business activities and technological aspects of the chlorate industry in the year 1980.

Colman has published a valuable review paper on the electrolytic production of sodium chlorate.[1]

15.1.1. Capacity, Production, and Market

The demand for sodium chlorate in pulp bleaching, which is the largest market of sodium chlorate, has strongly increased and is expected to rise 8% per year until 1985. High-quality pulp can be produced by using single-stage bleaching with sodium chlorate. Waste treatment with the new process using chlorate is easy, and is a major reason for the increase in demand. Demand for sodium chlorate in uranium milling is expected to rise 9–11% annually, although this market is not large at present: about 6% of total consumption. However, it is expected that the annual use of sodium chlorate will be 900,000 to 950,000 tons in this field by 1987. Sodium chlorate for herbicides has had a large share previously, but decreased to only 2% of the total market.

TABLE 15.1
Capacity and Production of Sodium Chlorate in the United States and Canada

Year	Capacity 10^3st/y	Production 10^3st/y	Production % increase	Price $/st
		United States		
1971	230.5	196.5	−1	134.0–134.5
1972	230.5	182.5	−6	134.5–135
1973	235.5	198.3	10	135
1974	235.5	201.5	2	135–165
1975	NA	167.1	−17	215–250
1976	253.5	196.4	17	250–310
1977	285	218.5	11	310)345
1978	415	255.6	15	325–345
1979	443	248.1	−3	345
1980	392	250	4	405
1981	389	NA	NA	420
		Canada		
1979	281.8			
1980	303			
1981	314			

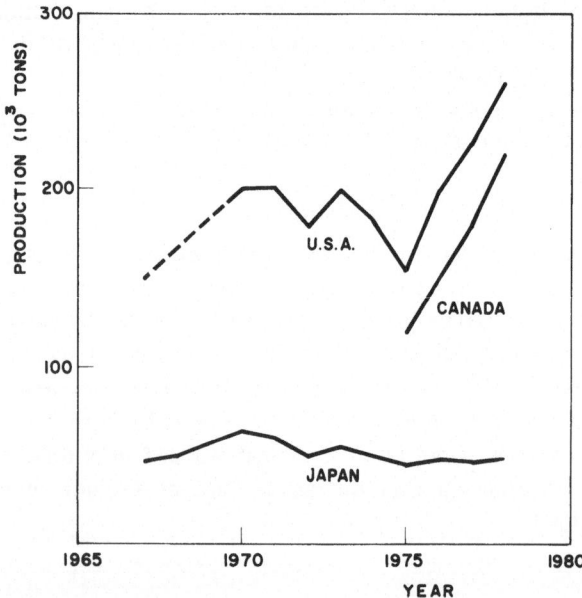

FIGURE 15.1. Production of sodium chlorate in the United States, Canada, and Japan.

The production of other chlorates and perchlorates, mostly ammonium perchlorate, used sodium chlorate in the amount of about 17,000 tons in 1979. Another market is expected if sodium chlorate is used in solid propellant boosters for the space shuttle program.

The U.S. capacity, production, and the price of sodium chlorate are increasing recently as shown in Table 15.1.[2] North American producers are expanding production capacity to match the bright future for business. Table 15.2 shows the capacity of the sodium chlorate production in the United States and Canada.[3]

In Japan, the pulp and paper industry consumes almost 85% of the total production of sodium chlorate, and other markets are small. At one time, herbicide was the largest market for sodium chlorate (64%), but is expected to be about 10% or less at present. The production of and demand

TABLE 15.2
Sodium Chlorate Producers in North America (10^3 t/y)

	1977	1979
U.S.A.		
Brunswick Chemical	8.0	8.0
Georgia-Pacific	3.8	30.8
Erco Industries		25.0
Hooker Chemical	119.0	139.0
Huron Chemicals	10.6	10.6
International Minerals & Chemical		40.0
Kerr-McGee	57.0	67.0
Olin	6.0	26.0
Pacific Engineering	6.0	6.0
Pennwalt	56.0	76.0
TOTAL	266.4	428.4
Canada		
B. C. Chemical	24.0	27.0
Canadian Occidental	21.0	24.0
Canso Chemicals	9.7	9.7
Erco Industries	105.0	145.0
Huron Chemicals	1.6	1.6
Prince Albert Pulp	1.0	1.0
QueNord		27.5
Stanchem (PPG)	30.0	33.0
Tidal Chemicals	13.0	13.0
TOTAL	205.3	281.8

for sodium chlorate in Japan have been almost unchanged for many years, as illustrated in Fig. 15.1.[4]

15.1.2. Technology

The use of chlorine dioxide, and hence sodium chlorate, is increasing due to the demand for brighter pulp and cleaner effluents. The technological developments with respect to the production of chlorate and chlorine dioxide for the last ten years are:

(1) process development for chlorine dioxide,
(2) new design for chlorate cells, and
(3) use of metal anodes.

1971–1972. Hooker Chemicals and Plastics Corporation has constructed a new facility for sodium chlorate with a capacity of 45,000 ton/year in Taft, Louisiana. The plant uses a new cell design with metal anodes supplied by Marston Excelsior Ltd., England. About 70% of chlorate is used to make chlorine dioxide bleach with the Hooker SVP (single-vessel process) chlorine dioxide generating system.

1973. Erco Industries Ltd. of Canada has expanded its chlorate production capacity using Krebs NC-12 electrolytic cells equipped with metal anodes. The cell allows for natural circulation of solution.

1973–1974. Standard Chemical Ltd., a subsidiary of PPG Industries, has increased chlorate production capacity in Beauharnois, Quebec. This plant uses a new cell design developed jointly with PPG. The cell is equipped with the Bimetal Electrode licensed by KemaNord, Sweden.

1974. Engelhard Mineral and Chemical Corporation has developed a new metal anode for producing sodium chlorate.

1975. Diamond Shamrock Corporation has developed a compact unit suitable for on-site use in the pulp mill. The cell is adaptable to load variation, tolerates a high level of brine impurities, and permits quick start-ups and shut-downs.

1977. The pulp industry is interested in chemical recycle processing because of the need for clean effluent. Great Lakes Paper at Thunder Bay, Ontario, has built the world's first closed-cycle bleaching kraft pulp mill. It is said that chlorine dioxide replaces 70% of the chlorine in the bleaching process.

1979. Brunswick Pulp and Paper Company has installed Krebs NC cells to upgrade its Brunswick, Georgia, plant to a 25,000 ton/year capacity.

There have been recent substantial improvements in chlorate cell technology. Older electrolytic cells have been operated at relatively low temperatures, since otherwise there is extensive consumption of the graphite anodes. In contrast, operation at high temperatures favors the chemical process for producing chlorate from hypochlorite.

The current efficiency of the chlorate cell equipped with graphite anodes is 82–88%, and the power consumption is in the range 5800–5900 kWh dc per ton of $NaClO_3$.[5]

The use of metal anodes permits the current density to be increased by three to four times as much as that used with graphite anodes. The current efficiency is increased to 95% or more under the conditions of high temperature operation. Also, the energy consumption has been decreased. There are two types of metal anodes now available: platinum–iridium alloys and ruthenium dioxide coated titanium anodes (DSA), as shown in Table 15.3.[5-9]

It is possible to operate the cell with a high rate of circulation of the electrolytic solution when the cell is equipped with metal anodes. This increased flow results in greatly improved current efficiency. It is also reported that the chromate content can be significantly reduced. Generally, the chromate concentration is in the range 4–5 g/liter, but Diamond Shamrock EH-120 cells contain only 350–400 ppm.[5-7] Usually, the electrolytic solution is fed to the cells in parallel, but the solution flow in the EH-120 cells is in the manner of a cascade in series by gravity. Also the EH-120 cell units are arranged in a single series electrical circuit. The cascade solution flow and the series connection of electrical circuit minimize the overall voltage drop of the system by allowing a higher concentration of sodium chloride. The cascaded flow in the EH-120 cell system may also be a factor in permitting the use of a smaller chromate additive concentration.

Some cells use pumping to circulate the electrolytic solution at a high rate. However, all the modern cells listed in Table 15.3 use the hydrogen gas lift action for solution circulation, known as the "inner circulation" or "self-circulation" system. Of these, the Krebs NC cell has a unique design, as shown in Fig. 15.2, i.e., the electrolytic cells are connected with a reactor to utilize the gas lift action for circulating solution efficiently.[7]

However, there is an optimum rate of solution circulation. The current efficiency increases with flow rate, reaches a maximum level at linear velocities of 12–15 cm/sec of solution flow, and then decreases quickly thereafter. The optimum Reynolds number is estimated to be about 1400.[10] According to experiment, the current concentration (amperes per unit volume of solution) also influences the current efficiency, with maximum current efficiency being achieved at about 25 A/liter. The flow rate of electrolyte is increased when the electrode gap is narrowed due to increased gas void fraction. Experimentally, maximum current efficiency was obtained with an electrode gap of 4 mm.[10] However, it is estimated that the optimum distance between two electrodes in practical cells is somewhat smaller, say

TABLE 15.3

Performance of Chlorate Cells (September 1978)

	Pennwalt P8-3		Diamond Shamrock EH-120	Pechiney Ugine Kuhlmann		Krebs NC
				TA 30	TA 2-60	
Current (kA)	15	30	2–10	30	60	30
Current density (A/dm^2)	16.8	33.7		25	20	25
NaClO$_3$ production (kg/day)	223	446	1800	450	900	450
Current efficiency (%)	93–95	93–95	91–96	>93	>94	95
Cell voltage including bus (V)	3.0	3.75	34.5–40.5[a]	3.00–3.15	2.90–3.05	3.0
DC kWh/ton NaClO$_3$	4810	6010	4650–5550	5120	4900	4770
Anode	Pt/Ir and DSA		DSA	Pt/Ir and DSA		Pt/Ir
Consumption per ton NaClO$_3$						
Cooling water (m^3)			166 @20°C			
NaCl (ton)			0.56			
HCl, 36 wt % (kg)			25			
Na$_2$Cr$_2$O$_7$ (kg)			1.7–3.0			
Feed brine						
NaCl (g/liter)	300–315					
Na$_2$SO$_4$ (g/liter)	0.1					<10 ppm as SO$_4^{2-}$
Hardness (Ca^{2+} + Mg^{2+}) (ppm)	10		ca. 300			<5
Na$_2$Cr$_2$O$_7$			350–400 ppm			4–5 g/liter
Cell liquor						
NaClO$_3$ (g/liter)	500–650		340–600			300–550
NaCl (g/liter)	70–110		100–200			100–150
NaClO (g/liter)	1.5–3.5 (reduced to zero in final product)					2
Na$_2$Cr$_2$O$_7$ (g/liter)				4–5		5
Cell gas						
H$_2$ (vol. %)	95.3–97.9			97.7		97.7
O$_2$ (vol. %)	1.5–3.5			2.0		1.5–2.0
Cl$_2$ (vol. %)	0.5–1.0			0.3		0.2–0.3
Cell operating conditions						
pH	6.0–7.0		6.5–7.5			6.5–6.7
Temperature (°C)	55–90		65	70–75		80

[a] This figure is the terminal voltage of the EH-120 cell system. If the system contains 10 unit cells, the cell voltage is estimated to be 3.45–4.05 V.

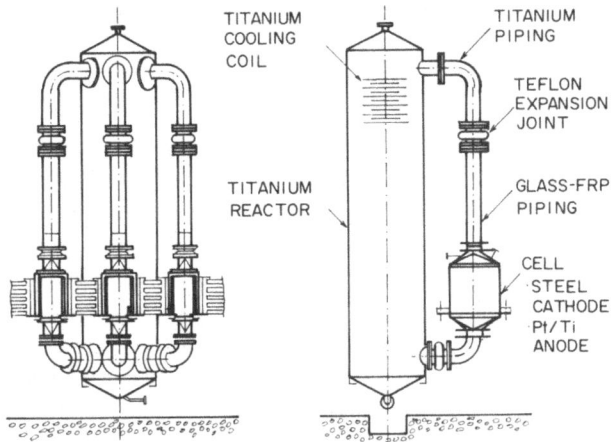

FIGURE 15.2. Krebs NC cell.

2–3 mm, because the *IR* drop in the solution, and hence, the cell voltage must be minimized. At high temperatures the chlorine content in hydrogen gas increases since Cl_2 releases from the solution. Also the oxygen content is high since reaction (15.10) is stimulated.

Although the Diamond Shamrock EH-120 cell is a bipolar-type cell, all others listed in Table 15.3 are of the monopolar type. Both types are used to an almost equal extent in the world at present, while U.S. producers operate more bipolar cells: 70% bipolar-type and 30% monopolar-type. The bipolar cells can be operated at low amperage and high terminal voltage with minimum investment for conductors, although the current efficiency is somewhat less in comparison with the monopolar cell because of difficulties in eliminating leakage of the electric current. Substitution of metal anodes for the graphite anodes in the bipolar cell is complicated due to the cell configuration and accounts for the slow pace of conversion to the use of metal anodes in the U.S., where more than 50% of chlorate is still being produced with graphite anodes.[6]

Titanium is a useful material for the chlorate industry because of its ability to withstand corrosion. Of particular interest is the use of titanium as the cathode material in Diamond Shamrock EH-120 cells. It is well known that the titanium cathode is not stable because of the formation of metal hydride which shortens the electrode life. However, the titanium cathode is corrosion resistant during periods of shut-down time, whereas the conventional steel cathode requires cathodic protection. During electrolysis, gradual spalling of the titanium hydride from the cathode surface rejects the scale deposited due to brine hardness, eliminating the need for periodic acid washing to remove deposits. The increase of cell voltage caused by scale deposition is also thus avoided.

Because brine hardness seriously affects cell performance, the modern chlorate cell requires a minimum concentration level of magnesium and calcium, say 5–10 ppm.[7.8] The Krebs NC cell requires the concentration of SO_4^{2-} to be limited to less than 10 ppm, while for the Pennwalt cell, the Na_2SO_4 content in the brine is prescribed to be 0.1 g/liter or less.

An example of the flowsheet for production of sodium chlorate is illustrated in Fig. 15.3. The saturated NaCl solution is purified and the solution pH is controlled prior to sending to the electrolytic cell. Since a small amount of chlorine escapes with hydrogen gas from the cell, the solution pH tends to rise with time of operation. Therefore, hydrochloric acid is added to the electrolyte to lower the solution pH. The depleted brine from the cell is sent back to the reservoir.

A part of the electrolytic solution is brought to the dechlorination tower to separate the active chlorine (mixture of Cl_2, HClO, and ClO^-), filtered to separate suspended solid particles, concentrated with the multi-effects evaporator, then crystallized to obtain solid $NaClO_3$. Dechlorination and filtration are quite important steps in order to prevent serious operating problems such as corrosion of the equipment and scaling. Sodium chlorate slurry is dehydrated with the centrifuge, dried with the rotary drier, and shipped as product.

High-purity sodium chlorate is white, but the regular grade product contains a small amount of chromate, and hence has a slightly yellow color.

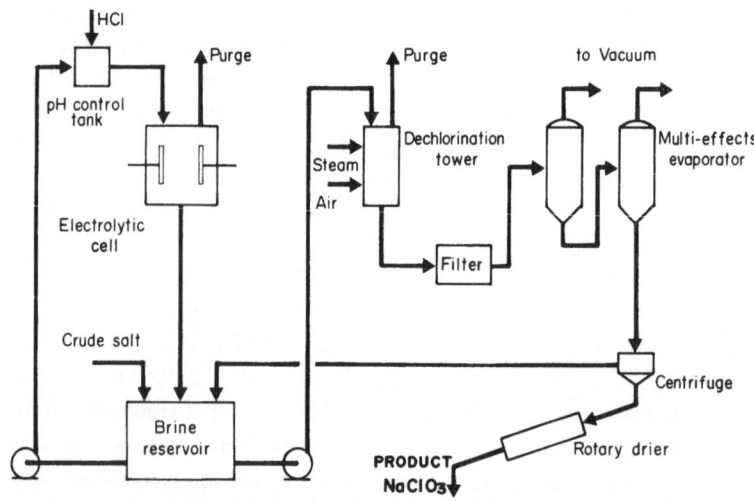

FIGURE 15.3. Flowsheet for electrolytic production of sodium chlorate.

15.1.3. Reaction Mechanisms for Electrolytic Production of Chlorate

About 80 and 60 years ago, respectively, Foerster and Mueller[11] and Foerster[12] published several excellent papers on the electrolytic production of chlorates. Their conclusions are still valid and form the basis of established theory.

Chlorine is generated at the anode and hydrogen and alkali are generated at the cathode when a neutral solution of NaCl is electrolyzed. When the electrolytic cell has no diaphragm between the two electrodes, dissolved chlorine reacts with alkali to form hypochlorite and then chlorate. That is, the main reactions in the cell are represented as follows:

At the anode:

$$2Cl^- \rightarrow Cl_2 + 2e \tag{15.1}$$

$$E^0 = 1.36 \text{ V vs. SHE at } 25°C$$

At the cathode:

$$2H_2O + 2e \rightarrow H_2 + 2OH^- \tag{15.2}$$

$$E^0 = -0.82 \text{ V vs. SHE at } 25°C$$

In the solution:

$$Cl_2 + H_2O = HClO + H^+ + Cl^- \tag{15.3}$$

$$\left. \begin{array}{l} HClO = H^+ + ClO^- \\ HClO + OH^- = H_2O + ClO^- \end{array} \right\} \tag{15.4}$$

$$\left. \begin{array}{l} 2HClO + ClO^- \rightarrow ClO_3^- + 2Cl^- + 2H^+ \\ 2HClO + ClO^- + 2OH^- \rightarrow ClO_3^- + 2Cl^- + 2H_2O \end{array} \right\} \tag{15.5}$$

The chlorine electrode process (15.1) is independent of the solution pH, whereas the thermodynamic potential for the hydrogen process (15.2) becomes more negative with an increase of the solution pH (see Fig. 15.4).

Jaksic et al. proposed a relevant equilibrium constant K^* for HClO in reaction (15.4) as follows[13-15]:

$$K^* = \frac{a_{H_3O^+} C_{ClO^-}}{C_{HClO}} = K_a \frac{\gamma_{HClO} a_{H_2O}}{\gamma_{ClO^-}} \simeq 10K_a \tag{15.6}$$

where a is the activity, γ is the activity coefficient, C is the concentration, and K_a is the thermodynamic dissociation constant of HClO.

They predicted the optimum pH for reaction (15.5) as

$$(\text{pH})_{opt} = pK^* - \log 2 \tag{15.7}$$

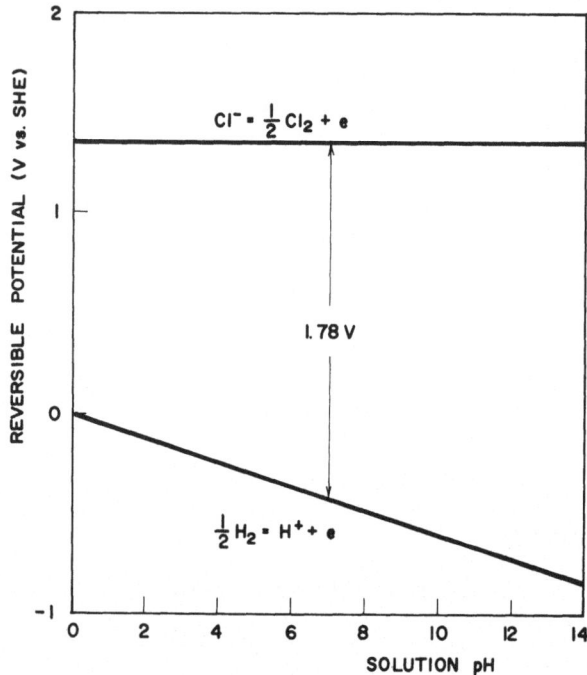

FIGURE 15.4. *Pourbaix diagram for the chlorine–hydrogen system at 25°C.*

From experiment, it was determined that

$$K^* = (3.0 \sim 6.8) \times 10^{-8}$$

and hence, $(\text{pH})_{\text{opt}}$ was estimated to be 7.3 at 25°C, and about 7.0 in the temperature range 40–50°C.[16] Thus, the decomposition voltage for chlorate production is evaluated to be about 1.78 V from Fig. 15.4.

Since the main reaction (15.5) depends on the concentration of HClO and ClO⁻, the solution pH is an important factor because reactions (15.3) and (15.4) depend on the solution pH (see Fig. 12.16).

There are several side reactions, and these cause current inefficiency.

(1) At the cathode:
   ● Reduction of ClO⁻

$$ClO^- + H_2O + 2e \rightarrow Cl^- + 2OH^- \tag{15.8}$$

$$E^0 = 0.89 \text{ V vs. SHE at } 25°C$$

   ● Reduction of ClO_3^-

$$ClO_3^- + 3H_2O + 6e \rightarrow Cl^- + 6OH^- \tag{15.9}$$

$$E^0 = 0.63 \text{ V vs. SHE at } 25°C$$

(2) At the anode:

● Anodic oxidation of ClO^- to ClO_3^-

$$6HClO + 3H_2O \rightarrow \tfrac{3}{2}O_2 + 4Cl^- + 2ClO_3^- + 12H^+ + 6e \quad (15.10)$$

$$E^0 = 1.14 \text{ V vs. SHE at } 25°C$$

● Oxygen generation

$$2H_2O \rightarrow O_2 + 4H^+ + 4e \quad (15.11)$$

$$E^0 = 1.23 \text{ V vs. SHE at } 25°C$$

● Anodic oxidation of ClO_3^- to ClO_4^-

$$ClO_3^- + H_2O \rightarrow ClO_4^- + 2H^+ + 2e \quad (15.12)$$

$$E^0 = 1.18 \text{ V vs. SHE at } 25°C$$

(3) Decomposition of HClO:

$$2HClO \rightarrow 2HCl + O_2 \quad (15.13)$$

(4) Loss of dissolved chlorine to atmosphere.

Reaction (15.8) is fast, and is controlled by diffusion.[17] It is delayed by $Cr_2O_7^{2-}$ added to the electrolytic solution due to an increase in the cathode overvoltage. The key discovery of the addition of chromate made possible electrochemical method of chlorate production.

Although ClO_3^- is a strong oxidizing agent, it is stable electrochemically, and hence reaction (15.9) is very slow, thus making only a small contribution to the current inefficiency.[17,18] However, ClO_3^- can be reduced cathodically when the solution is contaminated by a small amount of heavy metal ions such as Ti(IV), Mo(VI), and W(VI).[19-21]

Reaction (15.10) is the electrochemical process of chlorate formation, but is accompanied by oxygen generation, and thus, current inefficiency. With reactions (15.1), (15.3), (15.4), and (15.10), the overall electrochemical reaction from Cl^- to ClO_3^- is

$$2Cl^- + 9H_2O \rightarrow \tfrac{3}{2}O_2 + 18H^+ + 2ClO_3^- + 18e \quad (15.14)$$

That is, reaction (15.14) requires 9 Faradays to produce 1 mole of ClO_3^- instead of 6 Faradays as with the conventional route composed of reactions (15.1), (15.3), (15.4), and (15.5). Since reaction (15.10) is an electrode process, the reaction rate per unit volume of the solution is small compared to reaction (15.5), and is thus a weak point for the electrochemical route. Accordingly, the electrochemical formation of ClO_3^- should be minimized to be as small as possible, and it is preferable to enhance the chemical conversion of hypochlorite to chlorate in practical cells operated at high temperatures and at a suitable solution pH.

Oxygen evolution is competitive with chlorine generation, reaction (15.1), in that the thermodynamic potential for reaction (15.11) is less positive than that for reaction (15.1). Relatively high oxygen overvoltage is the only reason why preferential generation of chlorine is possible at the anode.[22] Consequently, these reactions are greatly affected by the electrode material and the operating conditions.

The anodic oxidation of ClO_3^- to ClO_4^- takes place on a platinum anode because the oxygen overvoltage is high enough. This reaction is also possible on a magnetite anode.[15] On the other hand, further oxidation of ClO_3^- is difficult to obtain on a graphite anode.

Decomposition of HClO, reaction (15.13), is accelerated cathodically by the presence of small amounts of metallic impurities in the solution. Loss of the active chlorine is insignificant because of the relatively high pH of solution, although it is still a factor which may contribute to reduced current efficiency.

It is clear that the chlorate formation reaction is related to many factors such as the NaCl concentration, solution pH, temperature, solution flow rate, current density, and the anode material. Of these, the solution pH seems to be the most important factor for keeping the current efficiency high.

A maximum current efficiency is obtained at a solution pH of about 7, in agreement with theory.[14]

Since the rate of the chemical process [reaction (15.3)] increases with temperature, the current efficiency also increases.

When the operating temperature is high enough, the solution conductivity increases and the overvoltage decreases, so that the cell voltage also decreases. All are desirable for production of chlorate. However, serious degradation of the graphite anode is a problem and the solution temperature of a cell equipped with graphite anodes is limited because the corrosion rate of the anode increases with temperature. With metal anodes or even the magnetite anode, operation at high temperatures is possible. Of course, there is an optimum condition because the loss of the active chlorine also increases at high temperatures.[23]

It is reported that current efficiency is almost unchanged when the current density is higher than 15 A/dm^2.[14,15,23,24]

It is considered that oxygen evolution takes place mainly with the electrochemical oxidation of ClO^- shown by reaction (15.10), and that the charge transfer reaction of H_2O, reaction (15.11), is negligible. Several processes for the anodic oxidation of ClO^- may be considered in addition to reaction (15.10), such as

$$5HClO + 8H_2O \rightarrow 2O_2 + 2Cl^- + 3ClO_3^- + 21H^+ + 16e \quad (15.15)$$

Hammer and Wranglén obtained the material balance in detail, and concluded that the Foerster–Müller mechanisms shown by reaction (15.10) was most feasible and preferential.[16]

Ibl and Landolt proposed that the mechanism for the anodic formation of chlorate in dilute NaCl solution differed from that in concentrated NaCl.[25,26] That is, the process of chlorate formation in dilute NaCl solution is controlled by hydrolysis of dissolved chlorine, whereas in concentrated NaCl solution, mass transfer of hypochlorite through the diffusion layer in the vicinity of the anode is rate-determining. It is clear that cell performance is affected by the solution flow. Thus, the current efficiency is related to the Nusselt number (mass transfer), the Reynolds number (solution flow), and the thickness of the diffusion layer.[14,25–28]

Figure 15.5 illustrates a profile of the diffusion layer near the working anode.[26] Hypochlorite ion coming toward the anode reacts with H^+ to form HClO. A part of ClO^- diffuses to the anode surface, and is oxidized. In the steady state, the diffusion rate of ClO^- toward the anode must be equal to the diffusion rate of HClO toward the bulk of solution. A very small amount of ClO^- is oxidized to ClO_3^- at the anode. Since the mass transfer process is rate-determining, there will be an optimum thickness of the diffusion layer, δ_M. Jaksic calculated the differential equations for the mass transfer to obtain δ_M, and compared experimental results.[28] The current efficiency is kept high when δ/δ_M is kept in the range 0.1–10, where δ is the thickness of the diffusion layer and is related to the flow rate of solution.

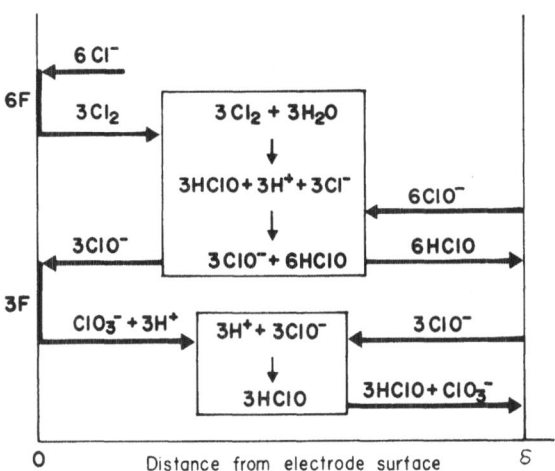

FIGURE 15.5. Model of the diffusion layer of the anode during electrolysis of concentrated NaCl solutions.

15.1.4. Anode Materials

A considerable amount of graphite is being used as anode in the chlorate cell even at present. Some electrolyzers are also equipped with magnetite and/or lead dioxide anodes.

The graphite anode resists the chlorine evolution reaction, but is attacked by the oxygen evolution reaction, especially at high temperatures. Degradation of the graphite anode in chlor-alkali cells has been investigated by many authors. Hine and co-workers studied the effect of the concentration of active chlorine and of the solution pH on the consumption of the graphite anode during electrolysis of NaCl solution, and concluded that the active chlorine in the solution is the key factor. The solution pH was a minor factor in the degradation of the graphite anode.[29] The concentration of active chlorine, especially $HClO$ and ClO^-, is related to the solution pH unless these concentrations are controlled intentionally. Consequently, the corrosion rate of the graphite anode is large at high pH. Degradation of the graphite anode was found to be almost independent of the ClO_3^- concentration. According to electrochemical studies, mostly by means of the potential sweep method, carbon oxide forms on the graphite surface, and the behavior of the oxide depends on the electrode potential. Formation and reduction of the lower oxide, $(C-O_L)$, which may take place at relatively low potentials, is almost reversible, while on the other hand, the higher oxide, $(C-O_H)$, formed at high positive potentials is difficult to reduce by cathodic polarization, and it becomes CO and/or CO_2 when the electrode is polarized further at high potentials.[22] Jaksic obtained the corrosion rate of the graphite anode in the chlorate cell as a function of the solution pH.[30] The corrosion rate increased quickly at pH higher than 7.

It is considered that the penetration of chlorate in the micropores of the graphite causes degradation of the anode. Thus, oil treatment is used to plug the pores and to prevent serious corrosion of the graphite anode during electrolysis.[31]

Matsumura et al. emphasized the effectiveness of the magnetite anode for producing chlorate in the laboratory and in practical cells.[32] Magnetite is a spinel, and is composed of an equal mole ratio of FeO and Fe_2O_3. The atomic ratio (AR) of $Fe(III)/Fe(II)$ is an important factor affecting the electrochemical characteristics of the anode.[33] That is, the electric conductivity changes greatly with the AR, and the corrosion rate increases extensively when the AR value deviates from 2. Hayes and Kuhn have reviewed methods of preparation and the physicochemical properties of the magnetite anode.[34]

Lead dioxide is also a candidate as the anode material. According to Izidinov and Veselovskii, the overvoltage of the lead dioxide anode was very high in $5.3\,N$ NaCl at pH $= 5.5$–6.0, but the anode was depolarized extensively by addition of $10^{-3} M$ CoCl$_2$. The overvoltage was almost the same as that of the platinum anode before its passivation.[35] Addition of CoCl$_2$ was also effective in reducing the corrosion rate of the PbO$_2$ anode. It was considered that addition of CoCl$_2$ reduced the oxygen supersaturation on the PbO$_2$ electrode, thus reducing the passivating action of oxygen on the electrochemical oxidation of Cl$^-$ to Cl$_2$.

Uzbekov *et al.* studied the corrosion behavior of the platinum anode by means of a high-sensitivity radiochemical method.[36] The corrosion rate of the platinum anode remained almost constant at 1.6×10^{-8} A/cm^2 (with an assumption of four-electron transfer) in the potential range 1.4–2.0 V, but increased greatly at higher potentials. It is considered that the platinum surface is covered completely by oxygen, and that transport of Cl$^-$ to the electrode surface to form Cl$_{ad}$ becomes difficult. This concept was supported by Djordjevic *et al.*, who called it "oxide-covered platinum".[37] The platinum anode in the chlorate cell would be covered by ClO$_3^-$ and/or ClO$_3$(ad), thus retarding the formation of oxygen.

It is well known that the dimensionally stable anode (DSA) has contributed to the modern development of the chlor-alkali industry. There is no doubt that the DSA is also useful as the anode of the chlorate cell and many modern cells are equipped with the DSA, as shown in Table 15.3. According to Jaksic *et al.*, oxygen generation on the DSA was somewhat less than on the platinum anode.[38]

The RuO$_x$-type anode is attacked by the oxygen evolution reaction, whereas it resists the chlorine evolution reaction.[39] The material can be modified by several methods such as addition of precious metals so as to improve its physical and chemical properties. Some adequate materials to serve as the working anode for chlorate production have become available, whereas only some research papers on production of chlorates with the metal anode can be found. Further investigation in this field must be conducted.

15.2. ELECTROLYTIC PRODUCTION OF ORGANIC COMPOUNDS

15.2.1. Introduction

Electrochemical processing of organic compounds has a long history. For example, the Kolbe reaction is well known, i.e., carboxylic acids are converted into hydrocarbons and carbon dioxide by anodic oxidation:

$$RCOO^- + R'COO^- \rightarrow R-R' + 2CO_2 + 2e \qquad (15.16)$$

Cathodic reduction of maleic acid to produce succinic acid is an example of an electroorganic process that has been operated successfully on a commercial basis for many years:

$$
\begin{array}{ccc}
& & \overset{\displaystyle H}{\underset{\displaystyle |}{}} \\
H-C-COOH & & H-C-COOH \\
\| & +2H^+ + 2e \rightarrow & | \\
H-C-COOH & & H-C-COOH \\
& & \underset{\displaystyle H}{\overset{\displaystyle |}{}}
\end{array}
\qquad (15.17)
$$

In general, however, a number of papers have described the scientific aspects of electroorganic reactions studied in small-scale electrolytic cells with currents in the microampere or milliampere range, but there are very few published reports of results obtained in commercial-scale electrolyzers.

Recently, commercialization of electrosynthesis of organic compounds has been considered to be attractive for several reasons:

(i) the yield of product, or the current efficiency, of the electrochemical process is higher than that of conventional nonelectrochemical methods;

(ii) both electrochemical procedure and operation are relatively simple;

(iii) the electrochemical method is capable of use of low-cost starting materials;

(iv) the formation of byproduct can be eliminated, or minimized at least, by control of the electrode potential;

(v) although the investment cost of the electrolytic system and the cost of electric energy are relatively expensive, the total manufacturing cost of electrosynthesis would be comparable, or even less, in comparison with the chemical route, which requires expensive raw materials and complicated processing, in some cases;

(vi) since the anode and cathode compartments are separated from each other in general, the electrochemical method is capable of production of useful materials simultaneously in both compartments,

(vii) the electrochemical synthesis is able to cut down on pollution problems.

The last reason is today's most attractive feature.[40-48]

The basic processes in electrochemical method are, of course, the anodic oxidation and the cathodic reduction without any chemical agent, but other types of reactions specific to organic chemistry can occur in parallel, such as the following examples:

Anodic oxidation:

$$C_6H_5-CH_2OH \rightarrow C_6H_5CHO + 2H^+ + 2e \atop C_6H_5CHO + H_2O \rightarrow C_6H_5COOH + 2H^+ + 2e$$ \quad (15.18)

Cathodic reduction:

$$C_6H_5NO_2 + 6H^+ + 6e \rightarrow C_6H_5NH_2 + 2H_2O \quad (15.19)$$

Substitution:

$$C_6H_5R + CN^- \rightarrow R-C_6H_4CN + H^+ + 2e \quad (15.20)$$

Addition:

$$CH_2=CH_2 + 2OH^- \rightarrow HO-(CH_2)_2-OH + 2e \quad (15.21)$$

Cleavage:

$$CH_3-S-S-CH_3 + 2H^+ + 2e \rightarrow 2CH_3SH \quad (15.22)$$

Coupling:

$$2(CH_3)_2C=O + 2H^+ + 2e \rightarrow (CH_3)_2-C-OH \atop | \atop (CH_3)_2-C-OH$$ \quad (15.23)

Cathodic reduction may be accompanied by dimerization. Cleavage, isomerization, cyclopolymerization, and other processes can also be conducted in electrochemical cells. Electrochemical reductions of nitrocompounds, nitriles, and carboxylic compounds are examples.

The well-studied Kolbe reaction is typical of anodic electroorganic reactions. It is of interest that the Kolbe reaction occurs on Pt and/or Ir anodes in appropriate aqueous solutions at high current efficiency, whereas in organic solutions, reaction proceeds efficiently on platinized platinum, gold, and carbon anodes.

Electroorganic halogenation, especially fluorination,[53] and the anodic oxidation of organometallic compounds such as Grignard reagent are of importance. The electroorganic processes now operated on a commercial scale are the cathodic hydrodimerization of acrylonitrile to produce adiponitrile, and the anodic oxidation of Grignard reagent on a lead anode to obtain alkyl lead (the Nalco process):

$$Pb + 4(C_2H_5) MgCl \rightarrow Pb(C_2H_5)_4 + 4Mg^{2+} + 4Cl^- + 4e \quad (15.24)$$

Some other processes under development for commercialization are listed below[42,45]:

Anodic oxidation:
- Propylene to propylene oxide (Bayer and Kellogg):

$$CH_3CH=CH_2 + H_2O \rightarrow CH_3CH-CH_2 + 2H^+ + 2e \quad (15.25)$$
$$\overset{\diagdown \diagup}{O}$$

- Monomethyladipate to sebacic acid:

$$\underset{\text{(monomethyladipate)}}{CH_3O\overset{O}{\overset{||}{C}}(CH_2)_4\overset{O}{\overset{||}{C}}OH} \rightarrow \underset{\text{(dimethylsebacate)}}{CH_3O\overset{O}{\overset{||}{C}}(CH_2)_8\overset{O}{\overset{||}{C}}OCH_3} + 2CO_2 + 2H^+ + 2e$$

$$\xrightarrow{\text{saponification}} \underset{\text{(sebacic acid)}}{HOOC(CH_2)_8COOH} \quad (15.26)$$

Cathodic reduction:
- Nitrobenzene to *p*-aminophenol (John Brown, England):

$$(15.27)$$

- *o*-Phthalic acid to 3, 5-cyclohexadiene-1, 2-dicarboxylic acid:

$$(15.28)$$

- Dimethyl ketone to pinacol

$$2(CH_3)_2 CO + 2H^+ + 2e \rightarrow (CH_3)_2(OH) C-C(OH)(CH_3)_2 \quad (15.29)$$

Combination of anodic oxidation and cathodic reduction:

● Benzene to hydroquinone via *p*-benzoquinone:

$+ 2H_2O \rightarrow$ $+ 6H^+ + 6e$ (anodic oxidation)   (15.30)

$+ 2H^+ + 2e \rightarrow$ (cathodic reduction) (15.31)

Although production by electrosynthesis of organic compounds is presently small in size (with some notable exceptions), there is no doubt about a bright future, chiefly because of the variety of possibilities presented for processing. The cost reduction possible with electrosynthesis is also a major factor in its favor. The large-scale production of adiponitrile by Monsanto using the electrochemical hydrodimerization process has stimulated R&D in electroorganic chemistry. According to Prescott, approximately 700 million pounds of nylon filament was produced in 1964 in the United States. Of this amount, almost 90% was nylon 66, which was obtained from adiponitrile as intermediate.[50]

15.2.2. Electrochemical Synthesis of Adiponitrile: An Example of Electroorganic Chemistry

Adiponitrile is an important intermediate for producing nylon 66, and is obtained by a catalytic reaction of adipic acid with ammonia. It is converted into hexamethylene diamine, and finally into nylon 66:

$$HOOC(CH_2)_4COOH + 2NH_3 \xrightarrow{catalysis} NC(CH_2)_4CN + 4H_2O \quad (15.32)$$

$$NC(CH_2)_4CN + 4H_2 \rightarrow NH_2-CH_2(CH_2)_4CH_2-NH_2 \quad (15.33)$$

$$NH_2-CH_2(CH_2)_4CH_2-NH_2 + HOOC(CH_2)_4COOH \rightarrow nylon\ 66 \quad (15.34)$$

Since adipic acid is expensive, a new route for producing adiponitrile was investigated by Baizer based on electrochemical dimerization of organic compounds.[51,52] The goal was to reduce acrylonitrile (AN) by cathodic reduction into adiponitrile (ADN), because AN feed was cheaper than adipic acid and was easy to obtain. The process is simple:

$$2CH_2=CH-CN + 2H^+ + 2e \rightarrow NC(CH_2)_4CN \quad (15.35)$$

Initial attempts to electrolyze 20% HCl solution containing AN with lead or alkali metal amalgam cathode to obtain ADN were not successful because of excessive propionitrile formation. After extensive experimentation, Baizer achieved success. He determined two important features, namely:

1. the process consists of electrochemical dimerization, and it is not necessary to contain free radicals as intermediate, and
2. the AN concentration in the catholyte must be kept high so as to stimulate the rate of dimerization.

The solubility of AN in an aqueous solution is 7%, but is increased to 10% by the use of a quaternary ammonium salt of an aryl sulfonic acid, named McKee's salt.

The unwanted side reaction of reduction to propionitrile was also controlled by the use of McKee's salt.

Formation of bis-cyanoethylether (another side reaction) was cut by lowering the solution pH. Use of quaternary ammonium salt is an important feature for the reduction of AN to ADN. The mechanism of the effect of quaternary salt on phase transfer reactions is a current topic of research in organic synthesis.[54-58]

The Monsanto electrolytic cell is a filter-press type equipped with cation exchange membrane separators, consisting of a sulfonated polystyrene resin.[50] The cell are made of polypropylene with neoprene gaskets. The effective area of the membrane is $10 \, ft^2$ $(0.93 \, m^2)$. The cathode is made of lead. The nature of the anode material has not be disclosed in detail, but it is said that a platinum-coated carbon anode gives good results.

The catholyte is circulated at high rate so as to prevent scale deposition on the cathode surface, and also to stimulate mass transfer of AN to the cathode surface. The solution temperature is kept at 85–160°F (30–70°C). The current efficiency is high, nearly 100%, and the consumption of electric energy is 7700 kWh/ton ADN.

A Belgian company, UCB, has developed two processes for producing ADN: one is by direct electrolysis, and the other is a potassium amalgam process.[59] The cell for direct electrolysis has no separator between the two electrodes, so that the cell voltage and power consumption (3300 kWh/ton ADN) are both significantly less than those of the Monsanto process. The electrolyte is an emulsion consisting of about a 2:1 mixture of aqueous and organic phases. The aqueous phase contains potassium phosphate, tetraethylammonium phosphate, and a small amount of sodium hexametaphosphate (corrosion inhibitor). The organic phase is recovered after electrolysis, and recycled to the cell. Acrylonitrile is fed to this flow. The cell is made of polyethylene-lined steel and is equipped with an iron oxide anode and a graphite cathode. The flow velocity of electrolyte is 1 ft/sec. Temperature control is provided by a cooling system. The current efficiency and the yield of product are 80–90% and 85%, respectively.

The UCB amalgam process is a dimerization of AN using potassium amalgam as a reducing agent:

$$2CH_2 = CH-CN + 2K(Hg) + H_2SO_4 \rightarrow NC(CH_2)_4CN + K_2SO_4 + (Hg) \tag{15.36}$$

The amalgam is produced by electrolysis of potassium salt with a mercury cathode, and hence chlorine is obtained as a byproduct. Byproduct potassium sulfate is also useful as fertilizer.

BASF has developed a method to produce ADN by electrolysis with an undivided cell. The cathode process is irreversible and AN is not oxidized at the anode, so that the diaphragm is not necessary.[60] The yield and the current efficiency can be increased when the salt concentration is kept at less than 12%. The solvent also contains acetonitrile. Since the solution conductivity is low, a unique electrolyzer, named the capillary gap cell (see Fig. 12.11), has been designed to minimize resistive potential drop. The anode-to-cathode gap is only 125 μm. The plate anode is made of electrodeposited PbO_2 and the cathode plate is graphite, each 40 cm diam. The cell is equipped with 100 electrode plates, and is operated at 10 A/dm^2. The plant has a capacity of 140 tons/month with 12 electrolyzers. The yield of ADN is said to be about 90%.

15.3. SAFETY PROBLEMS AND ENVIRONMENTAL PROTECTION IN ELECTROCHEMICAL INDUSTRIES

15.3.1. Safety Problems in Electrochemical Industries

Unfortunately, accidents are increasing with the increase of new methods and technologies being used in industry, whereas the ratio of accidents to the number of employees is decreasing gradually. While the labor accident rate in the chemical process industries is at the lowest level for all industries, continuing efforts must be made for improving safety measures for employees.[61,62]

There are three factors affecting labor safety:

1. the production capacity in a plant is increasing,
2. processes are becoming more complex, requiring a high technical level for operators although training and upgrading of labor skills are sometimes lagging, and
3. significant reduction of labor effort per unit of production.

For example, the average size of Japanese chlor-alkali plants in the 1950s was 20 tons/day, but the present size is 250 tons/day, with some plants being as large as 1000 tons/day. In the United States, several giant plants of 1000–1500 tons/day are being operated.

Scaling-up of the plant size is desirable from the viewpoint of safety protection, because more investment may be devoted to safety measures in a large plant than in a small operation. As the plant becomes more complicated, additional safety measures are required. Complication of the flowsheet and increased complexity of the technology are inevitable for the process industry because the factors may decrease production costs. Therefore, it is important that the employee have the ability for adapting himself to new procedures. Training for good work habits and education in the technology must be conducted extensively, but self-training is also of great importance for the individual worker.

The labor requirement for unit production in process industries has decreased when the flowsheet has been improved to increase productivity. Figure 14.1 shows the results of a survey on the labor productivity in Japanese chlor-alkali plants. The plot shows that the logarithm of man-hours vs. the logarithm of plant capacity is a straight line with a slope of about 1.5. For example, we may consider two plants of different sizes: the one having a capacity of 50 tons/day and the other 200 tons/day. The man-hours per unit production of these plants are 4.5 and 0.4, respectively. Therefore, the total man-hours per day in the small plant should be 4.5×50; that is, this plant requires 9 to 10 men per shift for operation, in contrast with 0.4×200 man-hours per day or only 3 to 4 men per shift in the large and new plant. However, the slope of the logarithm of man-hours vs. the logarithm of plant size is about unity when these plants have been constructed under the same design (see Section 14.1).

There are several potential causes for damage of chemical plants, such as

- equipment trouble,
- operator error,
- disasters such as storm, flood, and earthquake, and
- social causes such as interruption of electricity, suspension of water supply, civil unrest and strikes.

These factors may injure a number of employees as well as accidents to individuals due to electric shock, burns and scalding, explosions, and contact with corrosive and toxic chemicals.

For preventing these problems, various measures must be considered:

- improvement of the process flowsheet such as automation and mechanization,
- adequate system for safety management,

- adequate system for emergency situations,
- adequate system for emergency shelter and evacuation of personnel,
- regular training and education for improving safety attitudes of workers.

A good example of improvement of plant operation is the development of the tapping manipulator for the calcium carbide furnace operated at the Omi plant of the Denki Kagaku Kogyo Company. Tapping of calcium carbide from the electric furnace is a very dangerous job, and hence use of the manipulator is of great advantage for labor safety (see Section 9.5).

Emery and Currey[62] have identified some key requirements for safety programs in an electrochemical plant, as follows:

> First, a basic philosophy must be stated based on positive management attitudes and a solid management commitment. More than lip service is required. The dollar expenditures will be high and it will be easy to find reasons for not implementing expensive modifications. Secondly, the types of hazards and their injury potential must be identified. Thirdly, programs must be implemented in the design, operation, maintenance, and training areas to remove or reduce the injury potential in each hazard area.

Electric shock is a specific trouble in the electrochemical plant. According to Currey,[63] a circuit in the 400–500-V range is considerably safer than a 600–800-V circuit. In keeping with his philosophy, 400 tons/day circuits of H-4 cells recently installed at Hooker Chemical Corporation operate at circuit voltages between 300 and 350 V.

The OSHA of the United States has established regulations and inspection procedures for the electrochemical industry as follows:

1. Isolation of cell area—controlled access, signs.
2. Suitable training of personnel.
3. Protection of exposed bus in working areas.
4. Better access to cells in operating, renewal.
5. Suitable protective clothing.
6. Removal of mechanical hazards—lifting, ladders, cell tops.
7. Reduction of chemical hazards, due to spills, cell disassembly, etc.

15.3.2. Abatement of Mercury Discharge from Chlor-Alkali Plants

The Japanese chlor-alkali industry encountered serious pollution problems in 1973. Several plants using amalgam cells were switched-off for a week or so because sludge containing mercury was found near the plant sites.

The chlor-alkali industry has traditionally been the largest consumer of mercury, followed by electrical apparatus manufacturers,[64] so that reduction of the level of mercury discharge to the waterways and air has been a major social requirement of the chlorine industry. Many papers and patents on mercury problems have appeared.[64-76]

For reducing the mercury discharge from chlor-alkali plants to the environment, a task group was organized in the Mizushima plant of the Okayama Chemical Company and also in the Kokura plant of the Osaka Soda Company. The Mizushima plant is relatively new, and the employees are greatly concerned about mercury because the plant is located in the Inland Sea National Park. On the other hand, the Kokura plant is the oldest amalgam-cell plant, having a 60-year history, and is located in the industrial area of northern Kyushu.

The preliminary investigation showed that in a typical plant, the sources of mercury loss are:

1. brine yard,
2. drainage,
3. thick amalgam,
4. emission from cell room area, and
5. products—caustic and hydrogen, but not chlorine.

Accordingly, the following items have been emphasized:

1. treatment of sludge and waste water,
2. reduction of thick mercury,
3. preventing vaporization of mercury, and
4. polishing products.

Modernization of the plant flowsheet and layout has been carried out to achieve these requirements. Together with the technical activity, an improvement in the attitude of employees was discussed and tried. It is believed that reducing mercury losses depends largely on the attitude and overall morale of all employees, not only in the old plant but also in the new one. Lectures about mercury were given to plant employees, mostly office workers and laborers, and educational pamphlets were prepared and distributed. The mercury level in the urine of all employees is determined twice a year, and hence, every worker can compare himself to other people in the plant as well as to the general public.

The chief engineer of the plant and the general manager have been appointed as monitors to patrol the cell room area once a day, to check out any incident, and to report to the project office. All employees are instructed to inform the office if any release of mercury is found.

In the cell room, the operations group has been given the responsibility for prevention of mercury losses and for monitoring leakage. The maintenance group is responsible for keeping the work area clean. Inspection duties have been assigned to cell room personnel so as to check unusual behavior of each cell at an early stage.

Treatment of Sludge and Waste Effluent. Figure 15.6 shows the layout of the sewer system and arrangement for drainage in the Mizushima plant. The plant area has been paved with asphalt so as to recover rain water. Drainage around the plant (solid line) is separated from the general sewage (dotted line) and is treated completely prior to discharge or recycle. The cell room area, the brine treatment yard, the reservoir of brine, and the sludge treatment area are protected by a heavy wall (double line) so as to prevent any leakage of water containing mercury.

Figure 15.7 is the flowsheet for the recycle system. Sludge and drainage from the brine yard are filtered, and the filtrate is sent to a large reservoir (500 m^3). The cell is washed mainly with purified brine, and a minimum

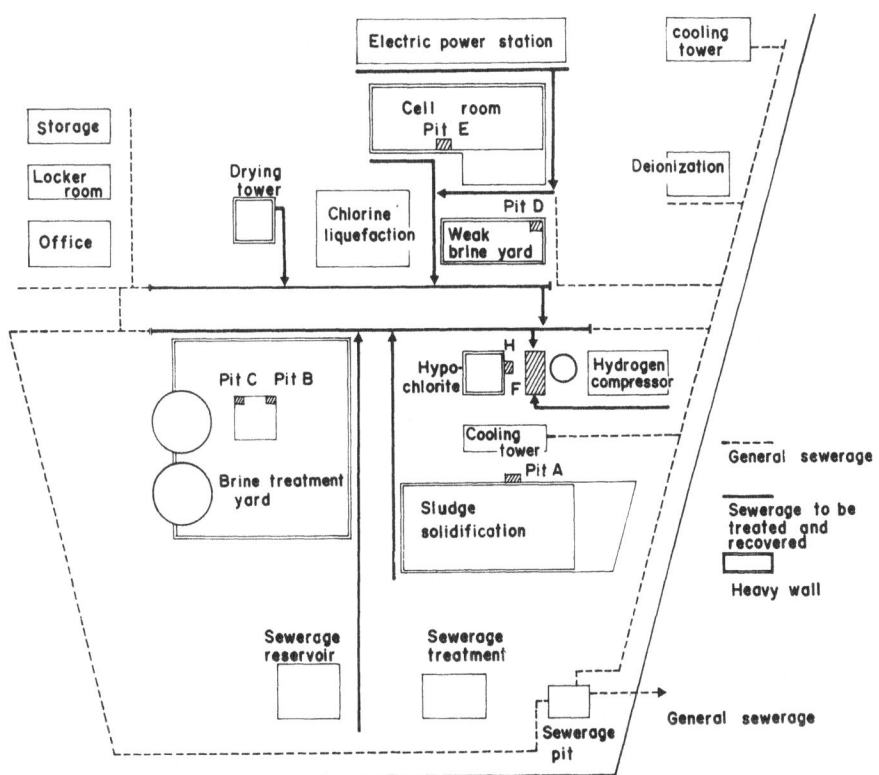

FIGURE 15.6. Layout of sewerage and arrangement for drainage in the Mizushima plant.

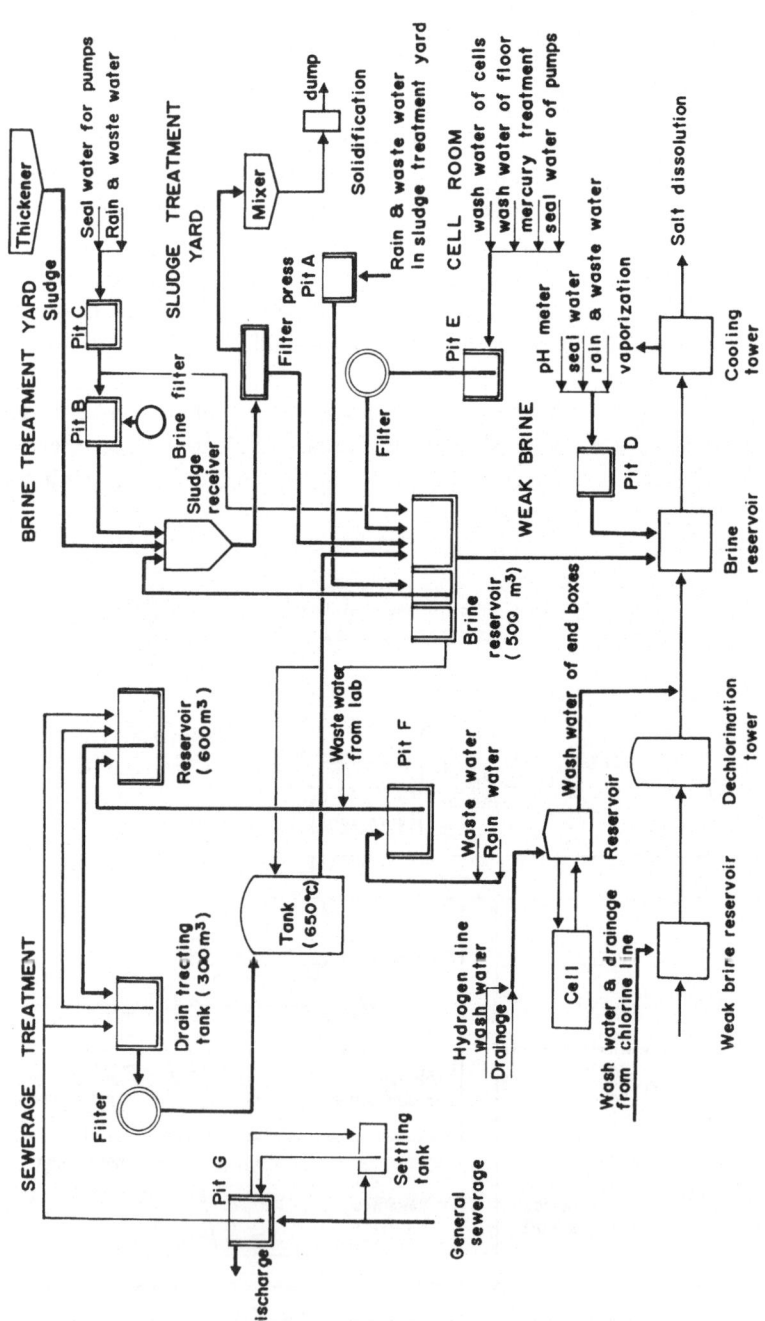

FIGURE 15.7. Flowsheet for recycle system of waste water.

amount of fresh water is finally used. The waste water is collected, then filtered.

The cell room floor is occasionally washed with a small amount of fresh water and water from Reservoir E.

Rain water in the brine yard and the sludge solidification area (ca. 2000 m^2) is sent to the weak brine reservoir, then recycled. Rain water in other parts (ca. 12,000 m^2) is collected in Reservoir F, filtered, then chlorinated. Mercuric chloro-complex is recovered with an anion exchange resin. Residual mercury is absorbed on a resin developed specially for removing mercury from the liquid phase. Purified water is stored and sent to the brine yard as process water. Some is used as wash water.

Water collected in a 500-m^3 reservoir is mixed with the weak brine from the cell. The brine is sent to the cooling tower, where some water is vaporized to maintain the water balance in the process (10 tons/hr in capacity). After the brine is well chlorinated, no discharge of mercury is measureable.

Figure 15.8 shows the distribution of mercuric ions in various forms. The coordination number n of mercuric chloro-complex, $HgCl_n^{-(n-2)}$, depends on the Cl^- activity in the solution. The average number of coordination $\bar{n}$ is usually nearly 4 for the tetrahedral anion, $HgCl_4^{2-}$, in highly concentrated chloride solutions as shown by the dotted line. On the other hand, $HgCl_2$ is more stable in dilute chloride solution. Therefore, excess chlorine (more than 28 mg/liter) is required to avoid a precipitate of the uncharged mercuric salt.

Rain water from outside the chlorine plant and general sewage are

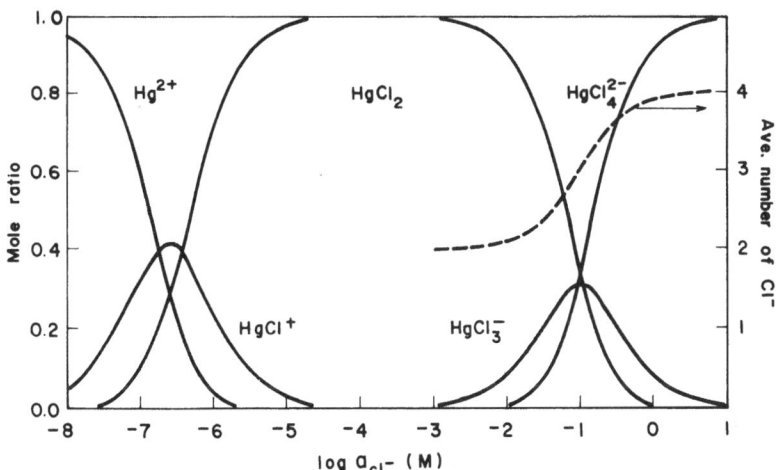

FIGURE 15.8. Distribution of mercuric ions in various forms of chlorocomplex as a function of the activity of chloride ions.

collected, and the mercury level is also continuously monitored. An automatic analyzer gives a warning if mercury in the effluent reaches 0.5 ppb. The flow rate and the pH of the stream are also controlled prior to discharge.

High-quality salt is used so as to minimize sludge from the thickener. Sludge is washed and filtered with an automatic filter press. According to present records, the formation of sludge (wet basis, 45% water content) is 12–15 kg/ton NaOH. The average content of mercury is 7 ppm, because the brine is sufficiently chlorinated. The filter cake is mixed with cement and sand in a 200-liter drum. The solid product is checked to be sure there is no dissolution of mercury, and is then dumped in the deep sea after three weeks of aging.

Sludge accumulated at the bottom of the brine resaturation yard is purged every six months, and is concreted after recovery of salt.

Filter backwash is combined with the sludge from the thickener, then treated. The concentration of active chlorine in the brine is kept constant at about 10 mg/liter. Thus, very little mercury precipitates from the brine.

Thick Mercury. Thick mercury, or mercury butter, is a complex mixture including carbon, calcium, and iron.[77] Strontium may be a serious impurity in thick mercury.[78] According to experience, the purer the feed brine at high temperature, the lower is the formation of thick mercury. Use of metal anodes (DSA) is a factor for reducing thick mercury, probably because carbon particles are absent.

Cell operation under stabilized conditions of brine composition, temperature, mercury flow, and current density is essential to reduce thick mercury. The NaCl concentration and temperature of feed brine are controlled exactly, which is useful to stabilize the flow in the thickener. The brine is polished by a leaf filter with activated carbon as filter aid. The filter requires a relatively small amount of purified brine and/or fresh water for backwash.

The temperature of the feed brine is 65–73°C, and the weak brine from the cell is at 75–85°C.

Only a small amount of thick mercury is recovered by regular maintenance every six mounths, and is distilled.

Reduction of Mercury Vaporization. Prevention of mercury loss in ventilation air from the cell room area is the most difficult task. For minimizing emission to the room atmosphere, the cell room was modernized.

The cell room floor was painted with an epoxy primer, covered with 5-mm-thick epoxy resin containing silica sand, then finished with a smooth layer of epoxy resin paint in order to recover mercury drops easily with a specially designed vacuum cleaner.

A conventional mercury pump was converted to the enclosed "canned pump" style, with carefully selected gasket material. The piping was annealed carefully after welding to avoid localized corrosion and/or cracking. Caulking was made with a room-temperature curing urethane to close pinholes, if necessary.

The inlet and outlet end-boxes of the cell were clamped with a cover of FRP (fiberglas-reinforced plastic). The caustic outlet of the amalgam decomposer and the caustic receiver were also covered and their vapor spaces connected to a slight vacuum. Individual orifices were used to equalize pressure (-10 to -20 mm Hg) at each location. The vacuum of the main pipe was kept at -100 to -150 mm H_2O. A small amount of fresh air was fed through a small hole at each location in order to stabilize the pressure and to prevent explosion of hydrogen.

The use of metal anodes with automatic brine gap adjustment and improved brine treatment permit a minimum of cell maintenance, with cleaning being required only twice a year or less.

The cell is cooled to room temperature before it is opened, and is then promptly sealed with a temporary cover.

Overhaul of the amalgam decomposer is carried out every three years. The tower is cooled and hydrogen is replaced with nitrogen. A temporary cover is used. Mercury and caustic are transferred to a sealed reservoir under vacuum. The graphite packing is renewed, and overhaul of mercury pump, valves, piping, and gaskets carried out to avoid undesirable troubles during normal operation.

Figure 15.9 shows the vacuum and treatment system for preventing mercury emission. Waste hydrogen is brought to an adsorption tower containing a resin for removing mercury from the gas phase.

Gases from the end-boxes of the cell, the wash tank for the end-box, the caustic outlet, and the receiver are washed with brine and fresh water, then sent to the adsorption tower. Mercury in the purge gas is controlled at less than 0.0005 mg/m^3, and drainage is sent to the brine treatment yard.

Maintenance of equipment containing mercury is carried out in a separate room. Thick mercury is treated in a closed chamber. Ventilation air from those rooms is brought to treatment under a slight vacuum, and is a useful procedure for preventing mercury emission. According to experience, the mercury concentration of the atmosphere 3 m and 10 m downward from the cell was 0.6–0.9 mg/m^3 and 0.2–0.4 mg/m^3, respectively, when the empty cell was opened. Although the air flow velocity affects the dissipation of the mercury, these data are typical of the plant record under normal conditions. It decreased to 0.04–0.06 mg/m^3 at 3 m from the cell when the cell was filled with water. The mercury content in the maintenance room increased quickly from 0.01 mg/m^3 to 0.2–0.6 mg/m^3 when a basket of graphite packing was brought from the decomposer. The open amalgam cells and decomposers under repair should be covered and connected to the exhaust duct.

The mercury content in the cell room air was $0.03 - 0.04$ mg/m^3 at the

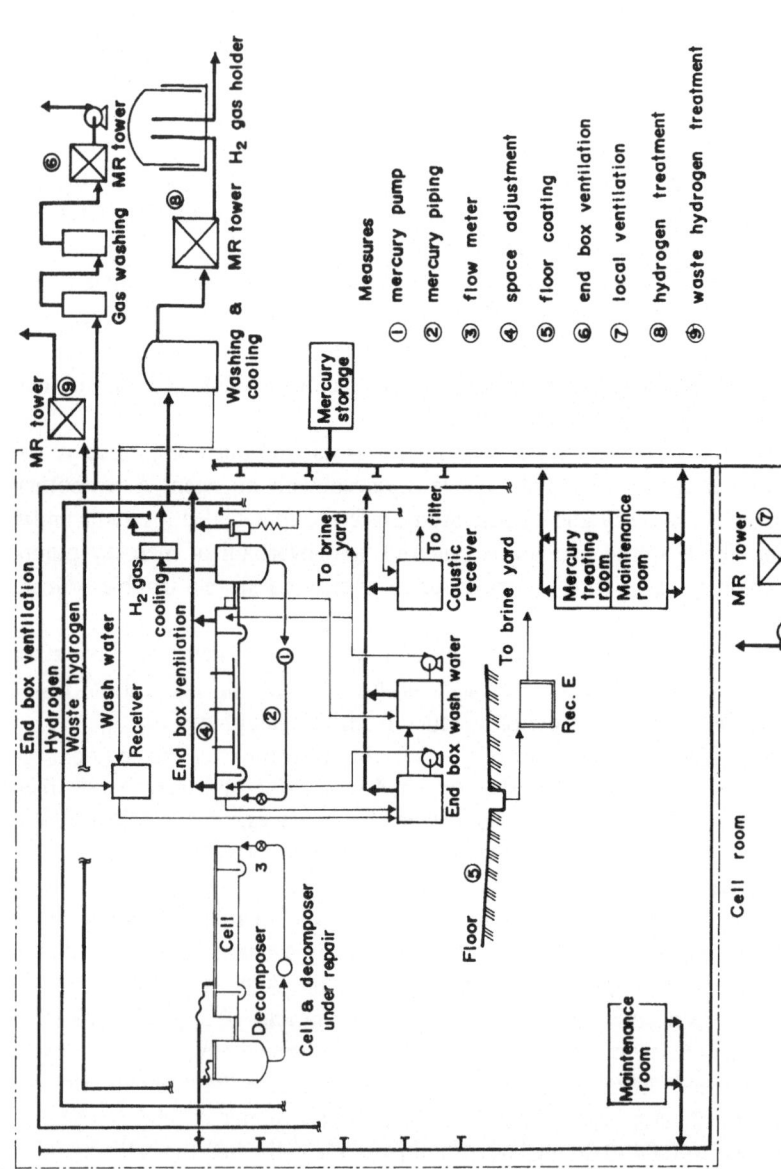

FIGURE 15.9. Vacuum and treatment system for preventing mercury emission.

start of the clean-up program. Sampling records indicate the steady improvement that has been made, with present data showing that mercury in the cell room is only 0.002–0.004 mg/m³, with an average of 0.0024 mg/m³. It is believed possible to achieve a level of 0.001 mg/m³. Although the Kokura plant is an old one, the mercury content is 0.008–0.03 mg/m³ at present, and efforts are aimed at bringing it to a goal of 0.005 mg/m³.

Polishing Products. The cell gas is doubly washed with a counter flow of fresh water at 15°C in packed towers in series to eliminate mist containing traces of mercury. Mercury is not detected in chlorine gas if the NaCl content in the second tower is kept at less than 0.1 g/liter. Chlorine is dried, then sent to liquefaction.

Hydrogen is cooled to 30–35°C with a shell-and-tube type cooler at the top of each amalgam decomposer. Condensed water and mercury drain back to the decomposer directly. The hydrogen gas is cooled further to 7–9°C to demist and reduce mercury levels, then brought to the adsorption tower where the mercury content is reduced to less than 0.0002 mg/m³.

Caustic soda from the amalgam decomposer is cooled to 60–65°C before sending to the storage reservoir. It is polished by a leaf-type filter with activated carbon filter aid. The mercury content in the polished caustic is 0.005–0.020 ppm or 0.010 ppm on the average. Two filters are sometimes used in series. The backwash water is sent to the sludge reservoir, then treated.

Results. Mercury in the canal near the plant was determined several times by the local governmental officers. The results showed that the mercury content in the sediments was about the same as the background reading.

The mercury content in the air near the cell room is determined on a regular basis. The mercury content is only 0.3 μg/m³ of air at a point 50 m downward from the cell room.

Table 15.4 shows the distribution of mercury consumption on a per ton of caustic produced as 100% NaOH basis at Mizushima. No mercury discharge has been detected since the start-up of the closed circulation system in both plants. The mercury loss into ventilation air at the Mizushima plant is small when compared with that of the Kokura plant, because cleaning of the cell room floor and prevention of mercury emission from plant equipment are relatively easier in the new plant.

Discharge of sludge from the brine treatment yard decreased by one-half or one-third of the previous record after high-quality salt was used. Recovery of mercury from brine sludge is carried out as well. The mercury loss with sludge, therefore, has decreased significantly. Attention is being given to the recovery of mercury from waste material such as filter cloth and filter aid, since this is the largest single source of mercury loss, followed by cell room ventilation, as shown in Table 15.4.

TABLE 15.4

Mercury Consumption at the Mizushima Plant (100% NaOH basis)

		Consumption g/ton
Effluent, closed-circulation system		0.000
Ventilation		
Cell room, avg. ventilation	4.72×10^5 m^3/hr	0.189
Max. content	0.005 mg/m^3	
Avg. content	0.0024 mg/m^3	
Hydrogen purged	400 m^3/hr	0.000
Avg. content	0.0002 mg/m^3	
Inlet and outlet end-boxes		0.000
Max. exhaust rate	1200 m^3/hr	
Avg. content	0.0005 mg/m^3	
Local treating		
Max. exhaust	3500 m^3/hr	0.000
Avg. content	0.0005 mg/m^3	
Products		
48.5% caustic, avg. content	0.01 mg/kg	0.021
Hydrogen, avg. content	0.0002 mg/m^3	0.000
Chlorine, mercury not determined		0.000
Sludge, discharge (wet basis, 45% water)	15 kg/ton	0.150
Max. content	10 mg/kg	
Avg. content	7 mg/kg	
Waste, caustic filter cloth, aid, resin, etc.		1.109
TOTAL		1.469

An excellent overall result has been achieved for a mercury consumption of less than 1.5 g/ton NaOH or only 450 g/day from a 300 tons/day plant. Mercury consumption has been recorded at less than 2.5 g/ton caustic even in an old plant. These results can be considered as acceptable from the viewpoint of environmental protection. An investment of about 2 million dollars was required to institute the clean-up program at the Mizushima plant with about the same amount of money spent for the Kokura plant. In addition to interest and depreciation charges, the cost for sludge treatment is a large factor.

It is concluded that abatement of mercury discharge from an amalgam-type chlorine plant can be economically and practically achieved, and the results show low discharge of mercury even at an old plant, which could be acceptable for environmental protection. High investment as well as the operating cost are problems. The manufacturing costs, and hence, the selling prices for chlorine and caustic soda are, of course, affected accordingly.

15.4. FEASIBILITY AND PROSPECTS
OF ELECTROCHEMICAL PROCESSES—CONCLUSION

Electrochemistry has an historical basis of well over a century since the early pioneering work of Faraday, Nernst, and others. Today, electrochemical science holds an important position as a basic science. On the practical side, electrochemical industry has a large share in the chemical process industry.

Although interest in electrochemical technology has fluctuated at various times because of reasons related to low productivity and economy, researchers and engineers have come back to use electrochemistry because of its unique features.

The purpose of the chemical process industry is to supply cheap and abundant products to the market, but the industry does not persist in using just one conventional route. Consequently, an electrochemical method may find a favorable opportunity to substitute for a conventional process when it is much more feasible than the old process. The feasibility of some electrochemical processes has been described in this book, but we should always be aware of the limitations of electrochemical processes.

Some electrochemical industries such as chlor-alkali and aluminum refining are being operated actively on a large scale, and will continue to have a bright future. New processes in molten salt electrolysis and electroorganic synthesis, such as adiponitrile, are expected to be of great advantage.

Primary cells and secondary batteries have been available for many years and new types are being developed. Whereas the price of individual cells may be small, the total sales of this important segment of electrochemical industry are not small. Recently, new and improved batteries of various types have attracted a great deal of attention from the viewpoint of providing new energy options. There is no doubt that energy conservation is today's most important subject worldwide. Effective use of existing energy production equipment as well as development of new sources must be conducted. For example, it is said that we could save about 20% of electricity if we could conduct a complete load-leveling for existing generation of electric power. The fuel cell and certain storage batteries may contribute to this concept if they are scaled up to meet the purpose. The philosophy of the storage battery must be changed extensively. A giant battery array will serve as a kind of energy conversion plant, and not simply as energy storage equipment.

Since the charge transfer during electrochemical processing is proportional to the mass transfer, electrochemical techniques are utilized in various fields for metering and control of processes. Practical applications of electrochemistry in metal surface treatment abound, such as plating, anodization, and electropolishing.

Corrosion of metallic species in electrolytic solutions is an electrochemical process, so that a variety of electrochemical methods are utilized in the scientific study, monitoring, and control of corrosion. The Wagner–Traud theory of the mixed potential has served as an important basis for the theory of corrosion. Researchers of catalytic chemistry have a keen interest in electrochemistry because the electrode process is a kind of heterogeneous catalytic reaction. Continuing interchange takes place between the disciplines of catalysis and electrochemistry with mutual benefits in understanding.

There is a growing awareness of the importance of electrochemistry in general. By examining the history of the science and engineering of electrochemistry, new pathways may be found to solve some of the important problems faced by mankind.

REFERENCES

1. J. E. Colman, Electrolytic production of sodium chlorate, in *Tutorial Lectures in Electrochemical Engineering and Technology*, edited by R. Alkire and T. Beck, AIChE Symposium Series, Vol. 204, p. 244 (1981).
2. Annual Report of the Electrolytic Industries, prepared by Industrial Electrolytic Division, Electrochemical Society (published every year).
3. Anonymous, *Chem. Week*, p. 19 (November 18, 1978).
4. G. Ishi, *Symposium on Electrochemical Industries*, Society of Chemical Engineers, Japan, No. 7, p. 167 (January 1980).
5. M. Bakes, J. Cormier, and R. Scalliet, Abstract No. 442, meeting of Electrochemical Society, Seattle, WA, May 1978.
6. T. J. Navin, Abstract No. 444, meeting of Electrochemical Society, Seattle, WA (May 1978).
7. F. Greiveldinger, Abstract No. 445, meeting of Electrochemical Society, Seattle, WA (May 1978).
8. J. R. Hodges, Abstract No. 447A, meeting of Electrochemical Society, Seattle, WA (May 1978).
9. N. W. Meyers, Abstract No. 461, meeting of Electrochemical Society, Seattle, WA (May 1978).
10. T. K. Suonperä and J. P. T. Karikko, Abstract No. 446, meeting of Electrochemical Society, Seattle, WA (May 1978).
11. F. Foerster and E. Müller, *Z. Elektrochem.* **8**, 8, 515, 633, 923 (1902); **9** 171 (1903); **10**, 781 (1904).
12. F. Foerster, *Trans. Electrochem. Soc.* **46**, 23 (1924).
13. M. M. Jaksic, B. Z. Nikolic, I. M. Csonka, and A. B. Djordjevic, *J. Electrochem. Soc.* **116**, 684 (1969).
14. M. M. Jaksic, A. R. Despic, and B. Z. Nikolic, *Sov. Electrochem.* **8**, 533 (1972).
15. M. M. Jaksic, *J. Electrochem. Soc.* **121**, 70 (1974).
16. L. Hammar and G. Wranglén, *Electrochim. Acta* **9**, 1 (1964).

17. F. Hine and M. Yasuda, *J. Electrochem. Soc.* **118**, 170, 182 (1971).
18. V. A. Shlyapnikov and T. S. Filippov, *Sov. Electrochem.* **5**, 806 (1969).
19. J. Koryta and J. Tenygl, *Coll. Czech. Chem. Commun.* **19**, 839 (1954).
20. I. M. Kolthoff and I. Hodara, *J. Electroanal. Chem.* **5**, 2 (1963).
21. I. Hodara and A. Glasner, *Electrochim. Acta* **15**, 923, 931 (1970).
22. F. Hine, M. Yasuda, and M. Iwata, *J. Electrochem. Soc.* **121**, 749 (1974).
23. V. A. Shlyapnikov, *Sov. Electrochem.* **7**, 1080 (1971).
24. M. M. Jaksic, A. R. Despic, I. M. Csonka, and B. Z. Nikolic, *J. Electrochem. Soc.* **116**, 1316 (1969).
25. N. Ibl and D. Landolt, *J. Electrochem. Soc.* **115**, 713 (1968).
26. D. Landolt and N. Ibl, *Electrochim. Acta* **15**, 1165 (1970).
27. A. R. Despic, M. M. Jaksic, and B. Z. Nikolic, *J. Appl. Electrochem.* **2**, 337 (1972).
28. M. M. Jaksic, *Electrochim. Acta* **21**, 1127 (1976).
29. F. Hine, M. Yasuda, I. Sugiura, and T. Noda, *J. Electrochem. Soc.* **121**, 220 (1974).
30. M. M. Jaksic, *J. Appl. Electrochem.* **3**, 219 (1973).
31. M. M. Jaksic and I. M. Csonka, *Electrochem. Technol.* **5**, 473 (1967).
32. T. Matsumura, R. Itai, M. Shibuya, and G. Ishi, *Electrochem. Technol.* **6**, 402 (1968).
33. R. Itai, M. Shibuya, T. Matsumura, and G. Ishi, *J. Electrochem. Soc.* **118**, 1709 (1971).
34. M. Hayes and A. T. Kuhn, *J. Applied Electrochem.* **8**, 327 (1978).
35. S. O. Izidinov and V. I. Veselovskii, *Sov. Electrochem.* **6**, 1552 (1970).
36. A. A. Uzbekov, V. V. Losev, and K. I. Nosova, *Sov. Electrochem.* **8**, 1812 (1972).
37. A. B. Djordjevic, B. Z. Nikolic, I. V. Kadija, A. R. Despic, and M. M. Jaksic, *Electrochim. Acta* **18**, 465 (1973).
38. M. M. Jaksic, A. R. Despic, B. Z. Nikolic, and S. M. Maksic, *Croatica Chemica Acta* **44**, 61 (1972).
39. F. Hine, M. Yasuda, T. Noda, T. Yoshida, and J. Okuda, *J. Electrochem. Soc.* **126**, 1439 (1979).
40. M. R. Rifi and F. H. Covitz, *Introduction to Organic Electrochemistry*, Marcel Dekker, New York (1974).
41. D. E. Dany, *Hydrocarbon Process.*, p. 159 (June 1969).
42. Anonymous, *Chem. Eng.*, p. 78B (May 17, 1971).
43. J. C. Davis, *Chem. Eng.* p. 44 (July 7, 1975).
44. J. L. Fitzjohn, *Chem. Eng. Prog.* **71**(2), 85 (1975).
45. M. Fleischmann and D. Pletcher, *Chem. Br.* **11**(2), 50 (1975).
46. R. B. MacMullin, *Electrochem. Technol.* **2**, 106 (1964).
47. R. B. MacMullin, *Progress and Problems in Electrochemical Engineering*, presented at Chicago Section, Electrochemical Society, January 13, 1966.
48. R. B. MacMullin, *J. Electrochem. Soc.* **120**, 135C (1973).
49. H. M. Fox, F. N. Ruehlen, and W. V. Childs, *J. Electrochem. Soc.* **118**, 1246 (1971).
50. J. H. Prescott, *Chem. Eng.* p. 238 (November 8, 1965).
51. M. M. Baizer, *J. Electrochem. Soc.* **111**, 215 (1964).
52. M. M. Baizer and J. D. Anderson, *J. Electrochem. Soc.* **111**, 223 (1964).
53. H. M. Fox, F. N. Ruehlen, and W. V. Childs, *J. Electrochem. Soc.* **118**, 1246 (1971).
54. E. V. Dehmlov, *Angew. Chem. Int. Ed.* **13**, 170 (1974).
55. H. Nozaki, *Kagaku to Kogyo* (Chemistry and Industry, Chem. Soc. Jpn.) **28**, 639 (1975).
56. E. V. Dehmlov, *Angew. Chem. Int. Ed.* **16**, 493 (1976).
57. W. P. Weber and G. W. Gokel, *Phase Transfer Catalysis in Organic Synthesis*, Springer-Verlag, New York (1977).
58. R. Oda, *Kagaku* (Chemistry) **32**, 179 (1977).
59. Anonymous, *Chem. Eng.*, p. 58B (April 20, 1970).
60. F. Beck, *J. Appl. Electrochem.* **2**, 59 (1972).

61. *Soda Handbook*, p. 411, edited by Japan Soda Industry Association (1975).
62. A. T. Emery and J. E. Currey, *Chlorine Bicentennial Symposium*, p. 235, Electrochemical Society (1974).
63. J. E. Currey, *Chlorine Bicentennial Symposium*, p. 187, Electrochemical Society (1974).
64. W. C. Gardiner, "Where Were the Mercury Pollution Problems in Pulp and the Paper Industry", private communication (1971).
65. W. C. Gardiner, paper presented at Niagara Falls Sections, Electrochemical Society, November 9, 1970.
66. W. C. Gardiner, paper presented at the Chlorine Institute, 14th Chlorine Plant Managers' Seminar, New Orleans, LA, February 3, 1971.
67. W. C. Gardiner and F. Munoz, *Chem. Eng.*, p. 57 (August 23, 1971).
68. W. C. Gardiner, *Electrochemical Contributions to Environmental Protection*, p. 16, Electrochemical Society (1972).
69. R. A. Perry, *Chem. Eng. Prog.* **70**(3), 73 (1974).
70. R. E. Ross, *Chlorine Bicentennial Symposium*, p. 120, Electrochemical Society (1974).
71. R. A. Perry, *Chlorine Bicentennial Symposium*, p. 131, Electrochemical Society (1974).
72. S. Payer and B. Strasser, *Chlorine Bicentennial Symposium*, p. 133, Electrochemical Society (1974).
73. H. O. Bouverg, *Industrial Waste Water, IUPAC*, p. 75, Butterworths, London (1972).
74. R. Johsen and E. Böhm, *J. Appl. Electrochem.* **1**, 163 (1971).
75. K. Shikaze, *Chem. Econ. Eng. Rev.* **6**(3), 17 (1974).
76. W. A. Lemmon, *Chem. Econ. Eng. Rev.* **6**(3), 32 (1974).
77. O. deNora, *Chem. Ing. Techn.* **43**(4), 182 (1971).
78. J. P. Guptill and C. J. Kaminski, *Chlorine Bicentennial Symposium*, p. 111, Electrochemical Society (1974).

INDEX